Aristotle's Ethics

Aristotle's Ethics

David Bostock

OXFORD
UNIVERSITY PRESS

OXFORD

UNIVERSITY PRESS

Great Clarendon Street, Oxford OX2 6DP

Oxford University Press is a department of the University of Oxford.
It furthers the University's objective of excellence in research, scholarship,
and education by publishing worldwide in

Oxford New York

Auckland Cape Town Dar es Salaam Hong Kong Karachi
Kuala Lumpur Madrid Melbourne Mexico City Nairobi
New Delhi Shanghai Taipei Toronto
With offices in
Argentina Austria Brazil Chile Czech Republic France Greece
Guatemala Hungary Italy Japan South Korea Poland Portugal
Singapore Switzerland Thailand Turkey Ukraine Vietnam

Oxford is a registered trade mark of Oxford University Press
in the UK and in certain other countries

Published in the United States
by Oxford University Press Inc., New York

ISBN 0-19-875265-2

To Jenny's memory

Contents

Chapter V **Responsibility (III.1, 5)**

Chapter VI **Self-control (VII.1–10)**

Chapter VII **Pleasure (VII.11–14, X.1–5)**

Introduction

FOUR works on ethics attributed to Aristotle have come down to us, namely two full-length works, the *Nicomachean Ethics* and the *Eudemian Ethics*, and two shorter works, the *Magna Moralia* (or 'Great Ethics') and *On Virtues and Vices*. There is nowadays universal agreement that the last of these is not by Aristotle, and may be ignored. There is dispute over the *Magna Moralia*. Some take it to be an early work by Aristotle himself, or perhaps a pupil's record of early lectures by him, but others (who, I think, are in the majority) take it to be a later compilation, reflecting thought on the topic within Aristotle's Lyceum, but probably written after Aristotle's death.[1] In any case, no one supposes that it has the authority of the two full-scale works, the *Nicomachean Ethics* (*EN*) and the *Eudemian Ethics* (*EE*).[2]

The relation between these two works is curious. The *Nicomachean Ethics* is a work in ten books,[3] and the *Eudemian* a work in eight books, but they *share* their central books. That is, *EN* V–VII are the same as *EE* IV–VI, so these books are often referred to as 'the common books'. It has been argued by Kenny (1978), mainly on stylistic grounds, that these 'common' books were originally composed as part of the *EE* rather than the *EN*, and this conclusion is now very generally accepted. The most natural explanation is that the *EE* represents an earlier course on ethics, and that later Aristotle determined to revise this course, but did not see any need for substantial revision of these 'common' books. They were therefore taken over into the *EN* without much (if any) change. By contrast, there are some quite noticeable changes between *EE* I–III and *EN* I–IV, and again between *EE* VII–VIII and *EN* VIII–X. I think that most scholars now accept this account of the matter, though there are those (including Kenny himself) who do not. In any case, I accept it, and I shall assume without further argument that the *EN* is the later work. It therefore gives us Aristotle's final views on the topics that it discusses.

Even if this chronological hypothesis is mistaken, it remains true that for many centuries the *Nicomachean Ethics* has been widely studied as giving us Aristotle's ethical opinions, while the *Eudemian Ethics* has received comparatively little attention. I think this is deserved. Though no doubt scholars cannot afford to ignore the *EE*, for most people wishing to be introduced to Aristotle's views on ethics the *EN* is

---

[1] For some discussion, see on one side Cooper (1973) and on the other Rowe (1975) and Kenny (1978: 215–30). (Bibliographical details of works thus briefly referred to are given in the List of References at the end of the book.)

[2] Whether one abbreviates the title to *EN* or *NE* is a matter of taste. The order *EN* reflects the Latin title (*Ethica Nicomachea*), and it used to be the case that Aristotle's works were always referred to by their Latin titles. But fashions are changing.

[3] In ancient texts what counts as a 'book' is partly governed by what could conveniently be fitted onto a single roll of papyrus. Often the division into 'books' corresponds to a natural division of the subject-matter, but sometimes it does not.

the place to go to. This is not because its thought is simpler (for in several ways it is not), but because its exposition is on the whole clearer, its arguments are often better, and it addresses a wider range of problems. In any case, this book is designed as a commentary on the *EN*, and though I do refer occasionally to the *EE*, it is the *EN* that is my central topic throughout.

Let us go back to a point mentioned three sentences ago: why *should* one wish to be introduced to Aristotle's views on ethics? After all, Aristotle lived well over 2,000 years ago (his dates are 384–322 BC), the society he lived in was not at all like ours, and he was not in any way aware either of the religious views that have influenced our moral thinking for centuries, or of the ethical theories of (say) Hume, or Kant, or Mill, which are now so much the philosopher's stock-in-trade. What does he still have to contribute? Well, I think that the first point to make is just that his treatment of the topic is *not* influenced by factors such as those just mentioned, and for that reason it is (from our perspective) a remarkably fresh approach, untrammelled by all the baggage that we bring with us. But, going on from this, it soon becomes clear that his approach has its own problems, and the attempt to think through those problems, starting from what he has to say, brings us up against what are still problems for us. But they certainly appear differently if we begin by looking at them through his eyes. One may, of course, take a merely 'antiquarian' attitude to Aristotle: one can be curious about what he thought, while not thinking of it as having any contemporary relevance. This is not my attitude. What makes him worth studying, especially on this topic, is largely that his problems are our problems too, and that is why his attempts to solve them are worth our close attention. For even if we cannot in the end agree with his solutions, still we shall learn a great deal by working out just why we disagree. That cannot fail to assist us in our attempt to find improved solutions.

This book is designed to be read alongside Aristotle's own *Nicomachean Ethics*. It is not in any way a substitute for that text. I assume that readers will have the text before them, and can look up passages that I refer to as they go along. I assume too that most readers will be looking at the text in an English translation, and I have not presupposed any knowledge of Greek. This creates a problem, since there are many English translations, and they do not always agree with one another on the treatment of key terms. The one which has the best claim to be 'the traditional translation' is that by Ross (1925), which formed volume IX of what is usually referred to as 'The Oxford Translation' of Aristotle's works. This was revised by Urmson in 1975 and his revisions were retained in *The Revised Oxford Translation* (Barnes 1984). Then some further revisions, made jointly by Ackrill and Urmson in 1980, were incorporated in the version published in the World's Classics series. These subsequent revisions have not been very extensive, and it is still entirely appropriate to think of this translation as Ross's. It retains his rendering of key terms at almost all important points. Many other translations are available,[4] but probably the best

---

[4] Of translations in popular series, I note: Rackham (Loeb, 1926; rev. 1934), Thomson (Penguin, 1955; rev. Tredennick, 1976, with new introd. by Barnes), Ostwald (Library of Liberal Arts, 1962). The most recent translation that I know of is that by R. Crisp (2000).

alternative to Ross is Irwin (1985a), which also contains some very useful notes. My guess is that the most widely used translations today are those of Ross (in one version or another) and of Irwin. Consequently I have tried to note their rendering of key terms where it differs from my own. The translations given in my text are mine, but are certainly influenced by Ross and Irwin.

Occasionally there are real problems of translation, which affect the interpretation of Aristotle's philosophical position. Where this is so, I have discussed the matter. But the occasions are rare, and for the most part it is fair to say that any translation will do, with this rider: it should give in the margin the page and line numbers of Bekker's edition of the Greek text (1831). This is because it is the standard way of referring to all of Aristotle's works, used universally by all scholars.[5] Unfortunately, many popular editions which are conveniently cheap do not satisfy this requirement very well.[6]

I add a small note on this matter of Greek-into-English, concerning the *transliteration* of Greek words. In older times the convention was that Greek words were transliterated into English via their transliteration into Latin. In consequence, Greek 'U' becomes English 'Y', Greek 'K' becomes English 'C', Greek 'X' becomes English 'CH', and Greek 'Φ' becomes English 'PH'. More recently, it has become usual to ignore the Latin intermediate, at least so far as to restore 'U' and 'K' in transliterations from Greek. If we follow this through consistently, then Greek 'X' should become English 'KH', since the Greek letter represents an aspirated 'K' sound (as in Scottish 'loch', not as in 'chop'), and Greek 'Φ' should become English 'F'. The result is that one loses the etymological connection between some Greek words and their familiar English derivatives. (For example, the Greek 'ψυχή' (conventionally translated as 'mind' or 'soul') was under the old style of transliteration the English *psychē*, but in the new style would be *psukhē*.) In this book I have mostly followed the modern convention for 'U' and 'K', though not for 'X' and 'Φ', but not entirely consistently. (For example, I do refer to our text as the 'Nicomachean' *Ethics*, and not the 'Nikomachean' or 'Nikomakhean'.) Readers should simply bear in mind that different modern writers follow different conventions, and on the whole the earlier writers follow the Latin convention while the later do not.

None of the Aristotelian writings that has come down to us was published by him. We know that he did publish some writings, mostly early in his career, but these have been lost. (They can to some extent be reconstructed from miscellaneous quotations and allusions in other authors.) What we think of as 'Aristotle's writings' is presumably the collection of papers that he left at his death, and they must have been to some extent edited by others before they assumed the form in which we now know them. It is natural to suppose that these papers formed the basis of

[5] Thus the first line of the *Nicomachean Ethics* is 1094ᵃ1, which means that it is line 1 of column *a* of page 1094 in Bekker's edition.

[6] The current World's Classics edition of Ross's translation gives Bekker references only at the top of each page, which is not really adequate. The same is true of the current Penguin translation.

the lecture courses that he would give, and perhaps he himself would one day have organized them into publishable form, if he had lived long enough. (He died of an illness, aged 62.) But there is also the possibility that he never thought of them as publishable works, but simply as a basis for teaching.

Some of these writings that have come down to us are, at least in parts, rather scrappy and disjointed. One might think of them as arising from Aristotle's successively revising his lectures, as the years went by, or perhaps as a collection of 'research papers', written up as 'work in progress', in the hope that later a final version could be distilled from them. In this respect, the *Nicomachean Ethics* is a more finished work than many others: there is good reason to suppose that it is planned as a whole, and there are not many places where one feels that a number of originally separate discussions have been simply tacked together, without much attempt to harmonize them with the main theme.[7] But there is, of course, one very obvious example where the problem created by 'the common books' shows through. There is a discussion of pleasure at the end of book VII (which is a common book), and another at the beginning of book X (which is special to the *EN*). If Aristotle himself had been revising the work for publication, he could not possibly have included both, as they now stand. (In fact the reasonable conjecture is that the account in book VII would simply have been dropped, and that that in book X would have taken its place.) We must presumably thank the piety of the ancient editors for preserving both versions. (There are a few other apparent discrepancies between the common books and those special to *EN*; I shall mention them as I come to them.) In any case, there is an overall moral to draw: the *EN* as we now have it was not prepared for publication by Aristotle himself, and we do not know what revisions he might have made if he had done the job himself. Indeed, we do not know how much his editors did for him, for they had the original manuscripts before them and we do not. (One is tempted to suppose that they did very little, and that that is why apparently inconsistent treatments of the same topic still persist. But perhaps they exercised their discretion more widely than we think.) Anyway, it should be remembered that Aristotle did not himself prepare the *EN* (or the *EE*) for publication in the form in which we now have it.

This leads me to a brief remark about the *titles* to the two works, 'Nicomachean' and 'Eudemian'. We do not know why these titles were given. We do know that Eudemus was a prominent member of Aristotle's Lyceum, who succeeded to its headship when Aristotle died. So it is a fair conjecture that the *Eudemian Ethics* is so called because it was he who put together the version that we have inherited. We also know that Aristotle had a son named Nicomachus, and so one might conjecture that in a similar way it was this Nicomachus who put together our *Nicomachean Ethics*. But, against this, we have no independent reason for supposing that this Nicomachus ever became interested in philosophy. So a different suggestion, which is embodied in the French title *L'Éthique à Nicomaque*, is that Aristotle wished to dedicate his latest treatment of ethics to his son. But the truth

---

[7] In my own opinion, the second half of book V does give this impression, but it is something of an exception to the general run of the *EN*.

is that we simply do not know why the *Nicomachean Ethics* has the title that it does.

I end this introduction with a brief word about what Aristotle himself took to be the purpose of his *Ethics*. In his scheme of things, ethics is a preliminary to politics, for—to speak very broadly—the aim of the politician is to secure the good of the whole community, and so as a preliminary one must know what counts as a good life for an individual. That is what the *Ethics* aims to determine, and Aristotle explicitly links its topic to his *Politics* at several places, most conspicuously in its final chapter. Those who wish to know how he proposed to put his ethical ideas into practice, at the level of constitution, law, and government, must therefore go on to his *Politics* (which I do not discuss). But he also thinks of his *Ethics* as 'practical', rather than 'theoretical', in another way: each of us wants to know how best to organize his own life, whatever political regime he may be subject to, and Aristotle intends what he says in this work to be of practical use for this purpose. For example, he says early on in I.3:

The aim [of our enquiry] is not knowledge but action; (1095^{a}5–6)

and more fully in II.2:

The present study is undertaken not for the sake of theoretical knowledge, as others are—for we are not enquiring in order to know what goodness[8] is, but in order to become good; otherwise nothing would be gained by the enquiry—so we must consider what concerns actions, and how they are to be done. For these also control what kind of dispositions we acquire, as I have said. (1103^{b}26–31)

Two consequences of this approach Aristotle elaborates in I.3. The audience that he desires for his lectures is a mature audience, who are concerned with how best to live their lives, and who can do something about it, since they are not (as young people are) at the mercy of any chance emotion. But also, it is impossible to lay down specific rules for action which will apply in every case, since the circumstances of action are so many and various. So 'What is said will be adequate if it is as clear as the subject-matter allows. For one should not expect the same exactness in all discussions (as one should not expect it in manufactured products)'[9] (1094^{b}11–14). There are some other occasions, apart from those just mentioned, where Aristotle indicates his concern for the practical relevance of what he has to say (e.g. 1096^{b}32–5, 1109^{a}30–^{b}26, 1109^{b}34–5). But despite the role that this practical purpose is given in his introduction, it in fact plays only a very small part in the ensuing discussion, and I shall pay it little attention. His *Ethics* is essentially an enquiry into what goodness *is*, and similarly for other concepts central to ethics, and that is the main focus of interest. The work is, in our terminology, as 'theoretical' as is any other philosopher's work on ethics, and I shall treat it as such.

---

[8] I here use 'goodness' to translate *aretē*. This is no doubt a mistranslation, though justified by the present context. For some discussion of how to translate *aretē*, see p. 20.

[9] Aristotle has in mind a contrast between the exactness expected of the geometer and of the carpenter. See 1098^{a}26–33.

Consequently I shall mostly ignore his claim that ethics cannot be 'exact', as other sciences are, for—as I see it—this is a point that is meant to concern ethical practice and not ethical theory.[10]

Further reading

At the end of each chapter I shall make some suggestions for further reading on the topics discussed in that chapter. I take the opportunity to note here some reading that concerns the *Ethics* as a whole, though I do not recommend you to work through it before beginning on the *Ethics* itself. Think of it more as background, to be consulted where necessary.

For those who can read the original Greek, I recommend the Oxford Classical Texts, abbreviated OCT (i.e. Bywater 1894, for the *EN*; Walzer and Mingay 1991, for the *EE*). I have already mentioned translations. For a full-scale commentary, which leaves no part of Aristotle's text undiscussed, I recommend Gauthier and Jolif (1958–9), which is in French. (This is the only reading that I shall suggest which is not in English. For those who can read French it is worth consulting on many obscure passages which other commentators pass over in silence.) There are some full-scale commentaries in English, but they are now somewhat out of date, and I do not recommend them.[11] It is usually better to consult the notes provided by Irwin (1985a), supplemented where possible by Ackrill (1973). In the Clarendon Aristotle series there are full-scale commentaries on two books of the *EN*, namely books VIII and IX (Pakaluk 1998), and on three books of the *EE*, namely books I, II, and VIII (Woods 1982).

There are many general introductions to Aristotle's thought which include substantial chapters on the *Ethics*. I select in particular Ross (1923), Ackrill (1981), and Lear (1988). One might also mention here, as a 'single chapter' on the *Ethics*, the introduction by Barnes (1976) to the revised Penguin translation.

Of works such as this one, which aim to discuss at book length all the main themes of the *Ethics*, but are nevertheless selective in what they cover, I strongly recommend Urmson (1988). This is short, easy to read, and very good value for its 130 pages. A book with much the same compass as this one is Hardie (1968), which often has useful things to say. A much longer and less straightforward treatment, which, however, covers fewer topics than Urmson or Hardie (or this book), is Broadie (1991). There are many books, and of course articles, on particular problems arising from this or that aspect of the *Ethics*, and I shall mention them as they become relevant. But I take the opportunity to notice here Annas (1993), which has quite a lot to say about Aristotle, but is by no means confined to him. It deals with the whole spectrum of Greek thinking on ethics, from Plato, through Aristotle, to the Epicureans, the Stoics, and beyond.

[10] I shall come back to this point about 'inexactness' much later, on pp. 223–4.

[11] Grant (1857); Stewart (1892); Burnet (1900); Joachim (published 1951; but written before 1917).

Chapter I

The good for man: first discussion (Book I)

1. Introduction

Iᴛ seems likely that it was Aristotle himself who first used the word 'ethics' as the name for a particular branch of philosophy,[1] but one may doubt whether he named it well. For the overall subject of this work would be better indicated by the title 'The Good Life for Man', whereas the Greek word 'ethics' means literally what pertains to a man's character, and Aristotle does not in fact think that all human goodness is goodness of character. On the contrary, he distinguishes excellences of character from intellectual excellences, and he discusses the part that *each* of these plays in contributing to the good life. We might nowadays expect that a work entitled 'Ethics', even if it does announce itself as concerned with the good life, must mean to be considering what we should call the *morally* good life; and we would not think of some of Aristotle's intellectual excellences as contributing in any way to moral goodness. But that is our perspective, and not Aristotle's. He does not operate with our distinction between moral goodness and other kinds of goodness, and nor did his contemporaries. In fact there is no Greek word that is naturally translated into modern English as 'moral'. We must try not to foist upon him an outlook which is ours but not his.

His question is simply 'What is the best way for a man to live?', and there is no presumption that what we call a 'moral' way of living will turn out to be the best. So one might naturally ask: if it is not specifically moral goodness that we are talking about, then what other kind of goodness is it? But here one must turn to Aristotle's own discussion to find an answer. The only thing that one can confidently say right at the start is that he is concerned here with the good *for man*.[2] While he is happy to affirm that the notion of goodness applies also to things other than men, that is not his topic in this work.

---

[1] One cannot here rely just on the title that has come down to us, for that might not have been Aristotle's own. But in fact in the *Politics* he frequently refers back to our *Ethics* as 'The Ethics' (1261ᵃ31, 1280ᵃ18, 1282ᵇ20, 1295ᵃ36, 1332ᵃ8, ᵃ22). One may also note a general reference to 'ethical philosophy' (*ēthikē theōria*) at *Posterior Analytics* 89ᵇ9, and similarly to 'ethics' as a topic (*ta ēthika*) at *Metaphysics* 987ᵇ1.

[2] The Greek word *anthrōpos*, which is standardly rendered 'man', undoubtedly means 'human being' rather than 'male (adult) human being', and it should be so understood. But at the same time one has to admit that almost everywhere it is in fact male adults that Aristotle is actually thinking of, and moreover only those who are free citizens, and in a position to live lives of their own choosing.

2. The good as the ultimate end

Aristotle's first move is to identify the good for man with what men aim for. As he says in his very first sentence,

Every expertise and method of investigation, and likewise every action and choice, seem to aim at some good; hence the good has been well described as that at which all things aim. (1094ª1–3)

This is a rough and ready generalization, and will be refined as we proceed. But one may well wish to object at once. Surely men often choose to act in ways that are bad? And surely there are branches of expertise too which are designed to achieve what is bad?[3] (To take a controversial example, there are some who would say this of the medical skill of abortion.) It is quite clear what Aristotle's first response to this objection will be: what men aim for is always something that at least *seems* good to them (e.g. 1114ª31–2). No doubt they may be mistaken, and the end pursued may not actually be good at all, but it must at least appear good to the agent, otherwise he would not pursue it. Similarly with skills and expertise: the skill would not be practised at all unless someone wanted what it aims to achieve, and what a man wants will at least appear good to him.

There is evidently scope to continue this debate. One might admit that what a man wants or chooses to do will always appear to him to be in *some* way good, though even he himself may admit that it is another way bad. Thus, I may choose to smoke because I gain pleasure from it, and that is some kind of a good to me. At the same time I may recognize that in another way it is bad for me, i.e. bad for my health. I may even admit that what I am doing does me more harm than good, but still do it all the same.[4] A different line of objection is that many things which we do, and can reasonably be said to choose to do, we do simply out of habit, without thinking at all of whether so acting is in any way good. But I shall not pursue this debate any further. The first objection and response is quite enough to bring out what is controversial in Aristotle's position.

He is going to argue that there is indeed some one thing 'at which all things aim', i.e. some one thing that is everyone's *ultimate* aim. Supposing that we agree with this, then we might also accept that this one thing is what appears to everyone to be the ultimate good. But from this Aristotle infers that it is actually the ultimate good. So one might say that he simply rules out the possibility of a universal mistake on what is ultimately good. (In fact he very explicitly does just this at 1172ᵇ35–1173ª4.) But is he entitled to do so? Well, for the present I must just leave that question hanging,[5] in order to return to his argument.

At first sight, it appears that there are very many different things that are aimed

---

[3] I shall oscillate between 'craft', 'skill', and 'expertise' as my translation of *technē*. (Irwin consistently uses 'craft' and Ross 'art', which is now a distinctly old-fashioned usage of this word.)

[4] I thus suffer from what Aristotle calls *akrasia*. This is the topic of Ch. VI.

[5] I come back to it in Ch. X.

for, and that this holds good even of the aims of a single individual, but still more so when we take into account all the different aims of different individuals. In the rest of Chapter 1 Aristotle shows that this is not by itself an objection to his position. For we may fairly say that some things are pursued simply for the sake of a further aim that depends on them, and what Aristotle is concerned with is *ultimate* aims, those that are pursued for their own sakes and not for anything further. His illustration of this point is perfectly straightforward, and needs no further comment at this stage.[6] This shows that it is possible both to admit that a person has many different aims and to claim that he has only one ultimate aim. But of course it does not prove that he has only one ultimate aim. Chapter 2 does appear to be designed to offer a proof for this claim. In fact it offers two proofs, one at a personal and one at an impersonal level.

The first argument (at 1094^{a}18–22) has been criticized by everyone as a *non sequitur*. Aristotle observes that if *everything* that we choose is chosen for the sake of something further, then there will be an infinite regress, and desire will be 'empty and futile'. (His argument here requires the stronger premiss: if everything that we choose is chosen *only* for the sake of something further.) Assuming that this is not the case—quite an interesting assumption, when you think about it—he is entitled to infer that at least some things are desired for their own sakes (at least partly, as we should add). But it evidently does not follow that there is just one end that we desire for its own sake, everything else being desired for the sake of this. Aristotle seems to claim that it does follow, and if so, then we can only say that his argument is a fallacy. Two responses may be made. First, it is not clear that Aristotle is claiming that this follows, since the whole passage is governed by an 'if' (or, in Irwin's translation, by a 'Suppose'). But in that case one can only conclude that the passage does little to advance his overall position.[7] Second, there is a way of continuing the argument which very much improves it. What Aristotle actually says may be taken as showing that there are at least some ultimate ends, pursued for their own sakes. We may then add that the sum of *all* these ends may therefore be regarded as a single overarching end, which alone is pursued for its own sake and not for anything further, for all other ultimate ends are pursued also as parts of this single overarching end. This continuation is very much in line with what Aristotle will say in chapter 7—at least as I interpret him—but one has to admit that he does not say it here. Besides, something more will still be needed to tie these separate ends together into a unity, and it is not clear that Aristotle ever provides that.

---

[6] Aristotle's illustration focuses mainly on the 'impersonal' aims of a skill or expertise, which may not be the same as the personal aims of those who engage in it. (For example, the aim of medicine is to cure the sick, but the individual doctor's personal aim may be something quite different. Perhaps he took up medicine only because it is reasonably well paid, or because he enjoys having power over other people, or maybe just to please his father.) But clearly the same general point could have been made about an individual's personal aims.

[7] It is ingeniously argued by Wedin (1981) that the passage *should* be construed simply as a conditional (with two antecedents that are to be established later), and that when it is so construed it can be seen as very relevant to the overall position.

The second argument (1094^{a}24–^{b}7) returns to the impersonal level, and claims that of all crafts, skills, or sciences—Aristotle is not drawing any distinction here—there is one that is the ruling one, namely the science of politics (of statesmanship). It counts as the ruling science, because it directs to what extent the other skills and sciences should be pursued,[8] and it also lays down laws for all citizens, stating what they should do and what they should abstain from.[9] Aristotle infers that its aim is therefore the supreme aim, and he leaves us to add that its aim is the good of all the citizens. So we are to conclude that, for each citizen, there is such a thing as his supreme good, for that is what the science of politics aims to promote. From a modern perspective, it is clear that this argument is unconvincing. We may agree, at least for the sake of argument, that there is something which may reasonably be called the science of politics, and that it has a particular aim, and that this aim is of benefit to the citizens. But one might add that what the politician aims (or should aim) to produce, for each citizen, is not his supreme good, but only a necessary condition of his achieving it (for example, a stable society in which crime and poverty are at a minimum, but also the restraints on individual freedom are equally at a minimum).

This view is, of course, a rather modern conception of the aims of politics, and is mostly espoused by those who do not share Aristotle's confidence that there is such a thing as *the* good for man. They hold that different people have different perceptions of the good, and that it is not the role of politics to adjudicate between these different conceptions, but rather to create the conditions in which each individual may pursue *his own* conception of the good. So politics does provide a benefit, but it is only a means to something else which each citizen will rank higher, which is for him a 'more ultimate' end. One could, of course, hold just this view about politics while still accepting Aristotle's own view that there is only one thing that really is 'the good' for man, and that if different people have different conceptions of this then they are just mistaken. For one might still hold that it is not the politician's job to educate people on what really is 'the good'; rather, this is the job of school-teachers, or bishops, or newspaper columnists, or whatever. But one must admit that the main reason why this view of the politician's role is nowadays so prevalent is that people are hesitant to say that only one conception of the good life is the correct one. In this, they disagree with Aristotle.

Let us come back to Aristotle, and summarize the position so far. He opens the work by identifying the good (for man) as what every man ultimately aims for. This apparently takes it for granted that there is some one thing which every man ultimately aims for, but in fact Aristotle is not taking it for granted. In chapter 2 he offers two arguments for it, but one has to say that the first, as stated, appears just to be a fallacy; and the second invokes a conception of the science of politics which we need not share. But it is clear that Aristotle himself does not regard these arguments as conclusive, for when the topic is resumed in chapter 7 he is still treating it as an open question whether there is this one ultimate aim that all men share. Before we

---

[8] One notes Aristotle's enthusiasm for a centrally directed economy.

[9] Aristotle assigns a greater role to the law than we do. See further pp. 56–7.

get to there, however, he introduces a further development which requires a new section.

3. The ultimate end as *eudaimonia*

Most of chapters 3 and 4 are taken up with methodological remarks, which I shall discuss later (Chapter X). But one important new theme is introduced: everyone agrees on what this ultimate end is to be called, namely *eudaimonia* (1095[a]14–20). The traditional translation (adopted by both Ross and Irwin) is 'happiness', but at the same time everyone agrees that this translation is misleading. I shall therefore leave the word untranslated throughout.[10] When the word is first introduced, we are immediately offered as synonyms 'living well and doing well', so one might well say that *eudaimonia* in Aristotle really means no more than 'a good life'. There is a lot of truth in this. But at the same time one must add that Aristotle did not invent the word; it is perfectly common in ordinary Greek. So it may be useful to start by asking how the word would usually be understood by 'the man on the Piraeus omnibus'.

First, let us say why 'happiness' cannot be the right translation. We are ready to talk of being happy just for a brief period (for example: 'I was happy when I got up, but even at breakfast I began to feel depressed'). *Eudaimonia*, however, is essentially something long-term; one cannot be *eudaimōn* just for an hour or so. A closer parallel might be a question such as 'Are you happy in your job?', which is clearly a question about the long term, and does not ask after euphoric feelings. But there are still two ways in which one could not ask the same question by using the word *eudaimonia*. First, the reply to the question about happiness might be 'Yes, I am perfectly contented with it. I'm not particularly good at it, but it satisfies me'. But the same reply would be hopelessly inappropriate if the question were about *eudaimonia*, for the Greek word connotes success and achievement, and requires more than mere contentment and satisfaction. Second, the same question could not be asked about *eudaimonia* anyway, for one cannot be *eudaimōn* in one respect (e.g. one's job) while not *eudaimōn* in another (say, one's marriage). One is either *eudaimōn* or not, absolutely. *Eudaimonia*, then, connotes overall success and prosperity and achievement, though it *also* connotes something that we may call happiness. For a life made miserable by psychological tensions, or by an inability to relate to other people, would not be *eudaimōn*, no matter how successful it was in other ways.

It is convenient here to look ahead to chapters 8–12 of book I, where Aristotle reviews the things commonly said about *eudaimonia*, and argues that they harmonize with his own account. It is not an external good, nor a good of the body, but a good of the soul, which is the highest kind of good (1098[b]12–18); it involves actions and activities of the soul (1098[b]18–20), which bring their own pleasure with them

---

[10] Etymologically, the word means (roughly) 'having a good guardian angel'. But evidently Aristotle does not take this etymology into account.

(1099^a7–21); it is thus simultaneously what is most good, most noble,[11] and most pleasant (1099^a22–31). At the same time some external goods and bodily goods are necessary conditions for it, for one cannot be *eudaimōn* without some wealth, nor without good looks, good birth, friends, political influence, and successful children (1099^a31–$^b8$). This, says Aristotle has led some to identify *eudaimonia* with good fortune, and it is true that one who suffers a notable misfortune cannot be counted as *eudaimōn* (1100^a5–9, 1100^b22–1101^a8). Nevertheless, the identification should be resisted, for while the *eudaimōn* does require some good fortune, it is more important that he cultivate the appropriate state of soul, which is something that many people can work at (1099^b18–28).[12] Finally, Aristotle accepts Solon's dictum 'call no man *eudaimōn* until he is dead', not because it is only when dead that he can be *eudaimōn*—that would be absurd (1100^a13, 1100^a34)—but because when one judges someone *eudaimōn* that is a judgement about his *whole* life, and his whole life can only be known when it has ended. He even feels some sympathy to the view that what happens to one's friends and children after one's death can affect the issue, but eventually he rejects this. (1100^a10–1101^b9)

While some of these claims about *eudaimonia* are more plausibly regarded not as common views but as Aristotle's own slant on those views—in particular, the claim that *eudaimonia* is closely associated with actions and activities of the soul—still we get a fair impression from this discussion of what the common views were.[13] And it seems to me that they strongly support the suggestion that the word should be understood as *meaning* just 'the good life'. Let us now return to where we broke off.

In chapter 4 Aristotle introduces it as agreed by almost everyone that the ultimate end (for man) is *eudaimonia*, but he at once observes that people have different opinions on what this comes to in practice (1095^a17–22). Ordinary people, he says, tend to think of it as something rather obvious, such as pleasure or wealth or honour, though they do not have a consistent view on the matter, and for that reason look up to those with a more sophisticated account to offer. But these people too hold different views (and Aristotle refers here to what was in fact Plato's view, which in chapter 6 he is going to reject) (1095^a22–30). Resuming the topic in chapter 5, Aristotle at once introduces the traditional 'three lives', i.e. the life of pleasure, the life of political activity, and the life of the intellect[14] (1095^b14–19). He quickly dismisses the first, which he evidently thinks of as a life devoted to what are called the 'bodily' pleasures, as merely fit for animals[15] (1095^b19–22). He then offers a brief discussion of the second, which he connects first with honour, and then with

[11] The Greek word here is *kalon*, which Irwin translates as 'fine'. For some discussion, see pp. 98–100.

[12] In VII.13, 1153^b21–5 Aristotle adds that too much good fortune can be an impediment to *eudaimonia*.

[13] At *Rhetoric* 1360^b4–29 Aristotle gives us what he explicitly calls the usual views on *eudaimonia*. His description is much the same as we find here, save that it does not put any particular stress on the goods of the soul. (For more on 'the soul', see pp. 18–19.)

[14] More accurately 'the life of *theōria*', which Ross translates as 'contemplation' and Irwin as 'study'. I discuss the question of translation on pp. 190 and 198–9. Meanwhile, I shall use the transliteration 'theoretical' in describing this life.

[15] The word used means 'grazing animals' (as Irwin translates), e.g. cows and sheep. It seems an odd suggestion that such animals devote their lives to what humans call 'bodily pleasures'.

virtue,[16] against which he offers the merely captious objection that one can be of a virtuous disposition even though one never *does* anything at all—an objection that he answers himself at 1098^{b}30–1—and the more interesting objection that virtue is compatible with great misfortune while *eudaimonia* is not (1095^{b}22–1096^{a}2). With this, he appears to dismiss the life of politics, or honour, or virtue, and moves on to the theoretical life, which, however, is reserved for later treatment (1096^{a}2–5). The later treatment promised is given in book X, chapters 6–8, where in fact we find a further treatment of the other two lives as well. So I reserve comment until we come to this (Chapter IX). Finally, he remarks that wealth clearly should not itself be regarded as an ultimate end, since it is useful only for what one can do with it (1096^{a}5–7).

The whole of this discussion should be regarded simply as making a prima-facie case for the opening claim that while people are agreed on what the ultimate end is to be called, namely *eudaimonia*, they are not agreed on what in practice this comes to. While it is true that Aristotle concludes the discussion by saying that *all* the traditional views on the topic have been refuted (1096^{a}9–10), he evidently cannot mean this seriously, for clearly he has not offered any objections to the theoretical life. And, as I mentioned, the other two lives will also receive a more extended treatment later. What one should infer is that, in Aristotle's view, one will not make adequate progress on this issue just by considering the traditional views. We need, then, a fresh approach, and this is given in the central chapter of the book, i.e. chapter 7.[17]

The chapter begins with a recapitulation. What we are seeking is the (ultimate) good (for man), which is what men (ultimately) aim for. Such an end must be 'complete',[18] but at present it is an open question whether there may be more than one 'complete' end. If so, then what we are seeking is the 'most complete' of these (1097^{a}15–30). Aristotle goes on to stipulate that one end counts as 'more complete' than another if either (*a*) it is pursued for its own sake, whereas the other is pursued (only) for the sake of something else, or (*b*) it is pursued *only* for its own sake, whereas the other is pursued both for its own sake and for the sake of it. Such an end is unconditionally 'complete', i.e. one that is pursued always for its own sake and never for the sake of something else (1097^{a}30–4). It should be noted that these definitions do not by themselves ensure that there can be only one end which is 'most complete' or 'unconditionally complete'. But Aristotle now argues directly that there is only one such, namely *eudaimonia*. For this, he says, is chosen always for its own sake, and never for the sake of anything else, whereas other candidates that one might suggest—and he mentions here honour, pleasure, understanding,[19]

---

[16] The Greek word is *aretē* and 'virtue' is the traditional translation, but it can be misleading. See p. 20.

[17] I relegate to an appendix the refutation of Plato's views in ch. 6.

[18] The Greek word is *teleion*; alternative translations would be 'perfect' or 'final'. Literally, the word just means 'end-like'.

[19] The word is *nous*. ('Understanding' is Irwin's translation; Ross here (but not elsewhere) uses 'reason'; for some discussion of this word, see pp. 76–7.) It is possible that it is here used to represent the goal of the theoretical life.

and every virtue—are chosen both for their own sakes and for the sake of *eudaimonia*. We choose them for their own sakes because 'we would choose each of them even if it had no further result', but we also choose them because we think that through them we will become *eudaimōn* (1097ᵃ34–ᵇ6). It is difficult to see what else he could mean by this than that we choose these things because they *contribute* to *eudaimonia*, and in that case it seems that *eudaimonia* must somehow *include* within itself all those other ends that are (correctly) pursued 'for their own sakes'.[20]

This interpretation is strongly confirmed by what immediately follows, where Aristotle adds that 'the complete good' must be something that is 'self-sufficient'. This means that all by itself it makes one's life choiceworthy, and lacking in nothing. It is not just one good among others, which could be improved by the simple addition of further goods, for if that were so, it could not rightly be called 'complete' (1097ᵇ6–20). It must, then, include within itself all other (genuine) goods, and that condition, says Aristotle, is satisfied by *eudaimonia*, and by nothing else. So it and it alone is the goal of all that we do (1097ᵇ20–1).

At first sight, this may seem merely platitudinous: it says only that the completely good life must simply be the best life, and therefore not capable of improvement. On a little reflection, the thought seems quite astonishing: could there be any life at all which *could not* be improved in any way whatever? To take a banal example, however much money I have, and to whatever excellent purposes I put it, could I not always have had just a bit more, enabling me to achieve just that bit more with it? (It is good, no doubt, to be able to donate £10 million to set up some charitable institution; but, if I had had twice that sum, I could have done the same thing twice over—in different countries, say—and would that not have been even better?) More generally, the opportunities in one's life will inevitably be limited in many ways, and often by factors one cannot control. Can we even envisage a life in which the alteration of these limits would allow of no room for improvement? The ideal of the *best possible* life (for man) seems merely quixotic, and yet if the text is read as I suggest, that must be what it implies.[21]

Anyway, whatever this 'complete good' is, Aristotle apparently claims that we do all aim for it: it is our ultimate goal. Let us understand this in the only way that seems reasonable, as meaning that we each aim for the best life that is possible for *us*, *given* our limited opportunities, our limited capacities, and our own particular circumstances.[22] Even here I think that one may reasonably question the idea that there is such a thing as a 'best' life, given these constraints, but—quite apart from this—the claim that we *do* aim for it, in all that we do, seems hopelessly optimistic. In the *Eudemian Ethics* we find a different claim:

[20] I take it that Aristotle means that we *simultaneously* pursue these goals for both reasons. Kenny supposes that his meaning is that on some occasions we pursue them for the one reason and on other occasions for the other (1965–6b: 96–7; 1992: 8). This seems to me implausible, for what would mark the distinction between the one kind of occasion and the other?

[21] That is a reason for trying to read the text in some other way. I take up this question in Sect. 5.

[22] As Heinaman (1993) stresses, I may rationally pursue the best life possible for me, even though I recognize that I shall never achieve what Aristotle calls *eudaimonia*.

Everyone who can live according to his own choice should adopt some goal for the noble life, whether it be honour or reputation or wealth or cultivation, an aim that he will have in view in all his actions; for, not to have ordered one's life in relation to some end is a mark of extreme folly. (1214^{b}6–11)[23]

Here Aristotle says not that we *do* adopt some one aim in all that we do, but that we *should*. Most of us, however, would come under his condemnation of 'extreme folly'. For though there are some important decisions about our lives that (usually) we take very consciously—for example, on a career, on whether (and whom) to marry, on whether to have children (and how many)—still most of our ordinary day-to-day decisions have no such large aim in view. They are taken simply with a view to how best to meet the immediate problem, and the effect on one's life overall is simply not considered.

I think we should conclude that although, in the *Nicomachean Ethics*, Aristotle apparently says that we all do adopt some one goal in all that we do, still he is best construed (as in the *Eudemian Ethics*) as recommending that we should. This goal is so far specified simply as 'the best life', but—to be more realistic—we should surely understand this as meaning, for each person, the best life available to *him*, given his particular circumstances, capacities, and opportunities. Aristotle will admit that for 'the best life' *absolutely* (if there is such a thing) many things are required which no one can adopt as a goal, e.g. good looks, good birth, and good fortune. He might have added some others, e.g. good teachers, and good intellectual capacities. But, however this list is filled in, he must presumably be taken as recommending that each of us should aim for what is the best life available to him, given all these constraints which he cannot alter. If the constraints are too severe, then such a life will not be *eudaimōn* however hard we try (1099^{a}31–^{b}7). But Aristotle does think that some of us are capable of achieving *eudaimonia*, so let us come back to the question of what it really is, for—as he has noted himself—there are many different opinions on the topic. He now goes on to give his own answer, at least in a preliminary outline. (As I have observed, there is further discussion to come in book X.)

4. The good for man and the 'function' of man

This brings us to the famous argument about the 'function' (*ergon*) of man. The general strategy of this argument is quite clear. Where a thing has a function, its good, i.e. its doing well, is to be found in that function. What this means is shown by the examples given. The flute-player has a function, namely to play the flute, and so the good flute-player is one who plays the flute well. Or again, the eye has a function, namely to see, and so the good eye is one that sees well. Aristotle is going

---

[23] The Greek text for the first line quoted here is uncertain, and the word 'should' arises from an emendation to a sentence which otherwise appears ungrammatical (cf. Woods 1982: 200). But in any case it is clear that Aristotle is *recommending* the adoption of some goal, since he goes on to imply that there are some fools who fail to do this.

to apply this line of thought to man. He will argue that man has a function, so the good man is the one who performs this function well, and by determining what this function is we will have our account of the good life for man (1097^{b}22–8).[24]

First, then, let us consider the claim that man has a 'function'. Aristotle makes little attempt to argue for this. Clearly one can admit that the various special skills he cites—i.e. flute-playing, sculpting, carpentry, leather-working—do have functions, so that those who practise them may also be said to have those functions, without supposing that the same applies to man as such. For being a man does not appear to be a similar and special kind of skill. Similarly, it would seem that we could admit that the various parts of the human body—the eye, the hand, the foot, and so on— have functions, without supposing that this applies to the human being as a whole. At any rate, most of us are prepared to accept the first claim while denying the second. But from Aristotle's own perspective there is more in this line of argument than at first meets the eye. It is part of his overall teleological approach to nature in general, and to biology in particular, so I here fill in a little background on this.

When we say that the eye has a function, we mean that there is something the eye is *for*, namely seeing. In his biology, Aristotle seldom uses the word 'function' (*ergon*), but usually speaks just of what something is for, or as we may also say its purpose.[25] Aristotle was the first to see that this notion is crucial in biology. (For example, why do we say that a human eye and an insect's eye are both *eyes*? After all, their anatomy is very different. The answer is evidently that each serves the same purpose.) He also saw that by saying what some part of the body was for one was (at least partially) explaining why the animal had that part. So this leads us to think of a chain of 'what is it for?' questions. For example:[26] what is the heart for? It is for pumping the blood, so that it circulates around the body. And what is the purpose of that? It is for distributing food and oxygen around the body. And why is that needed? Well, without it the muscles could not contract, and … As we go on pressing the question 'And what is *that* for?' we must apparently reach some *ultimate* purpose, which everything else is for. As Aristotle conceives it, the final stage comes when we ask about the body as a whole, with all its various capacities, what *it* is for, and the answer is that it is for living a certain kind of life. That is the ultimate goal. An animal's life is not *for* anything further,[27] but is the final reason why it has

---

[24] This account of Aristotle's overall strategy is almost universally accepted, but an exception is Everson (1998). He thinks that Aristotle wishes to include in the good life for man both external goods and theoretical activity, though *neither* of them is required by, or part of, man's 'function'. (But the former can surely be included as necessary conditions, and the latter is surely meant to be included by 1178^{a}2–8.)

[25] I suspect that the word 'function' is used here in the *Nicomachean Ethics* partly because this argument takes the place, in Aristotle's discussion, of a rather different argument about functions in the *Eudemian Ethics* (1218^{b}37–1219^{a}39). This different argument starts from the premiss that the function of the soul is to make one live, and it is a version of an argument found in Plato's *Republic* (352e–354a) which also speaks of the function of the soul.

[26] This is a non-Aristotelian example. (Aristotle did not know of the circulation of the blood.)

[27] In one quite untypical passage Aristotle suggests that other animals exist for the sake of man (*Politics* 1256^{b}15–22), but this is best viewed as an aberration from what is clearly his standard biological position.

all the various parts that it does have, arranged as they are, and functioning as they do.

One must at once add that different kinds of animals have different lifestyles, which explains why they have the different parts that they do. Thus the parts of a cow are as they are because that is what is suited to a cow's life; the very different parts of a bumble-bee are as they are because that is what is suited to a bumble-bee's life; and (Aristotle is very ready to add) the parts of a cherry tree are as they are because that is what is suited to a cherry tree's life. In general, then, each species of living thing has its own distinctive kind of life, which is the ultimate purpose of that species, and explains (at least partly)[28] all the various features of that species which differentiate it from other species. A further detail may be added. Aristotle holds that where a thing has a purpose, that purpose should figure in its definition. Thus one does not really understand what a knife is unless one knows that knives are for cutting with; one does not understand what an eye is unless one knows that eyes are for seeing; similarly, then, one does not really understand what a cow is, or a bumble-bee, or a cherry tree, unless one also knows the distinctive lifestyle that each has, for this is its purpose. Now let us apply these thoughts to the *Ethics*.

Man is a distinct species of animal, and there must therefore be a distinctive way of living which is the human way of living, for this is what distinguishes man from other species. It should form part of the definition of man, for one does not really understand what a man is unless one knows what the specially human way of life is. It may also be regarded as what man is for, the purpose or goal of man, and perhaps also man's 'function'. In consequence, the 'good man' is the one who well fulfils this particular kind of life. So, in a way, Aristotle's view is that the 'good man' is one who is 'good at being a man', and in this way we have an analogy to the way in which the 'good flute-player' is one who is 'good at playing the flute'. But there is also a slightly different analogy to draw on: the 'good man' may also be compared to the 'good cow', or the 'good cherry tree', each being a good specimen of its kind. But in each case what is required is not just a good physique (which is how we are likely to think of a good specimen), but a good life, a life which well exemplifies the particular kind of life appropriate to that species. With so much by way of elucidation of the claims that man has a function, and that the good man is the one who performs it well, let us now move on to Aristotle's account of what that function is.

His first move is surprising: we are looking, he says, for what is *peculiar* to man, and on this ground he sets aside as irrelevant those features of the human life which are shared with other animals, e.g. the fact that humans take nourishment and grow, and that they perceive (1097^b33–1098^a3). But, one at once objects, these are features of a human life, and any man who lacks them—for example, one who is blind, deaf, and paralysed—can hardly be called a 'good specimen' of the human species. However, the point is of no real importance, for we would not much

---

[28] In his biological practice, Aristotle is also prepared to assign some role to what we may call 'the laws of matter', for—in a metaphor which he often uses—nature can only do her best for the animal in question within limits imposed by the kind of matter available. But in his more theoretical discussions (notably *Physics* II.9 and *De Partibus Animalium* I.3–4) he tends to overlook this point.

disturb the argument if we insisted on including such features as these in the best life for man, so long as we also admit that they are not what is central to that life. (This, indeed, seems to be Aristotle's own attitude when he admits that *eudaimonia* requires such things as good looks (1099^{b}2–4) and good health (1178^{b}33–5).) As for what *is* central, and what alone Aristotle does include, this he says is 'some kind of life, involving action, of the part of us that has reason (*logos*)'[29] (1098^{a}3–4). This requires some elucidation.

The thing that 'has reason' is a particular part of 'the soul' (as is made explicit a few lines later). Now 'soul' is a merely conventional translation of the Greek word *psuchē* (or *psychē*), and it can be a very misleading one. This is not the place for an account of Aristotle's very puzzling views on the soul,[30] but one must say something, to guard against gross misconception. First, then, Aristotle takes it to be obvious that every living thing has a soul, and living things include not only the lower kinds of animals (e.g. shellfish) but also plants. He distinguishes different kinds (or 'parts') of soul, according to the different ways in which living things manifest life. Lowest in the scale is what he calls 'the nutritive soul' which is primarily correlated with being nourished by food, though he also associates it with growth and with reproduction. All living things have this kind of soul. Next comes 'the perceptive soul', and this is what distinguishes animals from plants, for animals all have some kind of perception, even if only the most primitive kind, which is touch. After that there is 'the locomotive soul', which distinguishes those animals that can move from one place to another, and finally 'the thinking soul', which (in Aristotle's view) belongs to man, but to no other kind of animal. Since the *Ethics* is wholly concerned with this part of the soul, we may henceforth ignore the others, and here if one wishes to translate the word as 'mind' rather than 'soul', no great harm is done. Aristotle's standard description of it is 'that which has reason (*logos*)'.

In I.13 Aristotle distinguishes two 'parts' of this. In fact he begins just by distinguishing a part that has reason from a part that does not (1102^{a}26–32), and then introduces a subdivision within the latter. Part of it is common to all living things, and we may set this aside as irrelevant to the specifically human excellences (1102^{a}32–^{b}12). But he contrasts this with another part which does partake in reason in a way, namely because it can 'listen to reason' (1102^{b}13–1103^{a}1). This part, as we shall see, contains desires and emotions, and because in human beings these can 'listen to reason' he counts it as a specifically human part. Finally, he says on the

---

[29] *Logos* is a word of many meanings. Ross frequently renders it as 'principle', or 'rational principle' (as here), or 'rule', or something of the kind. While no doubt the word can have this meaning, I think it would nowadays be generally agreed that this is not what Aristotle has in mind, and that in his case 'reason' is altogether preferable as a translation. (Thus Irwin.) Clearly the two translations carry very different suggestions. Ross's preferred rendering seems to arise, at least partly, because Ross *himself* was an 'intuitionist' in ethics, believing that the basis of ethics is our intuition of general rules and principles. But there is no reason to suppose that Aristotle too was an 'intuitionist' in this sense, and for him the word is much more closely associated with arguing, inferring, giving reasons, and so forth. In any case, as we shall see, in this discussion the notion of 'the part that has reason' should be widely construed.

[30] I have offered a brief account of the central problem in my (1994: 141–5).

same ground that it may be regarded as a part of the soul that *does* have reason, though not 'in itself' (1103ᵃ1–3). Within the specifically human part of the soul, then, the part that Aristotle broadly describes as 'what has reason', are included both the reasoning or thinking part proper, and the part containing desires and emotions that can be controlled by it. As soon as he has introduced the idea that the 'function' of man is a life of the part of the soul that has reason, he at once notes (in what is evidently a reference forward to I.13) that both of these parts are included (1098ᵃ4–5). This is certainly a very broad way of construing 'what has reason', and the conclusion to draw is that Aristotle is not intending, at this stage of his argument, to focus on some *one* feature that is specifically human; he means to include them all.

A similar remark should be made about his description of the life in question as one that 'involves action' (*praktikē tis*, 1098ᵃ3). We shall see that in VI.1 (1139ᵃ3–17) Aristotle will further subdivide the part of the soul that has reason 'in itself' into one part that engages in 'practical reasoning', i.e. reasoning that leads to action, and another that is confined to 'theoretical reasoning'. So it might be suggested that here in I.7 he means to include only the former and not the latter.[31] But this would surely be to misconstrue his intention. We can see this from the way he continues. He remarks that the notion of 'a life' can be taken in two ways, either as the disposition to do certain things (which one retains even when asleep—for a sleeper is still alive), or as the 'activity' (*energeia*) that manifests that disposition. He wishes to be understood in the second way: the life that is man's 'function' is to be understood as an activity, and not a mere disposition. Now Aristotle's use of the word 'activity' is sometimes puzzling,[32] but at least this much is clear: theoretical thinking is (for him) just as much an 'activity' as practical thinking is. So the fact that he is prepared to switch from 'action' to 'activity' is quite enough to show that he did not mean to be restrictive when characterizing the best life for man as one 'involving actions'. This is not intended to rule out thought of a purely theoretical kind, for that too is evidently a feature of human life that is not shared by other animals.[33]

Aristotle's claim, then, is that the life that is man's 'function', the life that a man is *for*, the distinctively human life, is an activity of the part of the soul that has reason, where 'reason' is to be broadly construed. (At 1098ᵃ7–8 he expresses himself even more vaguely: it is an activity of the soul that is in accordance with reason, or (anyway) not without reason.) Applying the main strategy of his argument, he infers that the good man is one who lives this life well, i.e. 'with excellence' (*aretē*). But we must pause again for a gloss on this word 'excellence'.

---

[31] The suggestion may be found in Joachim (1951: 50), in Broadie (1991: 36), and in Everson (1998: 94).

[32] For more on the word 'activity', see Ch. VII (*passim*). It may seem odd that Aristotle is prepared to describe living as itself *an* activity. But undoubtedly he does. Cf. X.4, 1175ᵃ12.

[33] Aristotle does occasionally call theoretical activity an 'action', e.g. *Politics* VII.3, 1325ᵇ16–23. The same is implied by *Metaphysics* Θ.6 (in a passage which I discuss below, pp. 152–4), where both thought and perception are given as examples of 'actions' which are 'activities' (1048ᵇ18–28).

The traditional translation is 'virtue',[34] but this so strongly suggests to us *moral* virtue that a note of caution is required. While the Greek word does of course cover what we regard as the moral virtues, it is not by any means confined to them. We must think of the word more broadly, so that we can say, for example, that the virtue of a knife is to cut well, the virtue of an eye is to see well, and the virtue of the reasoning part is to reason well. This does not *mean* to reason morally. As we shall see later (pp. 97–100), Aristotle does seem to be committed to the view that one who reasons well, whose reasoning contains no errors, will in fact reason morally. But that is not what his words mean here, when he characterizes the best life for man as one which involves reasoning 'well', or 'with excellence', or 'with virtue'. One can see this from the analogy, which he draws once more, with musical skills. The good harpist is one who plays the harp 'well', or 'with excellence', or 'with virtue', but clearly the 'excellence' here is a technical excellence rather than a moral excellence. So too, then, with the good man: it is an intellectual skill that is in question, and not—or not only—a moral disposition.

But does 'excellence' here at least *include* what we might view as moral excellence? Well, presumably it does. For Aristotle has said that he is including here the part of the soul which 'has reason' only in the sense that it can obey reason, and by this he means the part of the soul that contains desires and emotions. He goes on to describe the 'excellences' of this part of the soul in some detail in books II–V, and almost everywhere[35] he seems to be describing something that we would recognize as a moral virtue. There are, then, many 'human excellences', and some of them do coincide with what we might call moral virtues, but some of them certainly do not (e.g. the 'theoretical wisdom' discussed in book VI, and highly praised in X. 6–8). And apparently Aristotle's position here in I.7 is that *all* should be included in that life which is man's 'function'.

At any rate, this is what is required by his constant appeals to the analogy with the flute-player and the harpist, or indeed the eye and the hand. No doubt there are many excellences involved in being a good harpist—for example, one needs good sight-reading, good rhythm, good fingering, and above all this something that I shall simply call expressiveness. The good harpist, clearly, should have them all. So when Aristotle says of the good life for man that it is 'an activity of the soul in accordance with excellence, and if there are many excellences then in accordance with the best and most complete of them' (1098ª16–18), we should surely interpret this in the same sense. By 'best and most complete' he should mean an excellence which itself *includes* all the others. This may be because 'most complete' here just *means* 'all-inclusive', as it appeared to do earlier when he was speaking of an end or goal that is 'most complete' (1097ª28–ᵇ21). Or it may be because what we think of as one particular excellence (i.e. theoretical wisdom?) somehow brings all others with it. At any rate, the argument does require that all the specifically human excellences should be included in that life which is man's 'function'.

---

[34] Irwin uses 'virtue' as a translation throughout and so did Ross in his own translation. But the revision by Urmson has 'excellence' in this passage (though it soon slips into 'virtue' elsewhere).

[35] Not quite everywhere. See ch. II, note 11.

That completes Aristotle's first discussion of 'the good for man'. (In I.8–12 he attempts to show that his account harmonizes with the 'common opinions' on the matter, at least when these are sympathetically interpreted. While he himself regards this as an important part of his task (see Chapter X), I shall pay it no further attention here. In I.13 he opens his discussion of what, in detail, the specifically human excellences are.) But before I come to comment on this account, I must note that many would disagree with me on what that account is.

5. Rival interpretations

The interpretation that I have been giving is mainly due to Ackrill (1974), but it is far from being uncontroversial.

Much of today's debate begins from a distinction drawn by Hardie (1965) between what he called an 'inclusive' and a 'dominant' interpretation of one's ultimate end. On his account, an 'inclusive' interpretation allows that one's ultimate end may include a number of different things, each pursued for its own sake, and it should also include a long-term plan which ranks them in relative importance and attempts to secure the harmonious realization of all of them at once, so far as this is possible. By contrast, a 'dominant' interpretation picks one goal in particular as the main goal, and it downgrades or indeed rejects other aspirations, except in so far as they are seen as contributing to the main goal. Hardie's own position was that Aristotle fails to distinguish between these two conceptions of one's ultimate end, and that what he says requires us to take *eudaimonia* now in one way and now in the other. In subsequent discussion this notion of an 'inclusive' end has come to be understood simply as an ultimate end which includes within itself many different goals, each pursued for its own sake. Hardie's (obviously important) idea that it should also include a plan which assigns them relative weights, and works out how they are to be harmonized, has played little role in the discussion of Aristotle, for the evidently good reason that Aristotle himself never anywhere addresses the problem of how to resolve conflicts between two competing ends, each desired for its own sake. If we may continue to overlook this aspect, then the interpretation of I.7 that I have given does credit Aristotle with an 'inclusive' conception of man's ultimate end, for this is taken to include activities in accordance with each of the many human excellences, each pursued for its own sake (as well as for the sake of the whole to which they contribute). This, it seems to me, is what the argument about man's 'function' requires. But there is of course room for a yet more 'inclusive' interpretation, by which *eudaimonia* also includes *everything* that contributes to 'the best life', which may well seem to be more than what is either a necessary condition for the pursuit and practice of the specifically human excellences (virtues) or a necessary consequence of this.

If we consider just book I on its own, and pay no attention to the later discussion of *eudaimonia* in X.6–8, then I am sure that some 'inclusive' interpretation is the most probable. I remark also that it is (I believe) universally agreed that an 'inclusive' conception is just what we do find in the *Eudemian Ethics*. It is true that

1214^b6–11 (quoted earlier, p. 15) seems to suggest that each of us ought to adopt some one 'dominant' end—e.g. honour or reputation or wealth or cultivation—but later passages on *eudaimonia* make it clear that it is taken to include *all* the human excellences, both virtues of character and virtues of intellect.[36] However, here in the *Nicomachean Ethics* we do have to reckon with X.6–8, where it certainly appears that Aristotle wishes to identify *eudaimonia* with one activity in particular, namely the activity that expresses the highest human excellence, theoretical wisdom. This would then be a 'dominant' conception of man's ultimate end. I postpone to Chapter 9 my discussion of Aristotle's position in book X—for certainly the interpretation of that book is not altogether straightforward—but in any case the problem of harmonizing books I and X with one another gives us good reason to consider whether book I can, after all, be seen as proposing a 'dominant', rather than an 'inclusive', conception of *eudaimonia*.

There are four main passages in I.7 that are relevant to this issue. I take them one by one.

1. 1097^a30–4 explains what it is for an end to be 'most complete' (or 'most perfect', or 'most final'). The important part of the explanation is:

> *x* is more complete than *y* if
> (*a*) *x* is pursued for its own sake, and *y* is not, or
> (*b*) *x* is never pursued for the sake of anything else, whereas *y* is pursued both for its own sake and for the sake of *x*

It is quite tempting to suppose that in clause (*b*) Aristotle has not said quite what he means, for he means '. . . whereas *y* is pursued both for its own sake and for the sake of *something* else (not necessarily *x*)'.[37] Granted this, we may then note that in X.7 it is claimed that theoretical activity is never pursued for the sake of anything else, whereas the activities corresponding to other excellences—justice and courage are mentioned in particular—are pursued for something else, namely for the sake of the good outcomes that they aim to achieve (1177^b1–24). To be sure, X.7 does not say that justice and courage are pursued for the sake of theoretical activity, but (i) if we reconstrue the passage in I.7 as suggested, this does not matter, and one might add (ii) that the logic of Aristotle's position in X.7 does actually commit him to the view that justice and courage are pursued for the sake of theoretical activity (though whether he realized this himself is another question; after all, it is a distinctly unexpected claim).

2. 1097^a34–^b5 claims that we pursue honour, pleasure, understanding (*nous*), and every virtue both for their own sakes and for the sake of *eudaimonia*. If this means, as I suggested, that we pursue them as parts of *eudaimonia*, then apparently *eudaimonia* must include more than just (the activities expressing) the virtues, their

---

[36] See in particular 1219^a35–9, 1220^a2–4, 1248^b8–16, and for a general discussion Cooper (1975: 115–43). (I add that *EN* V.1, 1129^b18 casually implies that *eudaimonia* has 'parts', and I note that this is a 'common book'.)

[37] This is how Aristotle himself seems to treat the condition in the next line (1097^a33–4). Heinaman (1988) takes this interpretation.

necessary conditions, and their necessary consequences. For honour seems not to satisfy this description (though pleasure does).[38] But there is an alternative way of construing this passage. It is quite plausible to suppose that honour, pleasure, and understanding are mentioned here as the respective goals of the traditional 'three lives', the political life, the life of pleasure, and the theoretical life. Then we may suppose that 'every virtue' is added mainly because in Aristotle's view this is a better specification of the goal of the political life (1095^{b}22–31). (It would incidentally include the theoretical virtues too.) But now, if this is the right explanation of the goals listed here, then Aristotle need not be taken as saying that we—i.e. all of us—pursue them all. On the contrary, his thought may be that some pursue one while others pursue another, in each case thinking that the single goal they pursue is not a part of *eudaimonia* but the whole of it. So construed, the passage is not after all an argument in favour of the inclusive conception of *eudaimonia*.

It is a consequence of this construal that pursuit of the goals listed is mistaken. For the 'pleasure' that is here in question will then be the 'bodily pleasure' that is the goal of what Aristotle describes as the life of pleasure, and evidently he thinks that that is not what *eudaimonia* really is. He has also argued that neither honour nor virtue can be identified with *eudaimonia* (1095^{b}22–1096^{a}4), and though he was thinking at the time of virtues of character the same objection would apply too to the theoretical virtues. So on this account none of the goals mentioned is *correctly* pursued for its own sake, save possibly understanding. This seems to me to make the interpretation implausible. For the statement that we would choose each of these goals even if nothing further resulted from them does not at all naturally suggest that 'we', who would do this, are all mistaken. But perhaps the interpretation is not impossible.

3. 1097^{b}6–16 claims that the 'complete' (or 'final') good must be self-sufficient, and explains this as 'that which, on its own, makes life choiceworthy and lacking in nothing'. There is no ambiguity here, and it is very difficult to see how a single 'dominant' end, such as theoretical activity, could satisfy this description. But it may be held[39] that the succeeding sentence, which *is* somewhat ambiguous, is supposed to explain this condition. Since the sentence is ambiguous, I give a very literal translation:

We think *eudaimonia* is [self-sufficient], and further that it is most choiceworthy of all things, not being counted together [with other things]—but, counted together, evidently it [is/would be] made more choiceworthy when taken with the smallest of [other] goods. (1197^{b}15–18)

If *eudaimonia* is taken to be a fully inclusive end, including all goods within itself, then the point of this sentence must be that it *cannot* be 'counted together with other goods', since that would involve counting the same thing twice. But a rival interpretation[40] takes the point to be that *eudaimonia* does not already include all

[38] At IV.3, 1123^{b}20–1 honour is described as 'the greatest of the *external* goods', but it is inextricably bound up with the supposed 'virtue' that Aristotle discusses in that chapter.

[39] Heinaman (1988) apparently does hold this.

[40] Preferred by Heinaman (1988: 42–7), S. A. White (1990: 119–24), Kenny (1992: 24–9).

goods, but may be considered as one among many goods, and when so considered it is the most choiceworthy of each of them taken singly. But we *can* 'count it together' with those other goods, and in that case it must of course be admitted that *eudaimonia* + *x* is more choiceworthy than *eudaimonia* by itself, wherever *x* is a good (however small) that is not already included in *eudaimonia*.

Clearly, on the second interpretation *eudaimonia* can be taken to be a single 'dominant' end, or it can be taken to be an 'inclusive' end, including many activities (e.g. all those that manifest the specifically human excellences), so long as these do not include all goods. But I prefer to suppose that Aristotle is thinking as the first interpretation proposes, because (*a*) we otherwise have no explanation of why he should claim that *eudaimonia* by itself makes life 'lacking in nothing', and (*b*) elsewhere he argues himself that *x* cannot be 'the good' if *x* + *y* is better than *x* by itself,[41] and he does take *eudaimonia* to be 'the good' (for man). Admittedly, this does have the consequence that Aristotle is committed to the very implausible claim that some life possible for man is 'the best' life, and capable of no improvement (p. 14). It also points to a considerable gap in his account, in so far as he never even considers how we are to choose between distinct ends each desired for its own sake (p. 21). But what is the alternative? It can only be that when he says that *eudaimonia* makes life 'lacking in nothing' he means no more than that it makes life 'lacking in nothing that is essential to *eudaimonia*'. But that is a mere tautology: of course the presence of *x* entails the presence of everything essential to *x*, whatever *x* may be.[42]

4. At 1098ᵃ16–18 the 'function' argument concludes with 'the human good is an activity of the soul in accordance with excellence, and if there are many excellences then in accordance with the best and most "complete" of them'. Undoubtedly, we do not *have* to suppose that 'most complete' here implies 'most comprehensive', i.e. 'including all excellences'. (This clearly *is* what the phrase 'complete excellence' means in the *Eudemian Ethics*, at 1219ᵃ35–9, but that is not by itself a strong argument for saying that it means the same here.) The main reason for supposing that this is what it means is that this is what the argument requires, yet there is a nagging doubt: if Aristotle had meant 'all', why did he not simply say 'all', instead of using this ambiguous phrase 'best and most complete'? Moreover, there is a later passage in book I where he says that *eudaimonia* should be identified with 'the best activities—these *or* that one *of them* that is the best' (1099ᵃ29–31). This strongly suggests that we have two distinct alternatives to consider, for the one activity that is the best is not naturally seen as *including* all the others, if it is described as one *of them*. There are two similar passages in the later discussions of pleasure, where again Aristotle seems to hesitate between saying that *eudaimonia* is many activities,

---

[41] The argument is borrowed from Plato (*Philebus* 20d–22c), and turned against Eudoxus' claim that pleasure is 'the good' (1172ᵇ28–34). It is quite clear that Aristotle *endorses* this argument, as I show on p. 146.

[42] I add that self-sufficiency also recurs in X.7 (at 1177ᵃ27–ᵇ1) but it seems clear that it is there differently conceived. As Kenny has put it (1992: 36): 'Whereas in the early book the dominant sense of self-sufficiency was 'that which on its own makes a man *eudaimōn*', in the final book the dominant sense seems to be 'that which makes a man *eudaimōn* on his own'.'

or is one of those many (1153ᵇ9–12, 1176ᵃ26–9). This again is distinctly unnatural if 'one of them' is just another way of indicating 'all of them'.

I maintain my claim that, *if* book I were all that we had to consider, then an 'inclusive' interpretation of *eudaimonia* would be the more probable. (And we should set down 1099ᵃ29–31 as carelessly expressed.) But book I is not the full discussion, for we have more to come in book X, and several hints in between. So I am myself inclined to think that in book I Aristotle is quite deliberately using a phrasing which is less committed to the 'inclusive' interpretation than was his phrasing in the *Eudemian Ethics*, and is doing so in order to prepare the way for what will come later. But, of course, it will not come until *later*, and there is absolutely nothing in book I which limits the specifically human excellence to theoretical wisdom in particular. So for the time being it seems best to suppose that all the specifically human excellences are included, whether each in its own right or (implausibly?) because some one of them presupposes all the rest. I shall assume this when I come to comment on Aristotle's doctrine overall in the next section.

But one postscript should be added here. The 'inclusive' interpretation that I favour will say that *eudaimonia* includes all activities which manifest specifically *human* excellences. This does not include *all* the activities which we desire for their own sake, for a clear counter-example is the faculty of sight. On Aristotle's account, we do desire this for its own sake (1096ᵇ17), but it is excluded from human *eudaimonia* because other animals share it too (1098ᵃ2). There are also all those other goods which are necessary conditions of *eudaimonia* (for example, money), and which must therefore be included in the *eudaimōn* life. But here it seems best (*pace* Cooper 1985) to retain a distinction drawn in the *Eudemian Ethics*: something without which *eudaimonia* cannot occur need not, for that reason, be counted as a part of it (1214ᵇ24–7). And here we should add that some things—for example, sight?—facilitate *eudaimonia*, even if they are not (as Aristotle conceives it) strictly necessary conditions (1099ᵇ27–8), and some things—for example, honour?—are natural consequences, even if they are not strictly inevitable.

But let us now set aside these awkward questions of interpretation—which, as we have seen, inevitably bring in Aristotle's second discussion in book X—and come back just to the *main* idea of book I. This evidently is that if we wish to discover what really is the good life for man, then we must look to those activities of men that manifest a specifically human excellence. How should we react to this idea? How is it related to the considerations with which book I began?

6. Some reactions

I think the first thing to say is that there is really no relation, for the discussion divides into two quite distinct parts which are linked only by an ambiguity. The first runs (with many digressions) from the opening of book I to the first part of I.7, ending at 1097ᵇ21. Its main claims are that the good for man is what all men aim for, that there is indeed some one thing that is everyone's ultimate aim, and that we are all agreed on what to call it, namely *eudaimonia*. (But we are not agreed on what, in practice, it involves.) The second part is the argument about the 'function' of man, which occupies the middle part of I.7 (i.e. 1097^{b}22–1098^{a}20).[43] That there really is no connection between these two parts is easily seen. For Aristotle makes no attempt to show that what he calls the 'function' of man, i.e. the specifically human kind of life, *is* something that men aim for. And I would say that it is quite clear that they do not. At any rate people surely do not explicitly set before themselves the goal of being 'as human as possible'.

There is a purely *verbal* connection between the two parts, made by the words 'good' and 'man', but it is entirely spurious.[44] Aristotle begins by considering the good *for* man, which we may gloss as 'what most benefits a man'. With various qualifications (as indicated on p. 8), it is not too unreasonable to claim that men aim for what will most benefit them. But the argument about the 'function' of man considers not this notion but the quite different notion of 'the good man', which Aristotle construes as the one who is *good at* being a man (i.e. one who well exemplifies the specifically human kind of life). But why should one suppose that the good man (good at being a man) can be identified with one who secures the good *for* man (what most benefits a man)? There is evidently no reason at all, or if so then surely the reason must be given. But Aristotle, I think, fails to see that the notions are different. He constantly uses, and I believe thinks in terms of, a single phrase which is ambiguous between the two. (*to anthrōpinon agathon* may be translated either as 'human goodness'—i.e. what it is for a human to be good—or as 'human good'—i.e. what is good for humans.) So he does not even notice that, in juxtaposing the two parts of his discussion as he does, he is making a disputable assumption about how the two notions are connected. But it is quite clear that there *is* an assumption here, and it is absolutely basic to his overall position. I now make one brief remark about each part of the discussion separately, before bringing them together again.

Aristotle claims that each person seeks *eudaimonia* as his ultimate aim. This needs to be qualified, as we have said, so that it is confined to what one may call 'rational' actions, and even then it is perhaps better viewed not as a statement about what we do do, but as a recommendation about what we should do. But, considered either way, is it not open to this objection: Aristotle clearly means that each person seeks (or should seek) *his own eudaimonia*, but isn't that an essentially selfish view? I think the answer is basically 'yes', but there are some obvious points to be made first. Discussing the view that *eudaimonia* is supposed to be 'self-sufficient', Aristotle says that this does not mean that it excludes one's parents, children, and wife, or more generally one's friends and fellow citizens, for—in a famous phrase—'man is by nature a political animal' (1097^{b}8–11). (Ross rather nicely translates 'born for citizenship', which I think captures Aristotle's own meaning rather well. But we might perhaps broaden it a bit: it is man's nature to live in a society of other men.)

---

[43] Aristotle himself would count this second part as continuing to the end of I.12, but—as I have said—I shall ignore this continuation, which I think adds nothing of value.

[44] So far as I am aware, this point was first clearly made by Glassen (1957).

Aristotle clearly does not advocate the very narrow self-seeking that we usually mean by the word 'selfish'. Indeed, he frequently tells us that we should act 'for the sake of what is noble',[45] and the actions thus recommended are usually those that we would think of as benefiting others, rather than oneself. He also says, when discussing friendship, that a friend should be loved 'for his own sake'. So it is clear that he does not altogether deny the value of altruism, and he does keep a place for it in his 'good life'. But whether he gives it an adequate place, and whether his starting-point will allow him to do so, are serious questions. I do not pursue them here, since we must look first at these later discussions (particularly on friendship). All that I say now is that it is reasonable to *suspect* that his avowed starting-point—namely, that each man does, or should, pursue his own *eudaimonia*—will lead to a conception of the good life that is too egocentric.[46]

Even if the good life can be described as the pursuit of *eudaimonia*, it would not of course follow that it consists in fulfilling one's 'function' as a human being. Let us turn now to this second part of Aristotle's account.

We do not share Aristotle's thoroughgoing teleological approach to all aspects of biology, nor his presumption that species are fixed and eternal. From our own perspective within the theory of evolution, we can make sense of his claim that the various parts of an animal may each be regarded as serving a purpose (with a few exceptions, of the kind that he himself would have accepted), but we would not follow him in supposing that the 'ultimate purpose' was the kind of life that that animal leads. (If anything, our 'direction of explanation' would be the reverse of his: the animal lives as it does because of the way that its parts have evolved.) Let us, then, set Aristotle's biological theory on one side. Can we nevertheless extract some comprehensible claim from his argument about man's 'function'?

The best substitute that I can see is this: there is such a thing as 'human nature'. We may suppose that human nature has evolved from the nature of more primitive animals. We may add that in the millennia to come it may evolve yet further. But now, as things are, there is something that may be called 'human nature'. As to what this human nature is, it need not be so very different from what Aristotle describes as 'the function of man'. Of course human beings eat, and see, and walk. This *is* a part of human nature, even though it is a part which they share with other animals. But what is most distinctive about human nature is that humans have reason: they can *think*. (No doubt, as we now look at things, other animals can think too, at least to some extent. But it is enough to claim that human powers of thought far outweigh those of other animals, so that thinking is what most distinguishes humans from all other animals.) This, then, is the first premiss: there is such a thing as 'human nature', and it crucially involves thought.

We should add here a further rider, which our terminology has taken for granted,

---

[45] *dia to kalon.* I discuss this phrase further on pp. 98–100.

[46] As Annas (1993) demonstrates at some length, Aristotle's starting-point is by no means peculiar to him. It was shared by all the Greek moral philosophers of his period. But that is not a good reason for saying that we should share it too.

and which Aristotle's does too: *all* human beings share this same 'human nature'. Some, no doubt, are maimed, and cannot see, or cannot walk. These are unfortunate, for they cannot display their human nature to the full. More importantly some are 'intellectually maimed', for they cannot think very well. These too are unfortunate, and cannot display their human nature to the full. Much more commonly, many are disadvantaged, e.g. by being born into a poor environment, or by being in some other way denied the opportunities to develop as a human being could. For various reasons, then, many actual human beings cannot develop to the full the 'human nature' that they have. Nevertheless, they all do have a 'human nature', which could develop, if obstacles were removed.

On this version of the premiss that man has a 'function', we can no longer argue as Aristotle does. We cannot say that the life in accordance with human nature is what man *is for*, is the *purpose* of man, for we have no warrant for that. But that is just as well, for the argument would fall foul of the distinction we have noted between 'the good man' and 'the good for man'. So here we need to state openly as a second premiss: 'the good for man', i.e. the life that is best for a man, is the life that is most in accordance with his 'human nature'. No reason is given for this; it is just a basic assumption of the Aristotelian approach. Why, then, should one accept it? Well, as I have just said, there is no reason. The thought may be appealing, or it may fail to appeal, but in either case there is nothing more that can be said, either for it or against it.

Philosophers will, of course, protest that there is much more to be said. The hedonist will ask whether this life 'in accordance with human nature' will maximize the agent's pleasure; the utilitarian will ask whether, if all of us do try to live in this way, the consequence will be that the total sum of all human happiness will be maximized; the Kantian will ask whether we can will that acting on this maxim should become a universal law; and so on. But of course Aristotle is not a hedonist, or a utilitarian, or a Kantian, so from his point of view these questions are irrelevant. He is, if I may misuse a common word, a 'human-ist', for 'human nature' is what he starts from.

It is true that in his own version of this approach he does allow as relevant the further question 'Will this life be the best life for the agent?'; and he insists that the answer is 'yes'. But clearly there is room for a variation which claims (with Hume) that human beings are by nature altruistic as well as egoistic, and so it will not insist that the life in accordance with human nature always does prove best for the agent. Yet the 'human-ist' (as I am using the term) will still say that this is how human beings *ought* to live, for he takes human nature to be *the* basis for ethics. What should we say of this approach?

Well, first one may observe that its appeal will very much depend on how this 'human nature' is explained, and so far Aristotle has given us little by way of explanation, except in terms that may well be regarded as circular: 'human nature' is given by 'the specifically human excellences', whatever those may turn out to be. (Aristotle's own views on what counts as a 'human excellence' will emerge in the following books; I do not try to summarize them now.) But one might look at it very differently. For example, there is a well-known theological doctrine that man is born sinful, and consequently it urges us to transcend our human nature, not to conform to it. (A secular version of this approach urges us to transcend the natural 'selfishness' of our genes.)[47] What, then, *is* this supposed 'human nature', and how are we to discover it? But this leads naturally to a second question: is there really such a thing? I would say that there is not. No doubt there are purely biological descriptions of what is and what is not to count as a human being, but those by themselves give us little clue on what is to count as *acting* in the specifically human way. And how is this to be determined? Even today the lifestyles of different human races around the globe are so many and various that one would hesitate to ascribe a single 'nature' to us all. When we add a little historical perspective, and look back on, say, the Aztecs of Mexico, or the ancient Egyptians, or indeed the primitive hunter–gatherer people from whom we descend, does it not seem even less likely that there is a basic 'human nature' which we all share, and which is substantial enough for it to make sense to say: *that* is how you ought to live? Taking an even longer view, and looking at things from our own perspective within the theory of evolution (which, of course, Aristotle did not share), would one *expect* to find some shared 'human nature' in all who are counted as human beings?

As a last resort, the 'human-ist' may say: very well then, I accept that different human beings have different natures, but *each* should try to live in accordance with his or her *own* nature, for that is what is most 'natural'. But to this one can only respond that it is surely a recipe for chaos. There is no reason to suppose that what is 'natural' must also be good.

Of course, this leaves us with a question. If ethics cannot be based on something called 'human nature', which the biologist or anthropologist or sociologist might (in principle) reveal to us, then what can it be based on? But I must leave that question to you.

Appendix. Aristotle on the Platonic theory in I.6

Plato believed in objects which he called 'Forms'. These are intelligible but not perceptible, and he often calls them 'paradigms' (where we might rather say 'ideals'). As Aristotle understands this theory, it posits a Form of man as a paradigm (or ideal) man, perfectly exemplifying the definition of what it is to be a man, except that it is eternal and intelligible, whereas ordinary men are imperfect, perceptible, and transient (cf. 1096ᵃ34–ᵇ5). Aristotle often argues against this theory elsewhere, and I shall not repeat here the many points that he makes.[48] I shall concentrate upon what he has to say here about the Form of 'the Good' in particular.

The Platonic theory is that to call anything 'good' is to say that it stands in a certain relation—which Plato sometimes calls 'participation' and sometimes 'imitation'—to the Form of the Good, and he thus assigns the same meaning to the word 'good' in all its

---

[47] The classical exposition is, of course, Richard Dawkins (1976).

[48] His most consistent and prolonged attack was in his lost work *On the Forms*, recently reconstructed by Gail Fine (1993). But see also *Metaphysics* M.4–5.

occurrences. Aristotle claims that this is a mistake. His first argument[49] is that the word 'good' is said of many different types of thing, i.e. of things that are, in his own scheme, 'of different categories'. His examples are that a substance may be called good (such as God, or understanding),[50] and so may a quality (such as the virtues), and so may a quantity (such as what is moderate), and so on. Aristotle infers that since the word applies to things of such very different kinds it cannot mean the same in all cases (1096ᵃ23–9), but there is no obvious reason why we should accept this inference. (Aristotle, I suspect, argues *a fortiori*: it is certainly his view that the very general word 'being' (i.e. existence?) means different things for items in different categories (ᵃ23–4).[51] I believe he takes it for granted that if this holds of 'being', it must hold of every other word too, but it is clear that this does not in fact follow. There are many obvious counter-examples, most notably predicates of a more or less logical nature, such as 'has an opposite'.)[52] This first argument then, though it may be suggestive, is hardly compelling.

The second argument is that if 'good' had but one meaning, then there should be only one 'science' of it, but in fact many different sciences—e.g. generalship, medicine, gymnastics—are each in their different way concerned with what is good in their own sphere (1095ᵃ29–34). I remark first that as an argument against Plato this cuts no ice, for he had indeed supposed (*Republic* V–VII, *passim*) that there was one supreme knowledge, namely knowledge of the good, which the philosopher must attain, and that proper knowledge of *everything* else would depend upon this. But I remark also that in any case the argument is again a *non sequitur*. For example, from the fact that the dyer has one 'science' of making things look yellow, and the painter another, and the lighting designer yet a third, it evidently does not follow that 'yellow' means something different in each case.

After these two arguments Aristotle pauses for some quite general remarks on Platonic Forms (1096ᵃ34–ᵇ5), and then notices a pertinent objection. Someone might say that we should distinguish between things good in themselves and things merely instrumentally good, because they promote these things, or preserve them, or prevent their opposites, and so on. Then only in the former case will the word 'good' mean the same thing, i.e. participation in the Platonic Form of the Good (1096ᵇ8–16). Of course this suggestion will collapse if there is nothing that is good in itself except for the Form (ᵇ19–20), but we may reasonably suggest as different things good in themselves thought and sight and certain pleasures and honours, for we do pursue these for their own sakes, even if we also pursue them for

---

[49] I omit 1095ᵃ17–23, which opens with the claim that 'those who put forward this doctrine' do not posit a single Form for things which, though all called by the same name, exhibit among themselves an order of priority and posteriority; for example, they do not posit a form of number. This is a merely *ad hominem* argument, and the *homo* in question appears not to be Plato himself (or at least, not in his published dialogues). (For elucidation, see e.g. Woods 1982: 76–8, commenting on the somewhat different version of this argument in *EE* I.8, 1218ᵃ1–9.) In any case, the main point of this preliminary skirmish is repeated in what I do call Aristotle's 'first argument'.

[50] No doubt God is a substance, but understanding (*nous*) is an unexpected example here. (Ross once more has 'reason'.) One might have expected something more ordinary, such as 'some *men* are called good', but Aristotle appears to be trying to give examples of things called good universally, and without qualification. (We should perhaps recall that in the *De Anima* and the *Metaphysics* Aristotle counts the soul as a substance, because it is 'the substance of' a living body, and *nous* is of course the best 'part' of the soul.)

[51] This is a puzzling doctrine. I have discussed it at some length in my (1994: 45–52, 65–8).

[52] Aristotle himself asks of each of his categories whether the items in it 'have opposites' (*Categories* 3ᵇ24–32, 5ᵇ11–6ᵃ18, 6ᵇ15–19, 10ᵇ12–25, 11ᵇ1–4), and the phrase evidently does not change its meaning from one category to another.

something else (^{b}16–19). But to this suggestion Aristotle replies, in a very obscure phrase, that 'the account of goodness' should then be the same in all of them, whereas in fact it is not (^{b}21–5). What does he mean by this? For surely it is his *own* view that these things are good in themselves, and pursued both for their own sakes and for the sake of the *eudaimonia* to which they contribute (1097^{b}2–5). And he never gives any explanation of how the 'account of goodness' is supposed to differ from one to the other. Indeed, since he himself equates 'good in itself' with 'pursued for its own sake', it seems that his *own* view must be that they are each good *in the same way*, namely by being pursued for their own sakes (and for the sake of the *eudaimonia* to which they contribute, which alone is pursued only for its own sake).

One can only conclude, I think, that he has misstated his point. He wishes to say that there is no such thing as the Platonic 'Form of the Good', which exists quite independently of what men may pursue. He denies that ethics has this kind of 'objectivity'. For him, our only understanding of goodness is as what men do pursue, so there is simply no possibility of it turning out that we all pursue the wrong thing. Goodness is not in this way independent of man's actual aspirations, and he quite correctly sees that he and Plato are at odds on this question. But his argument goes awry at its final step. He should say that what thought, sight, certain pleasures, and so on do have in common is just that we do pursue them (for their own sakes), and that is why they count as good. They do not *also* have in common participation in some Platonic Form that has nothing to do with men's actual pursuits. For if they did, that would require them to have a common account of their goodness that made no mention of men's pursuits, and this they do not have. But instead of saying just this, Aristotle says rather that they *do* each have *different* accounts of their goodness—i.e. accounts which make no mention of men's pursuits?—and so he leaves us completely at a loss as to what these different accounts might be. As I say, he has misstated what he really means to claim.

I remark that there is a different way in which he might have made his case against Plato. In this work Aristotle is concerned with the good *for man*, and the examples I have just been discussing do fall under this general heading. But Aristotle does not in fact think that the general notion of goodness is so confined. For example, there is also what is good for fish (1141^{a}23), and in addition we may say that God is good (1096^{a}24), and that nature and all its works are good, in particular its primary elements (1141^{b}1). This does not mean that these things are good *for man*. So what does it mean? It is here that we might invoke his speculation that these uses of 'good' are to be understood 'by analogy' (1096^{b}27–31), though clearly he has not actually worked out this, or any other, theory of how the word 'good' is to be understood in general. But he could not unreasonably claim that the Platonic theory, by which it always indicates participation in some one form, simply cannot be made to fit these examples where human goodness is not in question. In any case, it is the good for man that Aristotle aims to focus on in this work, something that is achievable by man (1096^{b}34–5), and is—he thinks—something sufficiently specific not to fall foul of any ambiguities that affect the general notion of goodness.

Further reading

On the notion of *eudaimonia* ('happiness') as an ultimate end, some useful and comprehensible points are made in Kenny (1965–6b), and in N. P. White (1981). There is a much more sophisticated treatment in McDowell (1980), but I do not recommend this for those studying Aristotle's *Ethics* for the first time. (As an antidote to McDowell I recommend Kirwan 1990.)

On the argument about man's 'function', I recommend Glassen (1957) and Siegler (1967) as each drawing attention to simple and straightforward objections. For a vigorous defence of the argument against these objections, one might see Hutchinson (1986, ch. 3, sect. 2), and for a calmer assessment, Whiting (1988).

The 'inclusive–dominant' debate has received much attention in the literature. It is good to begin with Hardie (1965), who introduces the issue, and then my selection would be Ackrill (1974) and Devereux (1981) for inclusiveness; Heinaman (1988) and Kenny (1992, chs. 1–3, esp. 3) for dominance; and possibly S. A. White (1990) for a compromise position. There is a response to White in Crisp (1992). Others that could be further pursued are Cooper (1975, chs. II–III); Price (1980); Engberg-Pedersen (1983, ch. 1); Irwin (1985*b*); Cooper (1987), revoking at least some of his (1975); Kraut (1989, *passim*); Reeve (1992, ch. 3). See further the reading suggestions for X.6–8.

On the question whether 'external goods' (e.g. money) are to be counted as ingredients in *eudaimonia*, there is a provocative discussion in Cooper (1985) and a response in Kenny (1992: pp. 38–42).

For those who wish to follow up Aristotle's criticism of Plato in I.6, it is essential to compare *EE* I.8, on which there is a useful commentary by Woods (1992). This provides much illumination. Otherwise, to understand Aristotle's invocation of his theory of categories, one might first look at a general account of this theory, e.g. as given by Ross (1924, pp. lxxii–xc), and then Kosman (1968) and Ackrill (1972), who discuss the application of this theory in the present case. But really this topic leads on into very general questions about Aristotle's metaphysics (as evidenced by Owen 1960), which it would be inappropriate to pursue here. As for the substantive question that is raised, namely whether the word 'good' does have several meanings, that again is too large an issue to be pursued here, but I might perhaps mention as an introduction von Wright (1963, ch. 1, sects. 5–8), and perhaps the debate between Geach (1956–7) and Hare (1956–7).

Chapter II

Virtues of character (Book II)

1. The parts of the soul (I.13)

ARISTOTLE has claimed that the good life for man is 'an activity of the soul in accordance with excellence', so his next task is to discuss the various excellences of the human soul. Henceforth I shall use the translation 'virtue' rather than 'excellence', both because it is the traditional translation, and because it is a rather better approximation to what Aristotle actually has in mind in most of his discussion. His account begins in I.13 with the two relevant parts of the soul, where—as I have noted—he distinguishes a part that has reason 'in itself' and another part that can be said to 'partake in reason in a way' (1102ᵃ26–ᵇ14). The distinction needs some elucidation.

Aristotle introduces it by pointing to a certain kind of *conflict* in the soul, very roughly a conflict between 'reason' and 'desire', where one pulls in one way and the other in another.[1] In the man who is self-controlled (*enkratēs*) reason wins, but in the man who lacks self-control (*akratēs*) desire wins.[2] His ground for saying that the desiring part does 'partake in reason in a way' is just that it can oppose reason, and can also obey it, and can (in the virtuous man) be in harmony with it (1102ᵇ14–28). But it seems that Aristotle misdescribes this part when he goes on to call it 'the desiring part and in general the appetitive part' (*epithumētikon, kai holōs orektikon*)[3] 1102ᵇ30). The position may be clarified by contrasting Aristotle's account with Hume's.

For Hume, reason is entirely distinct from what he calls 'passion', a word which he uses widely to cover any kind of desire. Reason can calculate what will be the consequences of acting in one way or the other, but it cannot by itself motivate one to act in either way, just because it does not include any kind of desire for these consequences. So, for Hume, there cannot be such a thing as a *conflict* between

---

[1] He borrows this approach from Plato (*Republic* 436b–439d). But whereas Plato professes to obtain three parts of the soul in this way, Aristotle is here content with two.

[2] Both Ross and Irwin translate *enkratēs* and *akratēs* as 'continent' and 'incontinent'. This seems to me a poor translation. See further p. 123.

[3] Both Ross and Irwin translate the other way round, i.e. with *epithumētikon* as 'appetitive' and *orektikon* as 'desiring'. Against this, it is the first that corresponds to an ordinary Greek word, standardly translated as 'desire', whereas the second is a technical term of Aristotle's own invention, and so suggests a quasi-technical translation such as 'appetition'. On the other hand, Aristotle does use his technical term *orexis* in a wide sense, quite naturally equated with our 'desire', whereas he confines the ordinary word *epithumia* to an unusually narrow sense, for which there is no good English equivalent, though 'appetite' might serve (cf. n. 8). So either way of translating might be defended, and my own usage will not be entirely consistent, though I shall always (where relevant) indicate which Greek word is in question.

reason and desire (passion), and that is why he claims, in a famous phrase, that 'reason is and ought only to be the slave of the passions'.[4] But Aristotle takes it to be obvious that there are conflicts which can be described in this way, for example when I desire to eat or drink, but reason tells me that I should not (e.g. because it would be bad for me, or would be stealing, or on whatever ground). We must therefore say that Aristotle includes under 'reason' some of what Hume would count as 'passion' (i.e. desire), and it is easy to see how to harmonize this with what he says himself. His technical and general term 'appetition' (*orexis*) covers *both* what we may call the 'bodily' desires—e.g. desires for food or drink or sex—*and* what we may call 'rational' desires, e.g. the long-term desires for health, or honour, or virtue. And apparently we must think of the latter as belonging to that part of the soul that has reason 'in itself', not to that part which 'has reason' only in the sense that it can 'listen to reason', for otherwise—as Hume rightly observed—the two parts could not conflict.

In several places Aristotle recognizes three main varieties of 'appetition' (*orexis*), namely desire (*epithumia*), 'spirit' (*thumos*),[5] and wish (*boulēsis*) (as, for example, at 1111ᵇ11–30). Of these the first two must be credited to that part of the soul which can listen to reason, but the last to that part which has reason 'in itself'.[6] It is not worth our while to attempt any detailed exploration of this alleged trichotomy, for I do not think that it plays any important role in Aristotle's thought. But the main point is that Aristotle must here be attributing some forms of 'appetition' (*orexis*) to reason, and in that case he must be speaking somewhat carelessly when he characterizes the part of the soul that can conflict with reason as 'in general the appetitive part'; for it has no monopoly on 'appetition'. As for what it does contain, Aristotle's general word is 'feelings' (*pathē*)[7] and he gives a list of examples at 1105ᵇ21–23, namely 'desire (*epithumia*), anger, fear, confidence, envy, joy, love, hate, longing, jealousy, pity, and in general what is followed by pleasure or pain'. I shall call this, then, the part of the soul that has feelings, as opposed to the part that has reason (sc. 'in itself').

What is comprised under this heading 'feelings' is not entirely homogeneous. Simple bodily desires (*epithumiae*),[8] such as hunger and thirst, which arise simply

---

[4] Hume, *A Treatise of Human Nature*, II. iii. 3.

[5] Ross translates as 'anger' and Irwin as 'emotion'.

[6] One has to admit that the *EN* never explicitly says that wish (*boulēsis*) belongs to the reasoning part of the soul. But it is affirmed elsewhere (e.g. *Topics* 126ᵃ3–13, *De Anima* 432ᵇ5–6, *Rhetoric* 1369ᵃ1–3), and it would seem to be implied by such passages in the *EN* as 1095ᵃ10–11, 1166ᵇ7–8 with 19–22, 1169ᵃ3–6. I do not see how we can make sense of the *EN* without this supposition. (Cf. Broadie 1991: pp. 68–72). On the other side, *Politics* 1334ᵇ22–3 does explicitly assign wish to the unreasoning part. (I add that in Plato's division of the soul we must clearly understand the reasoning part as including at least some of what Aristotle calls 'wish'.)

[7] Ross translates 'passions', which shares the same etymology as Aristotle's word. For literally a *pathos*, like a 'passion', is something that happens to you as opposed to something that you do.

[8] The normal Greek usage of this word *epithumia* is almost as wide as that of the English word 'desire', but for the most part Aristotle tends to limit it to what we may call 'bodily' desires, and certainly he thinks of these as the paradigms. But he does occasionally slip into the wider usage—e.g. at 1111ᵃ31 he says that one ought to desire (*epithumein*) such things as health and learning—in which there is no clear distinction between 'desire' and 'wish'.

from bodily states, are quite appropriately thought of as 'feelings' of a certain kind, though one might wish to distinguish the feeling itself from the desire for food or drink that accompanies it. (It is presumably possible—e.g. for very small children—to have the feeling without knowing what will satisfy it, and hence without a desire that is directed to food or drink in particular.) But most of the items on Aristotle's list are more complex than this, for they arise in reaction to a thought of some kind (e.g. for anger, 'he has insulted me'), and they give rise to a desire for a specific kind of action (e.g. to punch him on the nose), and at the same time are accompanied by a 'feeling' which is associated with a state of the body (e.g. the heart beats faster). Aristotle is aware of this complexity in the emotions. For example, in the *De Anima* he says, concerning emotions in general,

The definitions should be of this kind, for example 'anger is a certain kind of movement of such a sort of body (or part or faculty of that body), brought about by this cause and for the sake of that goal . . .' The physicist and the dialectician would define each [emotion] differently. For example, when defining anger, one will say that it is a desire (*orexis*) to return pain for pain, or something similar, and the other will say that it is a boiling of the blood and hot substance around the heart [but we should include both] . . . (403^a25-^b1)

(The *Rhetoric* gives a slightly fuller version of the 'dialectician's' definition: anger is 'a desire for revenge on account of an unjustified slight to oneself or one's own', 1378^a30-2.) Modern analyses of the emotions often stress these 'cognitive aspects', and play down the role of 'feelings', on the ground that the very same feelings may accompany quite distinct emotions (say fear and hatred), and that in some cases 'feeling' may be more or less absent (say with envy or jealousy).[9] I do not imagine that Aristotle himself would have been happy to remove the element of 'feeling' altogether, but since he does clearly recognize that more than mere 'feeling' is involved, I shall sometimes paraphrase his word *pathos* as 'desire-or-emotion'.[10]

The relevance of all this to our main topic, namely the specifically human virtues (or excellences) of the soul, emerges at the end of I.13. There Aristotle tells us that the virtues are similarly divided. The intellectual virtues concern the part of the soul that has reason, and these will be treated in book VI, but virtues of character (in his own vocabulary, *ethical* virtues)[11] concern the part of the soul that has feelings (and which can listen to reason). These will be the topic of books II–V. In

[9] See e.g. Bedford (1956–7).

[10] It has been suggested by Fortenbaugh (1975, ch. 2, sect. 2) that, because Aristotle does recognize the cognitive aspects of emotions, he (*a*) credits this cognition to the non-rational part of the soul, and (*b*) holds that when reason 'controls emotion' it does so by influencing the associated cognition. But I see no firm ground for these claims, and the second strikes me as implausible. The truth is that Aristotle's division of the soul (in the *EN*) is not given in sufficient detail to allow us to answer many of the questions that one might wish to raise about it.

I note that usually Aristotle says that pleasure and pain 'follow upon' *pathē*, and does not say that they are themselves *pathē*. But occasionally he does appear to say this (1105^a3), though it surely conflicts with his later accounts of what pleasure is (which I discuss in my Ch. VII).

[11] Ross begs a question by translating *ethical* as 'moral'. *Most* of the virtues that Aristotle discusses under this heading do strike us as 'moral virtues', but a few clearly do not, e.g. 'ready wit' (IV.8).

book II we are offered a general definition, and in books III–V there is a discussion of particular examples.

2. Virtue as a disposition (II.5)

Aristotle begins his account in book II by asking how virtues of character are acquired. His main claims are that it is not 'by nature'—i.e. we are not born either virtuous or vicious—nor 'by teaching' (i.e. of an intellectual kind), but by habituation. And he (correctly) observes that the Greek word for 'character' (*ēthos*) is in fact derived from the word for 'habit' (*ethos*). The main outline of his doctrine here is, I think, sufficiently clear, and in any case I pass over it in order to come at once to his account of what virtue of character *is*, which begins in II.5.

Here he offers an argument by elimination. The things that occur in the soul, he says, are of three kinds: feelings, capacities, and states (dispositions) (1105^{b}19–21). But virtue is not either a feeling or a capacity, so it must be a state. There is evidently a premiss that he has failed to state: these are the things that occur in *that part* of the soul which has feelings, i.e. the part that can listen to reason but does not have reason 'in itself'. Of course, in other parts of the soul there are other things that occur; most obviously, in the part that has reason in itself there is also a grasp of premisses, deduction from those premisses, and contemplation of the results reached. None of these could reasonably be described as 'feelings' (*pathē*)[12] But Aristotle, without saying so, has already narrowed his attention to that part of the soul which he has earlier said (in I.13) is the part with which virtues of character are concerned. In *this* part, what actually happens can only be a feeling, though we may also say that this part has a *capacity* for certain feelings, or a *disposition* for certain feelings. And a virtue, says Aristotle, can only be the last of these three. Given the premiss, the conclusion is clearly correct, though the detailed arguments for it are not particularly convincing.

Aristotle argues that a virtue cannot *be* a feeling, because we are called good or bad people, are praised or blamed for our virtues or vices, but not for our feelings (1105^{b}28–1106^{a}2). This claim, however, clearly needs some qualification. (One can surely say: 'Although he is not in general a compassionate person, still he did feel compassion on that occasion, which is to his credit'.)[13] Again, he argues that we do not choose what feelings to have, sc. on a particular occasion, whereas the virtues 'are choices of a kind, or not without choice' (1106^{a}2–4). But, as we shall see later[14] (concerning III.5), Aristotle is in some difficulty over the sense in which 'we choose to be virtuous'. Finally, he notes that we are said to 'be moved' by our feelings, but not by our virtues (1106^{a}4–6), but this—it seems to me—is no more true as a

---

[12] The Greek word in question (*pathē*) can be construed very broadly, as covering anything that *happens* to one, as opposed to what one *does*. But still it seems to me that it cannot be taken to cover the activities of the reasoning part of the soul (especially in view of the list of *pathē* that follows).

[13] Cf. 1109^{b}30–1. (Similarly 1101^{b}14–18 apparently admits that we may praise people for their capacities.)

[14] Ch. V, Sect. 3(*b*).

comment on Greek usage than it would be as a comment on English usage: one can perfectly well speak of someone 'being moved' by generosity, just as much as of his 'being moved' by a feeling of generosity on a particular occasion. (But one might wish to say that the former usage is elliptical for the latter.) Turning, then, to the idea that virtue might be a mere capacity, he says first that all the same arguments apply (which, clearly, they do not; for example one is not 'moved' by a capacity), and adds that we have our capacities by nature, but not our virtues. But this apparently ignores the fact that we can, if we wish, set out to *develop* capacities.

The detailed arguments, then, are not particularly convincing. Nevertheless, the conclusion is clearly correct, given the premiss, for what Aristotle has in mind as 'virtues of character' evidently are dispositions. If one says of someone that he is brave, or temperate, or generous, or good-tempered, or honest, or just, in each case one is certainly not describing his feelings at the moment, or his mere capacities for feeling. If one is talking about feelings at all, it is a general disposition to feelings that is in question. For example, the brave man is one who is so disposed that, in general, he will not feel fear unless fear is merited; the temperate man is one who is so disposed that, in general, he does not have excessive desires for food or drink or whatever; and so on. The virtues of character, then, are certainly dispositions. But is Aristotle right to say that they are dispositions with regard to *feelings*? Might we not wish to respond that the brave or generous or honest man is one who is disposed to *act* in a certain way (no matter what he may feel)? Indeed, in the preceding chapters Aristotle has constantly spoken of virtue as concerned with actions, but in chapter 3 he has gone on to note the importance of taking pleasure in the right actions, and this is where feelings too become relevant.

Chapter 5 concentrates entirely upon feelings, and makes no mention of actions, but that is because it does not say all that Aristotle means. For feelings by them-selves are not enough. We can bring this out by recalling the 'self-controlled' man (*enkratēs*) whose feelings pull him in one direction while his reason pulls him in another. It must presumably be possible for such a man to be *well* disposed with regard to his feelings, i.e. he regularly feels, on each occasion, just the right kind of feeling that the situation demands. However, his reasoning may be wholly mis-guided, and so frequently pull against his feelings. Since he is self-controlled, his reason wins. As a result, he frequently ends up doing the wrong thing, though his feelings are invariably right.[15] Is such a man virtuous? If we take Aristotle's explicit definition in II.5 as his last word on the matter, then he must be, for his feelings are just as they should be, even though his actions are not. But this cannot be what Aristotle intends.

In fact there is no doubt that, in Aristotle's view, it is characteristic of the virtu-ous man that his feelings and his reason are in harmony. (That is why the virtuous man *enjoys* doing the virtuous thing, 1104^{b}3–11.) So the self-controlled man is not yet virtuous, even if he does always obey correct reasoning, though certainly he

---

[15] Example: in the case of bodily desires I may regularly have just the same feelings as the temperate man has, but wrongly believe that these feelings should be resisted, and so end up acting as an ascetic, i.e. one who Aristotle would say has the vice of insensibility (*anaisthēsia*).

ranks higher than one who lacks self-control, who in turn ranks higher than one who is vicious (1145^{a}15–^{b}2). It follows that when Aristotle defines the *genus* of virtue in II.5 as simply a disposition with regard to feelings, he has left something import-ant unsaid, for he also intends that these feelings should be in harmony with reason, and so lead naturally to the right actions. Let us think of his definition as emended in this way. Virtue of character is, then, a disposition to have the right feelings, and at the same time for those feelings to be in harmony with reason, so that they lead naturally to the right actions. This revised definition gives us one way of understanding why, in the next chapter, Aristotle so constantly speaks of virtue as concerned with both feelings and actions.[16] But whether that is the *right* way to understand this phrase 'feelings and actions' is a question we must take up again later.

3. Virtue as a 'mean' (II.6)

Having decided that virtue is a disposition, Aristotle goes on in chapter 6 to say what kind of disposition it is, and this introduces his famous 'doctrine of the mean': virtue is a kind of 'mean', that is to say something 'in the middle', between two extremes. (The doctrine has been anticipated in chapter 2, at 1104^{a}10–27, where he explains that virtue avoids the two opposing faults of excess and deficiency.) But one has to admit that his exposition of this doctrine is at several points obscure.

After some preliminaries (1106^{a}14–24), he introduces the doctrine thus: 'in every-thing that is continuous and divisible' (i.e. on every linear scale), 'it is possible to take a greater, smaller, or equal amount . . . what is equal is a kind of 'mean' (i.e. middle) between too much and too little' (1106^{a}26–9). The first point to note is that Aristotle never explicitly says, either here or later, just what 'continuous and divis-ible' scale he takes to be the relevant one where virtue is concerned. This is some-thing we must come back to. Instead he at once proceeds to a different point: a 'mean' (middle) may be reckoned either 'with regard to the thing itself' or 'rela-tively to us', and he makes it clear that he intends the second. But his elucidation of this point (1106^{a}29–^{b}7) is, I believe, seriously misleading. Before I comment on these issues, let us have his full definition before us:

Virtue is, then, a state involving choice, in a mean relative to us, determined by reason and by that by which the practically wise man would determine it. (1106^{b}36–1107^{a}2)

I take up one by one the clauses in this definition.

1. *Virtue involves choice.* This claim creeps into the definition without any explanation in this chapter. There are, I think, two suggestions worth considering. Aristotle *may* mean to point to the claim briefly noted in II.5, which he will attempt to defend in some detail in III.5, namely that we can reasonably be said to choose whether to be virtuous or not, and are therefore responsible for our virtue or lack

---

[16] The phrase has also been used earlier, at 1104^{b}14.

of virtue. (I discuss this claim in chapter V, section 3(ii)). But I think it is more probable that he means to point further back to his account in II.4 of what distinguishes the virtuous person from one who merely acts as virtue demands.[17] (This is in defence of his view that one becomes a virtuous person by being trained to act in the virtuous way; the trainee will act as virtue demands *before* he can be counted as a virtuous person.) Aristotle mentions three points which distinguish the virtuous person: (i) he knows what he is doing (i.e. his performance is not a mere accident), (ii) he chooses to act as he does, and chooses it 'for its own sake', and (iii) he acts in this way from a firm and settled disposition (1105^a28–33). As he goes on to comment, (i) is relatively trivial, but (ii) and (iii) are all-important (1105^a33–$^b9$). The virtuous person, then, will choose to do the virtuous thing, and will choose it 'for its own sake', i.e. simply because it is the virtuous thing, and not for any ulterior motive.[18] We could then understand the definition in this way: virtue is a disposition with respect to feelings (as II.5 has said), but one which issues in choice, and hence in action, which is—as still needs saying—in accordance with those feelings.[19]

It is useful to add a word here on Aristotle's use of the word 'choice' (*prohairesis*).[20] The literal meaning of the word is probably 'choice-beforehand',[21] and when he comes to explain the word in III.2 he says that it applies to what we may call a 'thought-out' choice, i.e. one that involves some deliberation. But in practice he quite often uses the word more widely, as we do, and will speak of something as 'chosen' even though nothing that could be called 'deliberation' was involved. The present passage is, I think, an example. The virtuous man surely will not always need to *deliberate* over what would be the virtuous way to act in his present situation; very often he will just 'see' it, and act at once. As Aristotle says himself, when discussing bravery,

To be fearless and unperturbed in sudden danger seems more the mark of a brave man than to be so in situations that are foreseen. For this is due to his state of character more than to preparation. In what is foreseen, one might choose how to act from reasoning and calculation, but when there is no warning one acts in accordance with one's state of character. (1117^a17–22).

The virtuous man will, in our sense of the word, choose to act as he does, and will choose this way just because that is the virtuous way to act, but we need not

---

[17] This is argued by Hutchinson (1986: 114).

[18] This gloss on 'for its own sake' is somewhat rough and ready, but I think it will do for present purposes. There is an interesting discussion of how the phrase should be more precisely understood in Williams (1995), Hursthouse (1995), and Wiggins (1995: 220–6).

[19] Contrast Kosman (1980: 110–3).

[20] Irwin translates 'decision'.

[21] Aristotle himself seems to explain the word as meaning 'choice-instead-of', i.e. choosing one thing in preference to another (*EN* 1112^a16–17, *EE* 1226^b5–9), and this is explicit in the *Magna Moralia* (1189^a12–14). But in that case (*a*) his etymology would seem to be mistaken, and (*b*) it does not altogether fit his own account of *prohairesis* in III.2, which allows for the consideration of alternatives but does not require it. A meaning which fits his own account would be 'choice-for-the-sake-of', i.e. a choice made with some definite end in view.

suppose that this always involves 'choice-beforehand' in the full sense, i.e. full deliberation perhaps including a weighing of alternatives.[22]

2. *A mean 'relative to us'.* Aristotle contrasts the two notions of a mean (or middle) 'with regard to the thing itself' and a mean that is 'relative to us'. The former, he says, is what is halfway between each of the two 'extremes', and is the same for everyone; but the latter is what is neither too much nor too little, and this is not a single thing, and not the same for everyone (1106[a]29–32). He then gives an illustration concerning the amount of food it is right to eat. Ten pounds is a lot, and two pounds a little, so the mean 'with regard to thing itself' is the arithmetic mean between ten and two, namely six pounds.[23] But it does not follow that each of us should consume six pounds, for this may be too little for the trained athlete in top condition, but too much for one who is just starting his athletic training. The expert coach will therefore prescribe different amounts for each (1106[a]33–[b]7). How is this illustration to be understood?

Well, quite a natural suggestion might be this.[24] What is making the difference in this example is that some people are more physically developed than others. So, to apply it to the case of virtue, we may note that some people are more 'morally developed' than others. There will be one way to act for the man who is 'fully developed' from the moral point of view, i.e. the man of 'practical wisdom' (the *phronimos*), and a different way for others, depending on how their state of moral development compares with his.[25] In a word, sinners are not expected to act as saints do; the standards required of them are lower. But if we do take Aristotle's doctrine in this way then we surely cannot accept it, for it would follow that if the sinner *does* behave as the saint would then he should be *censured*, since he has not hit the mean 'relative to him' but has 'gone over the top'. Besides, one must certainly observe that Aristotle himself never mentions the individual's 'moral development' as something that affects the mean 'relative to him'.

Let us look, then, at what he does tell us on this point. Virtue, he says, aims at the mean as also do crafts and skills (1106[b]8–16), and he adds,

I mean virtue of character, for this concerns feelings and actions, and in these there is excess and deficiency and a middle amount. For example, it is possible to feel both too much and too little fear, confidence, desire, anger, or pity, and in general to be pleased or pained too much or too little, and in each case this is not good. But to feel them *at the right times, for the right objects, towards the right people, for the right reason, and in the right way,* that is both middle and best, and is the part of virtue. In a similar way there is also with actions an excess, a deficiency, and a middle amount. (1106[b]16–24)

Here it is the words I have italicized which list the factors which affect what is the

[22] Cf. Fortenbaugh (1975: 70–5); Broadie (1991: 79).

[23] In what sense are ten and two the 'extremes'? (Perhaps because one who eats more than ten, or less than two, will not survive?)

[24] This is Joachim's suggestion, in his (1951: 88–9).

[25] In support of this interpretation, where Aristotle says that the mean relative to us is 'determined by *logos*, and by that by which the man of practical wisdom would determine it', Joachim takes *logos* to mean 'ratio' or 'proportion' (which is indeed one of the meanings of this many-faceted word). So the mean for *x* is given by the ratio in which *x* stands *to* the man of practical wisdom.

right, and therefore 'middle', way of feeling and acting. There is no mention of the 'moral development' of the agent concerned, nor indeed of anything else which concerns *his* particular state. It is *other* features of the situation in which he finds himself that are stressed as relevant. (For example, is it right to feel anger at what someone has just said? Well, this will depend upon whether the occasion is one on which anger is better controlled, on who the speaker was, on what his topic was, on just why his remark was provoking, and so on.) Clearly Aristotle's thought is that one cannot lay down any general rules about when a feeling would or would not be appropriate, not because *agents* are so different from one another, but because *situations* are so different. It was therefore somewhat misleading when he character-ised the 'mean' as 'relative to us' and 'not the same for everyone', and even more misleading when he went on to illustrate this with the case of the athlete's food intake. For what he actually has in mind is much better described not as a 'relativity to the agent' but as a 'relativity to the circumstances of the action'. (These may, of course, include the particular way in which the agent is related to other parties involved in the situation.)

3. *Middle of what?* As Aristotle says when introducing his doctrine, one can fairly use quantitative terminology ('too much', 'too little', 'the right amount') when one is dealing with a *scale* of some kind, i.e. in his words something 'continuous and divisible' ($1106^a26–7$). But just what scale is he thinking of as the one on which virtue 'aims for the middle amount' ($1106^b14–16$)? There are many suggestions that one might make. I shall discuss just two.

Virtue was defined in II.5 as a disposition with respect to feelings. I have noted that this is only part of the story, for the virtuous man's feelings must also be in harmony with his reason, and so must lead to the appropriate action. But since it is the part that Aristotle himself concentrates on in II.5, it is natural to suppose that it is what is uppermost in his mind. It is also quite natural to say that there is a scale associated with each feeling, namely a scale of the *intensity* with which it is felt. For example, one can feel very angry, slightly angry, or 'middlingly' angry; and so too with fear, compassion, desire (say, for sex), and many others. So this suggests that what is characteristic of the virtuous person is that he always, on each occasion, has the right amount of feeling, the degree of intensity of that feeling that is appropri-ate to the particular situation in which he finds himself. This, indeed, seems to be exactly what Aristotle is thinking of when in II.5 he describes the states of character he has in mind. They are

States in virtue of which we stand well or badly with respect to the feelings, for example, we stand badly with respect to anger if we feel it strongly or weakly, but well if we feel it middlingly. ($1105^b25–8$)

There is, however, an evident objection to his doctrine if we take it in this way. First, we can hardly understand it as meaning that in every situation whatever, one should always feel a 'middling' amount of absolutely every desire-and-emotion, for evidently in most situations most emotions are entirely out of place (e.g. when it is time to brush one's teeth). So perhaps we might say that in all situations where some people do feel a given emotion—say fear or anger—the right amount of that emotion to feel is always middling. But this is not much better, for some people do fear in situations where the right amount of fear to have is zero (e.g. when there is a mouse loose on the floor). We might, then, try to limit the doctrine in this way: in all situations where it is *appropriate* to feel a given emotion, the right amount to feel is always a middling amount. This formulation automatically excludes counter-examples where the right amount to feel is the minimum, i.e. zero, but still permits counterexamples where the right amount to feel is the maximum.[26] Did Aristotle really want to say that there is *no* situation which deserves a maximum reaction? Well, perhaps he did, but it does not seem particularly plausible. We could rescue him from this objection if he was prepared to say—against the natural meaning of the word—that sometimes the 'middle' and the 'extreme' could coincide. But it appears that he is not prepared to say this. For example, he tells us that there is no 'middle amount' of certain feelings, such as spite, shamelessness, or envy (or of certain actions, such as adultery, theft, murder), because it is built into these words that what they name is bad (1107ª8–14). But of course he would not put it this way if he were prepared to count zero itself as (in some cases) a 'middle'. We may con-clude that this attempted rescue cannot be sustained, so the doctrine that results is indeed an implausible one. One looks, therefore, for a different way of explaining what is 'middling' about virtue.

The best suggestion seems to be that it is not the virtuous action, on each occasion, that has something 'middling' about it, but rather the general disposition from which it flows. It is, as Aristotle says, this *disposition* that is 'in the middle', for each virtue lies between two vices, one of excess, and one of deficiency. What, then, is the *scale* involved? A simple suggestion is that it is the number of occasions, or better the number of types of occasion, on which the emotion in question is felt.[27] As Aristotle has said earlier, when anticipating his doctrine in chapter 2,

The man who flees from everything, fears everything, and withstands nothing, becomes a coward; the man who fears nothing at all, but goes to meet everything, becomes rash. Similarly the man who indulges every pleasure and refrains from none becomes intemper-ate, while he who shuns all pleasure, like a boor, becomes a kind of insensitive person. So temperance and courage are destroyed both by excess and by deficiency, but are preserved by the mean. (1104ª20–7)

Undoubtedly, Aristotle is thinking here not of the intensity of the emotion on a given occasion, but of the way that it may be displayed on too many, or too few, occasions (or types of occasion). Moreover, this way of taking the doctrine clearly avoids the previous objection, for now the claim that the virtuous disposition is always 'in the middle' means only that the emotion with which it is concerned is always one which can be felt too often, and can be felt too seldom, and that seems

---

[26] Admittedly, it is not very clear what the 'maximum' amount of fear or anger or gratitude might be, but (on this account) Aristotle is apparently committed to supposing that there is one, especially when he speaks of the middle 'with regard to the thing itself', i.e. what is halfway between the two 'extremes' (1106ª28–31). If this is to make sense, there must *be* 'extremes'.

[27] Strictly speaking, a scale of (whole) numbers is not something that counts, in Aristotle's vocabu-lary, as 'continuous and [sc. infinitely] divisible'. But this is a complication we may reasonably ignore.

unobjectionable. Unfortunately one cannot claim that this is all that Aristotle himself means, both for the reasons already given and because he twice says that virtue *is* a kind of mean (i.e. a middling disposition) *because* it aims at a mean (i.e. a middling emotion, on particular occasions) (1106^b27–8, 1109^a20–4). So in his own mind the two suggested interpretations are closely connected with one another. But, to be charitable, we may suggest that what is uppermost in his thought is that virtue is a middling disposition.[28]

In that case, all that is necessary, to guard against misunderstanding, is to say what Aristotle himself says, that not every emotion admits of a mean, for there are some that one cannot feel too seldom, i.e. those that one should never feel at all. As examples he gives spite, shamelessness, and envy, and in the sphere of actions adultery, theft, and murder (1107^a8–14). His discussion of these shows that he wishes to explain the point by saying that these words already have built into them a 'too much' or a 'too little' (1107^a14–27). The point is easily illustrated in the case of murder. Murder is a special case of killing a person, and Aristotle will surely think that there are occasions (e.g. in battle) when it is right to kill. But murder is by definition wrongful killing. Similarly, adultery may be taken as a special case of having sex, and theft perhaps as a special case of taking something for one's own use (in particular, a case where what is taken belongs to someone else, and is taken without his consent). The same account may be applied to the emotions that he mentions (and is so applied by him at 1108^a30–$^b6$). But the general moral to draw is simply this: not everything that could be called a 'middle disposition' is a virtue, and Aristotle is not committed to claiming that it is; he claims only the converse, that every virtue is a 'middle disposition'.

This point may usefully be elaborated. I suspect that in Aristotle's own thinking the two different scales that we have mentioned tend to be run together. For example, the coward is *both* someone who feels fear too often *and* someone who, when he does feel it, tends to feel it too strongly. But we can easily imagine someone who gets the *amount* of fear right, in both these respects, but who still lacks the relevant virtue because he does not satisfy Aristotle's other conditions. For example, he may feel fear for the wrong objects (e.g. for spiders—which, in this country, are harmless—but not for snakes), or on the wrong occasions (he tends, say, to be too fearful when sober, but too fearless when drunk, and these balance out), or for the wrong reasons, and so on. Only a little thought is needed to see that simply getting the amount right cannot be enough, and Aristotle's own discussion makes this clear.[29]

Why, then, does he so emphasize the point that virtue is a 'middle', and always lies between a possible 'too much' and a possible 'too little'? It may be that he was over-influenced by what one might call 'conventional wisdom'; there was, for example, a famous inscription at Delphi which proclaimed 'nothing to excess'. It

---

[28] This point is nicely argued by Urmson (1973: 160–3), and I think that it is usually accepted. But some still take it that he really means to argue for moderation on *every* occasion (e.g. Kraut (1989: 339–41).

[29] This point is stressed by Hursthouse (1980–1), surely rightly. (Curzer 1996 attempts to disarm it, but I think unsuccessfully.)

may be that he was over-influenced by some of the analogies that he mentions himself, for example that it is possible to exercise too much or too little, or to eat too much or too little ($1104^a11–19$, $1106^a36–^b7$), or again that we say of a well-made article that nothing could be added to it or subtracted from it ($1106^b8–14$). But I imagine that what mainly influenced him was that he thought he could discern a general pattern that applied to all virtues: a virtue can always be regarded as something that lies *between* two opposing vices. This is easy to explain if, as he has already said, each virtue is concerned with a particular desire-or-emotion. For then there will be one common way of going wrong which is to feel that emotion too much (either too strongly, or too often, or—more probably—both), and another which is to feel it too little. This is what generates the triadic scheme of one virtue between two vices. Of course, there will be *other* ways of going wrong too, as I have just noted, but these are so many and so various that they have not usually been named.

4. *Determined by reason.* If not every 'middle disposition' is a virtue, then we naturally ask: which of them are? Aristotle's answer is that the relevant 'middle' is the one which is 'determined by reason, and by that [method] by which the man of practical wisdom would determine it'. (Here, as often, it seems best to take 'and' as meaning 'i.e.': the man of practical wisdom (the *phronimos*) will determine 'by reason' which dispositions are to be counted as virtues.) Since practical wisdom is to be discussed later in book VI, I shall postpone until then any further comment on just how 'reason' is supposed to reach its decisions on this topic. (We shall find that Aristotle is vexingly unforthcoming on the matter.) But here I wish to raise just this question: is Aristotle meaning to imply that, in order to be virtuous, one must oneself have this 'practical wisdom'?

The right answer, it seems to me, is 'no'. His account in this book of how virtue is acquired is that it is acquired by training and habituation: one is brought up to act correctly, 'steered by pleasure and pain' as he later says ($1172^a20–1$), i.e. by a system of rewards and punishments. If the training is successful, one becomes accustomed to acting in the right way, and to enjoy acting in that way (II.3 *passim*). One chooses that way of acting, and chooses it 'for its own sake', i.e. simply because it is the virtuous way. One acquires a firm and settled disposition to do this, and thereby one becomes virtuous ($1105^a28–33$). There is no mention, in this book, of any reasoning ability that may be required in addition. The situation as I conceive it could be illustrated in this way. We are brought up to be honest, fair to others, unafraid where there is no real danger, and so on. The normal person, brought up successfully, then *wants* to be honest and fair to others, and acquires a settled disposition to be so. They then qualify as having the virtues of honesty and fairness. But if you were to ask them *why* it is good to be honest, you would probably find that their answers would soon dry up. In a word, people are brought up to know that this or that is the virtuous way of acting, and they do as a result develop a disposition to act in this way, but they usually cannot supply the 'reasoning' which—it is alleged—would show why this way of acting is indeed the right way. Most of us are in this position, and it is—I believe—the position that Aristotle means to describe in this book.

It is true that later, in VI.12–13, Aristotle will introduce a 'higher' conception of virtue, which he calls 'full virtue' (*kuria aretē*), and which he evidently does construe as including the 'reasoning' which is characteristic of practical wisdom. But it seems to me that here in book II that is not what he is trying to describe. I offer two indications. (i) When in II.4 he is discussing what distinguishes the virtuous person from one who merely performs virtuous actions, he does include the condition that the virtuous person acts 'knowingly' (1105[a]31). But he goes on to say that this is an altogether trivial condition, and he puts much more weight on the other two, i.e. that the virtuous person chooses the virtuous action 'for its own sake',[30] and that he acts from a firm and settled disposition (1105[b]2–5). If he were seriously thinking that, in order to be virtuous, one requires the *knowledge* that characterizes the practically wise man, he could hardly have spoken as he does in this passage.[31] (ii) His definition of virtue of character, in book II, is that it is a middle disposition 'determined by reason, i.e. in the way that the practically wise man *would* determine it'. He surely would not have said 'would' if he had meant that in any virtuous person his practical wisdom *is* determining it. His overall position must, then, be this: it is practical wisdom which determines which dispositions are in fact virtuous dispositions, but nevertheless one can have such a disposition without oneself being practically wise. It is usually enough that one has been well brought up, and has responded to this upbringing in the right way.[32]

4. Feelings and actions

In II.5 Aristotle defines the virtues as dispositions with respect to feelings. As I have noted (in Section 2), he cannot quite mean this, for one who has the right feelings but fails to act on them surely does not count as virtuous. Indeed throughout II.1–4 his discussion focuses on the relation between virtue and actions, and feelings are scarcely mentioned.[33] In II.6 we appear to have a compromise position, for there Aristotle constantly speaks of virtue as concerned with both feelings and actions (1106[b]16–18, [b]24–7, 1107[a]3–6, [a]8–12). As I have suggested, the most natural way to interpret these apparent changes of position is to say that the full doctrine is as follows. Each virtue is associated with a particular feeling or emotion, and this indeed is what distinguishes one virtue from another. But the virtuous disposition is one which involves a harmony between emotion and reason: both pull in the same direction. (This is what distinguishes the virtuous person from one who is

---

[30] Admittedly, choice in the full sense requires deliberation (III.2), and good deliberation certainly falls under practical wisdom (VI.9). But I shall argue in my Ch. IV that ordinary everyday deliberation is only a small part of practical wisdom, and I have already argued that Aristotle in this passage is not—or should not be—requiring choice in the full sense (pp. 38–9).

[31] Cf. Williams (1995: 14–15).

[32] Here I broadly agree with Fortenbaugh (1975, ch. 4, sect. 2), Burnyeat (1980), Broadie (1991, ch. 2, sect. 10). But the contrary view has often been defended.

[33] Unless one wishes to count pleasure and pain, the main topic of II.3, as feelings. (On this, see n. 10 above).

merely self-controlled (*enkratēs*) the self-controlled man *acts* in the right way, but this is because his reason overcomes his feeling; he is subject to an internal struggle, which the virtuous man is not.) Consequently the virtuous man will act as his emotion dictates; he will always pursue the right course of action, and will do so because that is what he wants to do. That is why he will enjoy so acting. I think it is fair to call this the 'standard' account of Aristotle's position on virtue of character. But is it really Aristotle's own position?

One reason for hesitation emerges in II.6 itself: on two occasions Aristotle speaks separately of a mean in emotions and a mean in actions. At 1106^{b}16–24 he says,

I mean virtue of character, for this concerns feelings *and* actions, and in *these* there is excess and deficiency and a middle amount. For example, it is possible to *feel* both too much and too little fear, confidence, desire, anger, or pity . . . In a similar way, there is also with *actions* an excess, a deficiency, and a middle amount.

Again, at 1107^{a}8–12 he says,

But not every action, or every feeling, admits of a mean, for in some cases the name already incorporates badness, for example spite, shamelessness, envy, and in the case of actions adultery, theft, murder.

These passages suggest that the doctrine of the mean is to be applied both to feelings *and independently* to actions, and so they lead one to ask whether Aristotle's thought might perhaps be that it is only *some* virtues that concern feelings, whereas *others* concern actions.

Moreover, this thought appears to be strengthened when we move on to II.7, where Aristotle lists a number of virtues, with their associated vices, in order to indicate how the doctrine applies in their case. (These virtues are then discussed one by one, in more detail, in III.6–V.) Let us just review the list, which runs thus:

(No heading)
    (1) Courage (1107^{a}33–^{b}4; III.6–9)
    (2) Temperance[34] (1107^{b}4–8; III.10–12)
Concerned with money
    (3) Generosity[35] (1107^{b}8–16; IV.1)
    (4) 'Magnificence' (1107^{b}16–21; IV.2)
Concerned with honour
    (5) Pride[36] (1107^{b}21–3; IV.3)
    (6) Nameless; (5) on a smaller scale (1107^{b}23–1108^{a}3; IV.4)
(No heading)
    (7) Good temper[37] (1108^{a}4–9; IV.5)

---

[34] This is the traditional translation of *sōphrosunē*, which I use for want of a better. In ordinary Greek usage the word has a wide range of application, and often approximates to our 'sensible'. But Aristotle narrows it to moderation in respect of the bodily pleasures of touch and taste.

[35] Ross translates 'liberality' (which is closer to the etymology of Aristotle's word, i.e. *eleutheriotēs*).

[36] This (or 'proper pride') is Ross's translation. Aristotle's word, *megalopsucheia*, means literally 'being great-souled', and Irwin's translation, i.e. 'magnanimity', preserves the etymology but not the sense intended.

[37] Irwin translates 'mildness'.

Social virtues

    (8) Honesty (1108^{a}19–23; IV.7)

    (9) Ready wit (1108^{a}23–6; IV.8)

    (10) Friendliness (1108^{a}26–30; IV.6)

Feelings that are not virtues

    (11) Shame (1108^{a}30–5; IV.9)

    (12) Righteous indignation (1108^{a}35–^{b}6)

(No heading)

    (13) Justice (1108^{b}6–9; V)

As one would expect, Aristotle easily supplies suitable feelings for courage, temperance, and good temper (namely fear,[38] desire (of a certain kind), and anger). In the case of shame and righteous indignation, he comments first that these are feelings rather than dispositions, and then that—even construed as dispositions—they are not virtues;[39] but he nevertheless wishes to claim that they fit his triadic scheme of one condition that is praised being 'between' two which are blamed. (Thus shame—or its corresponding disposition[40]—is taken to be a mean between shamelessness and bashfulness, and righteous indignation to be a mean between envy and spite.)[41] So far, then, i.e. with items (1), (2), (7), (11), (12) on Aristotle's list, we are clearly concerned with feelings (*pathē*). But what of the other cases?

In connection with items (5) and (6) Aristotle does supply what we might be ready to call an 'emotion', namely the desire for honour. But here one must ask whether this 'emotion' is properly credited to that 'semi-rational' part of the soul where feelings proper are found, and which 'partakes in reason' only in the sense that it can listen to reason. For, as we have observed, Aristotle does not actually suppose that all 'appetitions' (*orexeis*) belong to this part, and one would certainly think that ambition, love of honour, and claims to respect for great things done belong more to the fully rational part, which has reason 'in itself'. But I am content to leave this merely as a question. The case with the *other* virtues here listed is altogether more compelling. For with these examples Aristotle makes *no* attempt to point to a relevant feeling, either in the summary at II.7 or in the more detailed accounts in book IV. He is concerned simply with the *actions* that are appropriate.

For example, generosity and 'magnificence' (*megaloprepeia*) are introduced simply as concerning the giving of money. The generous man does not give away too much or too little, but the right amount, to the right people, at the right time,

---

[38] In fact Aristotle supplies two distinct feelings for courage, namely fear and confidence, which confuses his account. There is a very nice and succinct discussion in Urmson (1988: 63–7), and something more complex in Pears (1980).

[39] He says in IV.9 that shame (or its disposition) is not a virtue because it is displayed appropriately only when one has done something wrong. But the virtuous man does not do what is wrong. (Our text contains no detailed discussion of righteous indignation.) Commentators have suggested a further reason, namely that there is no particular type of action to which either shame or righteous indignation lead naturally. Cf. *EE* 1234^{a}23–34; *Rhetoric* 1385^{b}12–14; Fortenbaugh (1975: 82); Urmson (1973: 168–9).

[40] Ross translates 'modesty'; Irwin merely offers 'proneness to shame'.

[41] It was noted by Ross (trans., *ad loc.*) that this account of how righteous indignation is a mean will not do. The point is elaborated by Urmson (1973: 166–7).

and so on. The 'magnificent' man is simply one who has the means to be, and is, generous on the grand scale. Neither of them are characterized in terms of any emotion, but simply with respect to their actions (and similarly with their opposites, the miser and the prodigal, the niggard and the vulgar). While no doubt Aristotle *could* have pointed to one, or (more probably) several, emotions that may lie behind their behaviour, it is significant that he does not actually make any attempt to do so. Finally, I add that in the case of the three 'social' virtues—honesty, ready wit, and friendliness—there is again no attempt to show that each involves a particular feeling; and while there does appear to be such an attempt in the case of justice (which I discuss in the next chapter), it is notoriously unsuccessful.

I am inclined to think, then, that what I have called the 'standard account' of Aristotle's position on virtue of character is, in a sense, mistaken. It is true that in I.13 Aristotle claims that virtues of character belong to the part of the soul that has feelings. It is also true that in II.5 he defines these virtues as dispositions with respect to such feelings. But in II.6 he appears to be moving towards the view that it is only in some cases that feelings are important, and that in others we should speak more directly of dispositions to act in this or that way. And in the summary account of particular virtues in II.7, as well as in the more detailed discussions in later books, this appears to be borne out. Quite often Aristotle does in fact characterize a virtue as a disposition to act in a certain way, and pays little attention to the feelings that may be involved. Interestingly, there appears to be a *recognition* of this point at the start of III.10, where Aristotle has finished discussing courage and is proposing to move on to temperance, for, he says, these two 'appear to be *the* virtues of the non-rational part of the soul' (1117^{b}23–4). This surely indicates, contrary to I.13 and II.5, that the remaining virtues to be discussed are *not* virtues of this non-rational part. (But I would wish to make an exception for the virtue of good temper.)[42]

How, then, does the doctrine of the mean apply in these other cases? When we are concerned not with feelings but with actions, what is to count as 'too much', 'too little', and 'the right amount'? Aristotle's response is to find a particular range of actions where these notions make sense. To take a simple example, consider the actions which are cases of giving money to others. Some people give too much, and they are prodigal (or profligate); some people give too little, and they are mean (or miserly); but the generous (or liberal) man gives just the right amount, taking into account what he can afford (1120^{b}7–11). Of course he also gives to the right people, at the right time, for the right reason, and so on (1120^{a}23–26, ^{b}3–4), so once again the mere amount is not by itself enough. But at least we can talk in terms of amounts in this case, and it is not too forced to think of there being two opposing vices, one characterized as the disposition to give too little, and the other as the disposition to give too much.[43] But in some other cases the account becomes very forced indeed. Consider, for example, Aristotle's account of honesty in IV.7.

---

[42] I shall later (in Ch. III, Sect. 6) come back to the relationship between what Aristotle calls 'virtues of character' and what he recognises as desires-or-emotions.

[43] I have here simplified Aristotle's account in IV.1, for he himself talks both of giving *and of taking* money (or other goods).

Tacitly, he restricts the sphere of this virtue to the case when one is describing one's own achievements. The extreme in one direction is boastfulness, when one overstates them; the extreme in the other direction is self-deprecation (*eirōneia*)[44] when one understates them; but the relevant virtue is to state them as they actually are. As in many other cases, Aristotle does not try to claim that there is some one emotion that leads people to be boastful; indeed he admits that the boaster may have many different motives (1127^b9–22). So his 'too much' and 'too little' do not concern any emotion, but simply the content of what one says. But this clearly will not do. First, even confining attention as Aristotle does to what I say about my own achievements, it is easy to see that some *mis*statements are not naturally classified either as overstatements or as understatements. (For example, I may say 'I have never learned German, because my school didn't teach it', when the truth is that my school did teach it, but I opted out of lessons there, did learn some German subsequently, but have now forgotten it.) But second, and more importantly, it is clear that the virtue of honesty is not in fact confined to statements about one's own achievements, and that in the general case there is no prospect of dividing deliberate misstatements into those that are overstatements and those that are understatements. (Suppose, for example, that a perfect stranger asks me the way to the station, and out of pure maliciousness I give wholly wrong directions. That is an example of dishonesty, but not because it is 'too much' or 'too little' of something, nor because it was addressed 'to the wrong person, at the wrong time, in the wrong manner, and so on'.) It is clear that there is no good way of applying the Aristotelian scheme to the virtue of honesty, and a sign of this is that we do *not* think of honesty as lying between two opposing vices. The vice opposed to honesty is simply dishonesty, and there is not also another vice, which diverges from the virtue in the opposite direction. I conclude that Aristotle's general scheme fails in this case.[45] It fails too in some other cases, and most importantly in the case of justice, but that is the topic of the next chapter. Meanwhile, I would summarize the position as follows.

Aristotle's own account seems to waver. It begins in I.13 (and continues in II.5) by focusing entirely on feelings (*pathē*), but to this we must add 'and actions', as Aristotle himself elsewhere does. The best way to resolve this problem would seem to be what I have called the 'standard account' of Aristotle's position, whereby it is still insisted that each virtue (of character) is associated with a particular kind of feeling, but added that this feeling must lead naturally to action, since it is not

---

[44] This is Irwin's translation. Ross prefers 'mock-modesty'. (It is a trait that Socrates is often accused of, in Plato's early dialogues.)

[45] I should perhaps note that the *EE* does not count honesty as a virtue, but that is not for the reason that I have raised here. On the contrary *EE* III.7 lumps together what are called in *EN* 'social virtues' (i.e. honesty, ready wit, and friendliness) and what are called in *EN* 'feelings that are not virtues' (i.e. shame, and righteous indignation, to which it also adds 'dignity'), and it says that *all* of them are 'middle states' which do not count as virtues because, though they do involve emotions, they do not involve choice (1233^b16–18 and 1234^a23–7). It is difficult to make any sense of this, whatever meaning we assign to the vague phrase 'not involving choice'. The classification in *EN* appears to be better on this score.

opposed either by 'reason' or by other contrary feelings. But in fact Aristotle's subsequent discussion cannot easily be squared with this 'standard account', for in some cases he makes no attempt to diagnose a particular feeling that characterizes a given virtue. On the contrary, he conceives of that virtue simply as a disposition to *act* in a certain way (e.g. to tell the truth), and is ready to admit that departures from the right course of action may be motivated in all kinds of different ways. In this, he is entirely correct, for certainly *some* virtues (and I would say *all*) are better characterized as dispositions to act in a certain way, rather than as dispositions to feel. We may—or may not[46]—agree with him that the truly virtuous man is one who feels no temptation to act otherwise than virtuously. But it is not contrary temptations that define what the virtue is.

In consequence, the 'doctrine of the mean' cannot be upheld. While we still think of each virtue as concerned with a particular emotion, then we can make reasonable sense of Aristotle's 'too much' and 'too little' as meaning too much of that emotion, or too little of it, in ways already explored. But, when we focus directly on actions, the 'too much' and 'too little' lose their significance. For example, there is no suitable scale on which the assertions of the inveterate liar can be ranked as saying either 'too much' or 'too little'. As Aristotle himself quotes: 'We are good in just one way, but bad in many' (1106^{b}35). There is often no obvious scale on which the many bad ways of acting may be ranked as 'too much' or 'too little'.[47]

5. Concluding remarks

Since, as we have seen, Aristotle's own account of virtue is scarcely consistent, there is little more that one can offer by way of further comment and criticism at this stage. But one point worth observing is this. Aristotle shows no awareness that the concept of a virtue is a 'contested' concept, in that different people have different views on what counts as virtuous. As we read through his own list of virtues, we are likely to respond in several cases that *that* way of acting does not seem to us virtuous at all. And I do not mean merely that we do not see it as *morally* virtuous; I mean more strongly that we see *nothing* admirable in the conduct described.[48] Obviously there have been over the centuries many changes of view on what are to be counted as virtues. For example, Christians extol humility as a virtue, whereas

[46] In several instances, for example courage, we do in fact admire the self-controlled man (who overcomes his fear), as much as, if not more than, Aristotle's virtuous man (who has no unworthy fears to overcome). But in other instances one sympathizes with Aristotle's perspective. I would much rather deal with someone who is 'by his nature' honest, than with someone who is constantly tempted to dishonesty, but overcomes the temptation.

[47] You *might* say: even if the many bad ways of acting cannot all be ranked on some *one* scale, still they can be ranked on a number of different scales, diverging in different directions from the one central point which is the good way of acting. So the good is still a 'middle', though a middle of many scales and not just one. But I defy you to specify the different 'directions'.

[48] Hardie (1968: 119) cites from an unnamed 'Oxford lecturer' this verdict on Aristotle's man of 'proper pride' or 'magnanimity' (the *megalopsuchos*): 'a prig, with the conceit and bad manners of a prig'.

for Aristotle it would appear to be a vice (1123^b9–15, 1125^a16–27). Again, Christians have taken chastity to be a virtue, though there are many nowadays who would disagree, and one suspects Aristotle would join them. (He might perhaps regard chastity as a kind of 'insensibility', 1119^a5–11.) But, if we set aside changes of view over time, and try merely to see things from Aristotle's own perspective, still his complacency seems strange. Only one generation earlier Plato was finding it necessary to *argue* that justice is indeed a virtue, in reaction to those who claimed that 'might is right'. And only a generation later Epicurus was to praise above all peace of mind (*ataraxia*), though it is quite clear that Aristotle would not do so.[49] One can hardly believe, then, that in his own day either the philosophers, or the majority of people, were of one mind on what character traits to count as virtues. Yet he writes as though this is a question on which there can be no dispute. One can only note the point, and pass on. It is but one of many indications that Aristotle is not prepared to admit that good people, equally well brought up, may nevertheless form different opinions on what is to be valued.[50]

A more central question is whether Aristotle is right to place the concept of virtue at the heart of his morality, as he so evidently does. There are two aspects to this. First, when reading I.7 one might quite naturally suppose that Aristotle wished us to focus on the actions (or activities) which good reasoning leads to, and that when he says that these are the ones done 'with excellence' he just means that they are actions which are well done, which are (in the circumstances) the right actions. But we now see that this is not what he means by 'with excellence'. For he construes 'excellence' ('virtue') as a disposition of a person, rather than a feature of this or that particular action, and so the focus has shifted from actions to people. His recommendation is not just that we should act well, but that we should aim to become the kind of person to whom it is 'second nature' to act well. So one might notice here a difference of focus from most modern ethical theories: the interest is not so much on what one ought to do as on what kind of person one ought to be, though of course the two are closely connected. But second, Aristotle goes on to develop this (entirely in accordance with Greek tradition) by talking not of one virtuous disposition but of many. One might think that a single virtuous disposition—always to do the right thing, for the right reason, in the right spirit, and so on—is all that he should require, and in a way I think that it is what he does require. But in fact he conforms to his tradition and speaks of many separate virtues—courage, temperance, good temper, justice, and so on. Clearly the main reason for separating these as distinct virtues is the thought that a person may have some of these but lack others, and this Aristotle would explain by saying that some of his desires-and-emotions have been well trained while others are still wayward. That seems very reasonable, but it leads to a problem.

---

[49] It would seem that Speusippus propounded a similar view, which Aristotle was evidently aware of (see Ch. VII n. 9). He probably means to refer to it in II.3, where he says that some people define the virtues as a certain way of not being affected (sc. by pleasure and pain) (1104^b24–5).

[50] The contrast with Plato is obvious: his early dialogues are full of characters who do not share his own values.

For suppose now that each of the various emotions is separately well trained in its own way, so that we have someone who has acquired each of the separate virtues, and values the actions that accord with each of them. Then it seems that he must often be caught in a situation where two distinct virtues conflict with one another. For example, it seems that honesty can conflict with compassion (e.g. where telling the truth would be hurtful), that justice can conflict with prudence (e.g. where punishment is strictly deserved, but would in this case have quite the wrong effect), or with generosity (e.g. when I wish to treat you more generously than you deserve), and so on. Similarly courage, or one's sense of honour, may insist upon issuing a challenge, whereas good temper would make light of the insult; politeness may counsel silence where sincerity demands a protest; and so too with many other examples. Those who give an independent value to each separate virtue must surely admit that such conflicts can occur, and then they are faced with the question of how they are to be resolved.

But, on Aristotle's own account, such conflicts are impossible. For he defines a virtue simply as a disposition to do, in each situation, whatever is right in that situation. So if, in a particular situation, it is right not to tell the truth, then his honest man will not tell the truth, and will not even feel any desire to do so. This may seem unrealistic, but it is what his doctrine demands.[51] So he does in effect define just one single virtue of character, and the various separate virtues that are conventionally recognized should be thought of as mere stages on the way to this. Indeed, I think he would himself agree that the picture of several different virtues is only appropriate for the early stages of training and habituation, for he will himself later say (in VI.13) that in 'full' virtue all are unified, and one cannot have any one of them 'fully' without having all the others too. But one has to add that this is not the impression that he gives in books II–V, where it does appear that each virtue has its own separate value.

One final comment: the virtues are dispositions to do what is *right*. In book II Aristotle has given us virtually no lead on how to discover what is right, except by saying that it is determined by practical wisdom. Since that is something that he will discuss in book VI, further comment must wait until then. But I have noted that he does not seem to expect disagreement on this topic, which certainly strikes us (and I imagine his contemporaries too) as an over-optimistic attitude.

---

[51] Hardie (1965–6: 44–5) protests that even the virtuous may sometimes find good actions difficult, not because they have bad desires to overcome (as the *enkratēs* has) but because they have competing desires which are each equally good. In support he refers mainly to Aristotle's discussion of 'mixed actions' in III.1, which does indeed draw attention to such conflicts. But I think he is stepping outside his text when he infers that even the *virtuous* man will be torn by such a conflict.

Further reading

On the division of the soul into one part that has reason 'in itself' and another (containing desires-and-emotions) which can 'listen to reason', I recommend in particular Broadie (1991, ch. 2, sects. 2–3). There is an extended (and often controversial) account of Aristotle's concept of an emotion in Fortenbaugh (1975); for present purposes one might limit attention to sections 1–2 of his chapter 1, and chapter 4. On the general connection between virtues and desires-and-emotions the discussion in Kosman (1980) raises some pertinent questions, and the topic is pursued at length in Hutchinson (1986).

On the doctrine of the mean important reading is Urmson (1973), together with the objections in Hursthouse (1980–1). There are also useful points made in Hardie (1965/6).

On the general topic of 'learning to be good' see Burnyeat (1980). This begins from book II but also ranges quite widely over the rest of the *Ethics*. Sorabji (1973–4) is also relevant here, but perhaps more appropriately consulted under book VI.

Modern treatments of an ethics based on 'virtue' as the fundamental notion (which often claim descent from Aristotle) are many and various, and I cannot cover the subject here. But one might perhaps begin with von Wright (1963, ch. 7), and Foot (1978, title essay), who both pay some attention to Aristotle. Then one might go to the useful collection of readings in Crisp (1996).

Chapter III

Justice (Book V)

I N book II Aristotle has given his general account of virtues of character. This general account continues into III.1–5, with a discussion of responsibility, choice, and related notions. But it is more useful to regard this section of the *Ethics* as opening a different topic, which one may call Aristotle's 'theory of action', and I therefore postpone it for later treatment (in Chapter V). In III.6–V he gives his account of some particular virtues of character. There is no space here to comment on each of these, and I shall take up only the last, the account of justice in book V. Evidently, Aristotle takes this to be a virtue of supreme importance, since he gives it much more space than any other. But, at the same time, one has to add that book V, taken as a whole, is an unsatisfying piece of composition, and here, more than anywhere else in the *Ethics*, one has to bear in mind that what we are dealing with is not a finished work, prepared by Aristotle himself for publication. It is rather a compilation of several different essays, written sometimes from rather different perspectives, and not always brought into harmony with one another.[1]

As a matter of fact the overall structure of V.1–5 is reasonably clear and coherent, and I shall therefore devote most attention to this part of the book. But the remaining chapters are disjointed, and follow no clear overall plan, so I treat them only briefly in an appendix.

1. Universal and particular justice (V.1–2)

After some brief opening remarks of a rather general nature (1129^{a}7–17), Aristotle begins at 1129^{a}17 to develop the main topic of chapters 1–2, that the notion of justice is used in two rather different ways. Since his own way of leading up to this distinction is less clear (to the modern reader) than it could be,[2] I shall first introduce it in my own way.

One of the things that may make his account confusing to us is that the ambiguity he wishes to diagnose really does affect the Greek words 'just' and 'unjust' (*dikaios*, *adikos*) in a way that hardly applies to their English analogues. In many contexts the standard translations are perfectly adequate, for the Greek words in question do apply to those actions (or people) that we too would call just or unjust. But the Greek words *also* have a wider use, in which they are naturally translated simply as 'right' and 'wrong'. So in Greek it is not too easy to say something like

---

[1] To rectify matters commentators have often proposed reorganizations of the text as we have it (e.g. Jackson 1879; Gauthier and Jolif 1958–9; Irwin 1985a), but I do not find such reorganizations successful. Cf. Hardie (1968: 184).

[2] Aristotle is following a method advocated in his *Topics* (106^{a}1–22).

'that would be the *just* thing to do, but—in the circumstances—it would not be the *right* thing to do', for one would very naturally wish to use the same word in each case.[3] This is essentially the point that Aristotle wishes to make. Justice in the special sense is, on Aristotle's account, one virtue among others, and when he wishes to focus on this sense he speaks of 'particular justice'. Hence it is customary to label the more general sense, in which the word means simply what is right, 'universal justice'.[4] (Aristotle does not use this label himself, but he does not consistently use any other label either.) In this sense, he claims that, in a way, justice encompasses the *whole* of virtue.

Aristotle associates universal justice with what is lawful, and particular justice with what is fair (or equal).[5] That is why he leads up to his distinction by claiming that both the lawless and the unfair man are called 'unjust', wishing us to infer from this that 'unjust' has two meanings, and so correspondingly does 'just' (1129^a17-^b1). He admits, of course, that the two meanings are not unconnected, and explains that that is why they are not always distinguished from one another (1129^a27-31). Nevertheless, he insists that there is a distinction, and—as I say—he is basically right about this. But one may still doubt whether he has drawn this distinction in quite the right way.

Particular justice Aristotle associates right from the beginning (1129^a32) not only with conduct that is fair or unfair but also with a particular motive, *pleonexia*, which is—translated literally—the desire to have more than the next man.[6] This, I think, was simply a mistake on his part. It may be reasonable to say that one who acts on this motive will always be aiming for a result that is unfair, but it clearly is not true to say that one who acts unfairly is always, or usually, motivated in this way. I return to this in Section 5. Meanwhile, let us just set it aside.

Universal justice he introduces as lawfulness, but goes on later to characterize as 'complete virtue in relation to other people' (1129^b15-17, 1130^a8-13). As his discussion makes clear, he believes that one who displays all the virtues in relation to himself need not do the same in relation to others, whereas the converse does not happen. Consequently the man who is just in this universal sense will actually have complete virtue without qualification, but still all that is *meant* by saying that he is just (in this universal sense) is that he behaves to *others* with complete virtue ($1129^b15-1130^a13$). Now this second characterization of universal justice seems to me essentially right, but it then follows that the first is not. Aristotle thinks that he can move from the one to the other, because he claims that the law itself requires us to practise all the virtues (at least, in relation to other people) (1129^b19-24, 1130^b20-4). But this is a mistake, for two reasons.

First, it presumes that the laws in question are *good* laws, and so only do prescribe

---

[3] I am not saying that it would be impossible to make such a claim, for Greek—like English—also has other ways of expressing what it is right to do (e.g. as what one *ought* to do).

[4] 'Particular' is the standard translation of Aristotle's phrase *en merei* (literally 'in part'). But Irwin prefers 'special' throughout, and so supplies as a contrast 'general' rather than 'universal'.

[5] It is the same Greek word, *ison*, that one translates sometimes as 'fair' and sometimes as 'equal'.

[6] Ross translates 'being grasping', and Irwin 'being greedy'. Each captures one aspect of the notion, but omits the thought that one activated by this motive is trying to do better *than others*.

conduct that is in fact virtuous. But this is a presumption one cannot grant, and at two points Aristotle himself indicates as much. At 1129^{b}14–17 he tells us that the laws aim *either* at the common advantage of all, *or* at that of the best people, *or* of those in power, *or* something similar. (As 'something similar' one naturally suggests 'those who are wealthy'.)[7] In each case, the proponents will claim that the system aims for the good (*eudaimonia*) of the community as such (1129^{b}17–19), but presumably some must be mistaken. It is not clear how the different systems of law resulting can each be equally good. On a different point, Aristotle goes on to acknowledge that some laws may be correctly framed (given the aim), but others are 'hastily conceived' (1129^{b}25). So presumably he will admit himself that the actual laws of a particular community may go astray, and fail to prescribe conduct that is in fact virtuous. We may elaborate on this thought. Suppose that the laws demand that I should report to the authorities the slightest remark by anyone—including my closest family and my dearest friends—that is to the smallest degree in disagreement with the official government propaganda; then would it really be *right* for me to comply? Surely it is possible for laws to enjoin conduct that is not virtuous? Now it is *possible* that Aristotle really did think that, however bad the law is, it is always right to obey it,[8] but it is much more probable that he had not thought this point through. Let us qualify his position, then: when the laws are *good* laws, the conduct that they enjoin is in accordance with all the virtues.

But what of the converse? Should we accept that, at least where the laws are good laws, they will forbid *all* conduct that conflicts with the virtues? Aristotle clearly thinks so. The law, he says, instructs us to act bravely, temperately, and with good temper (1129^{b}19–24); again, whatever is unfair is against the law (1130^{b}12); again, and very generally, the law bids us live in accordance with every virtue, and avoiding every vice (1130^{b}23–4). But, one protests, am I really disobeying the law if I am more angered by your behaviour than I ought to be, if I divide sweets unfairly between my children, or if I fail to be as generous to charity as I should? Aristotle is apparently committed to saying 'yes'; in his view the law does—or anyway should—govern every one of my actions.

It is relevant here to note his discussion of 'equity' (*epieikeia*)[9] in V.10. His general conception of equity is that it supplies *corrections* to law, when a law is brought to bear on situations which the legislator had not foreseen. His point is that a law has to be framed universally, so that there will be an unlimited number of situations that fall under it, but one cannot expect them all to be anticipated in full

---

[7] There is a clear echo here of the position espoused by Thrasymachus in Plato's *Republic* I.338c–339a. Justice is obedience to the law, which is laid down in the interest of the rulers, whether the rulers be the people as a whole (democracy), or those who are high-born (aristocracy), or those with wealth (oligarchy), or whatever. (Compare also Aristotle's reference, at 1130^{a}3–5, to Thrasymachus' claim that justice is 'another's good', *Republic* 343c. It would appear that Aristotle thought better of Plato's Thrasymachus than Plato himself did.)

[8] The thought would not be without precedent. It is apparently endorsed in Plato's early dialogue the *Crito*.

[9] Irwin translates 'decency'. (This translation does better fit the use of the same word in other contexts, but in this context 'equity' is traditional, so I retain it.)

detail (1137ª31-ᵇ27). He then comments: 'This is the reason why not all things are governed by law, namely because in some cases it is impossible to lay down a law, so that a decree is needed' (1137ᵇ27–9). (The point about a decree is that it is a one-off ruling to fit a one-off situation.) We have an admission here that it is not practical for the law to regulate every minutest detail of our lives, but this is consistent with what I take to be Aristotle's perfectly genuine opinion that the more you can regulate by law the better.[10]

There is a clear contrast here between Aristotle's attitude and our own. We think that what is governed by law should be restricted, not only by practical considerations, but also for another reason: we value individual liberty. While of course we recognize that such liberty must be curtailed to some extent, if we are to have a stable society, and possibly for some other reasons (e.g. to promote fairness in society), still in general we insist that some individual liberty should be preserved. But I think it is fair to say that Aristotle does not recognize that individual liberty is of value for its own sake.[11] This is a significant divergence of views.

Two points may be urged in his defence. First, since his argument here concerns virtue *in our dealings with others*, one may say that this still leaves a sphere of 'private morality' that may remain immune from legal interference. While I do not think that this does actually represent his own opinion, still the point may be granted *argumenti causa*; it still leaves a considerable gap between his attitude and ours. Second, it may be said that the Greek word (*nomos*) here translated 'law' need not mean anything so formal as a written law, or even a rule which one can be tried in the courts for breaking; it is often to be translated simply as 'custom'. Then one might add that it is not particularly unreasonable to say that *custom* bids one to share fairly between one's children, to give generously to charity (if one has the resources), and so on. Now even here one might stickle, especially given Aristotle's own list of the virtues. (For example, neither law nor custom bids us have 'a ready wit'.) But the more important point is that it is clearly not mere custom that Aristotle is talking of, for if so he would hardly speak as he does of lawgivers ('custom'-givers?), or of some laws (customs?) being hastily conceived (1129ᵇ13, 25). He clearly does mean that the *laws* do—or ideally should—enjoin every virtuous action and forbid every vicious one. But in fact they never have done, and it is not sensible even to hold this as an ideal.

It was a mistake, then, to equate universal justice with lawfulness. If lawfulness is a virtue—i.e. always treating the law with respect, just because it is the law, and conforming to it unless there are really strong reasons not to—then it is surely one of the particular virtues, and not the whole of virtue. (And this remains true even if we say that it is not just the laws of the state that are concerned, but also the rules

---

[10] Compare I.2, 1094ª24–ᵇ7, where it said that political science (pursuing *eudaimonia*) directs what other sciences are to be pursued in the state, and to what extent, and that it lays down laws on what should be done and what avoided.

[11] In the *Ethics* one might say just that Aristotle ignores the topic. But in the *Politics* he treats explicitly of individual liberty, associates it with democracy (e.g. 1317ª40–ᵇ17), and does not approve of it (e.g. 1310ª28–38). (In this he resembles Plato, and rejects what we are apt to think of as a typically Greek ideal.) For some discussion, see e.g. Barnes (1990).

and regulations of any other institution one may belong to.) It may even be regarded as a particular form of justice (or anyway, of what *dikaiosunē* means in Greek), but, if so, it would not seem to be specially connected with fairness.[12] However, I turn now to what Aristotle does recognize as particular forms of justice, and which he does connect with fairness. At the end of chapter 2 he announces two such forms, namely justice in distribution and justice in rectification (1130^{b}30–1131^{a}9). It is clear that these are discussed in the ensuing chapters 3 and 4 respectively. But in chapter 5 we appear to be offered yet another form of justice as fairness.

2. Justice in distribution (V.3)

Aristotle's account of a fair distribution is essentially very simple: it is one in which the goods to be distributed are divided in proportion to the worth (*axia*)[13] of the recipients. Thus those of equal worth should receive equal shares, those of greater worth greater shares, and those of lesser worth smaller shares. He adds rather a lot of mathematical detail, which is not really of any importance, and the actual description of a fair distribution occupies just five lines (1131^{a}25–9). Moreover, most of these five lines are devoted to an interesting caveat: while all agree, he says, that a fair distribution is in accordance with 'worth of some kind', still there is no agreement on what is to count as worth; in a democracy each free citizen is counted as of equal worth, in an oligarchy worth is measured by wealth, or by high birth, and in an aristocracy by virtue (excellence).

Now the main and most obvious criticism is that Aristotle is evidently thinking too narrowly. As is shown by his initial description at 1130^{b}32, and by his caveat here, he has in mind distributions made by the state, e.g. distributions of honours, and of state offices (which are normally worth money). He does at one point mention distribution not of goods but of evils (1131^{b}20–3), but apparently he takes it to fall under the same account, so that the more worthy should receive less of the evil, and the less worthy should receive more, just because less of an evil may be equated with more of a good. We have only to think about this for a moment to realize how odd it seems from our point of view. But before I develop this, let me first state the general point: there are many *more* kinds of distribution than Aristotle is thinking of, and the concept of fairness applies to them all, but one cannot suppose that the criterion of fairness is the same in all cases.

We may begin by staying with the case in which it is the state that is the distributor. No doubt Aristotle would be astonished at the very wide range of benefits and burdens that the state distributes today; e.g. among benefits we may mention health care, education, and various financial subventions such as a children's allowance, or the dole; among burdens we may mention most obviously taxation in all its forms, but also such things as liability to jury service, or to military service, and so on. If

---

[12] To take a modern example, one who—when driving—always obeys the speed limits is showing (one aspect of) this virtue of lawfulness. But it is hardly natural to describe this as 'behaving fairly', or to describe as 'unfair' one who readily exceeds the limits, when there is clearly no danger in doing so.

[13] Ross translates 'merit', which I think is less close to what Aristotle is actually thinking of.

there really were some one notion of the 'worth' of a person, then apparently Aristotle would be committed to saying that the same people (the 'more worthy' ones) should get more of *each* of the benefits, and less of *each* of the burdens, but this is obviously absurd. To bring out the point, let us notice a quite different principle, famously endorsed by Karl Marx: 'from each according to his ability [i.e. his ability to bear whatever burden is in question]; to each according to his need [i.e. his need for whatever benefit is in question]'. This at least has the merit of recognizing that the appropriate criteria for benefits and for burdens are different, and (tacitly) that the criteria for each are relative to the particular burden or benefit in question. For example, those who 'need' education are not to be identified with those who need the dole. One thing Aristotle clearly got right: in cases where the state is the distributor the question of the right criterion to adopt, in any given case, is often a matter of political dispute. (For example, to what extent should taxation be 'progressive', aiming to take from the rich and give to the poor in such a way as to reduce the gap between the two?) But what he got wrong is more important: *different* criteria are appealed to in different cases, and there is often no agreement even about which criteria are the appropriate ones.[14]

The point becomes even more obvious when we note, as we should, that there are all kinds of other distributions, not done by the state at all, as for example when I am dividing a cake between several children. Should I take account of their different 'worth' (whatever that is), or of their different needs (e.g. if some are under-nourished), or of their different desires (e.g. if some are more hungry, or if some don't care much for strawberry icing), or of their different deserts (e.g. if one has been naughty), or what? Or again, consider a school prize-giving. Some may argue that prizes should be given to those who would benefit most from receiving them, but a more usual view would be that they should be given to those who have deserved them most. Yet how are we to reckon this? For example, desert may be reckoned by achievement, or by improvement, or by effort, and these need not coincide. So, again, one asks: what is the *fair* way of doing things? But it can hardly be assumed that there will be a unanimous answer.

Thus I think that the main criticism of Aristotle on distributive justice is that the problem is more complex than he allows. He mentions one criterion of fairness, and admits that there is dispute over how to apply it. But the truth is that there are lots of different criteria that we appeal to, and there is often dispute, not only on how to apply a given criterion, but also on which criterion is appropriate for the particular case. There is also another aspect to this question of fair distributions which should be mentioned. Sometimes one may count a distribution as 'fair' not because the *outcome* it delivers is in some suitable way fair, but because the *method* employed was fair. For example, a non-divisible good (or evil) may be distributed by lot,[15] and when all candidates have an equal claim this may be counted as fair, not

---

[14] In the *Politics* Aristotle has more to say on distributions by the state, and he not infrequently invokes what seem to us to be different and competing criteria, though he seems not to notice this point himself. For a useful summary, see Irwin (1988a: 427–8).

[15] In Greek democracies political office was often distributed in this way.

because it distributes equally but because it gives everyone an equal chance. This thought has played some role in modern accounts, though clearly it is not recognized in Aristotle's discussion here.[16]

But I end with a word of approval. Fairness is always fairness *as between* two or more people (one of whom may be oneself), and while it may sometimes seem a little forced to think of this in terms of 'distribution', still in fact the notion generalizes quite naturally. For example, consider the employer who discriminates against women, or Jews, or Japanese, or whatever it may be. He acts unfairly, and that is because his distribution of what it is in his power to give—e.g. employment, increased salary, greater responsibility; or on the other side rebuke and penalty—is an unfair distribution. Or again, consider the parent or the teacher who has a favourite child. This too is unfair to the others, and again it is a matter of distribution, i.e. distribution of love, or of care and attention, if not of more concrete things. Aristotle's idea of connecting fairness with distribution was, I think, a good one.

3. Justice in rectification (V.4)

Justice in rectification, says Aristotle, applies to transactions between one person and another, both voluntary and involuntary[17] ($1131^{b}25-6$). He has earlier listed some of these 'transactions'. Among 'voluntary' transactions he mentions buying and selling, lending and pledging, renting and depositing and hiring out; among 'involuntary' transactions he further distinguishes those that are done in secret, and those that are done by force; the former include theft, adultery, poisoning, pimping, enticement of slaves, treacherous killing, false witness; the latter include assault, imprisonment, murder, robbery with violence, mutilation, slander, insult ($1131^{a}1-9$). What is at first sight puzzling is that while the so-called 'involuntary transactions'—which hardly deserve to be called 'transactions' (*sunallagmata*) at all—do clearly introduce a situation that needs rectifying, it is not immediately clear why a 'voluntary' transaction, say of buying or selling, should need any 'rectification' at all. Indeed, no such transaction is given anywhere as an example in the discussion of V.4. But the answer is contained in a little parenthesis at $1131^{a}4-5$: 'they are called voluntary because the *origin* of these transactions is voluntary'. The implication is that what Aristotle is concerned with is what happens *subsequently*, and the universally accepted interpretation is that he is thinking of cases where one party *defaults* on the agreement that was originally reached voluntarily; he fails to fulfil his part of the bargain.

Aristotle's 'voluntary' cases, then, have become 'involuntary' (for one party involved) by the time they need rectification. In fact they seem to correspond well enough with one branch of what we call 'civil law', namely that dealing with

---

[16] In the *Politics* Aristotle does, in effect, recognize this reason for distributing by lot (e.g. $1317^{b}20-1$). The relevance to modern theories of justice is noted in Williams (1980: 195–6).

[17] *Hekousion* and *akousion*. I discuss the meaning of these words in Ch. V; meanwhile I use the conventional translations.

contract. Some of the cases that he calls 'involuntary' would fall under our other branch of 'civil law', dealing with tort (e.g. slander, assault, and in some cases theft), but most of them would come under what we call 'criminal law' (and we see little significance in the distinction between crimes committed in secret and those done by force). I shall revert to this distinction later.

Despite the wide variety of offences calling for rectification, Aristotle offers the same account for them all. In each case one party (the victim) has suffered a loss, and the other (the offender) has made a corresponding gain. So the fair way of rectifying the situation is to restore the status quo; the offender should be made to give up his ill-gotten gain, which should then be restored to the victim, so that the result is 'equality' once more. (In this case, then, the right mathematical description involves not a 'geometrical' but an 'arithmetical' proportion. That is, we take no account of the relative 'worth' of the parties involved, but treat each equally, 1131^b32–1132^a6. Once more, Aristotle goes into needless detail on the very simple mathematics involved, 1132^a24–$^b20$.) That is Aristotle's account of how *all* the various misdemeanours he has mentioned should be 'fairly rectified', but he does at least notice one problem.

He begins by mentioning theft (1132^a2–7),[18] where one can at least see how his leading principle should be applied: the stolen goods should be restored to the victim. But he goes on to mention wounding and murder (1132^a7–19), where evidently things are not so straightforward. He himself admits that the terminology of 'gain' and 'loss' is not clearly appropriate to these cases. So far as wounding is concerned, it does seem to us to make sense to set a financial 'price' on a wound, and to require that the offender pay that price to the victim in compensation. But when it comes to murder things are hardly so straightforward, for even if a price could be set, it cannot be paid 'in compensation' to the victim, for he is no longer there to receive it. If Aristotle had thought rather more about his longish list of 'involuntary transactions', he would surely have seen that his rather simple scheme for restitution cannot always be applied.

But let us go back to the case he mentioned first, namely theft, where at least it seems easy to apply his principle. If I steal your wallet, and am then apprehended, then on Aristotle's account what justice demands is simply that I return the wallet, with its contents intact. In that way I have forfeited my gain, your loss is restored to you, and we are once more back to where we were. So what do I do next? Well, of course, I look for another wallet to steal. For if I get away with it then I have gained, and if I do not then still I have not lost, but am simply back where I was. There is, in his account, nothing to *deter* the wrongdoer, for it contains no element of *punishment*. That is surely a rather serious omission.

Now one might well say that the two notions of punishment and restitution should indeed be distinguished from one another, and kept apart from one another. But one might wish to add that for each offence *both* notions should be considered.

---

[18] And, very briefly, adultery (1132^a3). Presumably he thinks that in this case one *man* has gained over another *man* by having sexual intercourse with his wife. But he does not tell us how this gain is to be restored to the injured party.

There is something very unsatisfying about the official theory behind English law, which is that in civil cases one considers only compensation and restitution, but not punishment, whereas in criminal cases one considers only punishment, and not compensation or restitution. For one wants to say that in many cases both notions have a role to play, and it is wholly artificial to separate the two considerations as English legal theory does.[19] But, to return to Aristotle, the main criticism that we must make is obvious: in his account in V.4 of how the law should operate he considers only restitution and not punishment, but the latter is quite clearly a particular area in which the notion of justice is applied. Indeed, if one looks at the chapter headed 'On Justice' in J. S. Mill's *Utilitarianism*, one finds that it is almost entirely concerned with the question of what counts as a just punishment, and why. Nor can it be said, in Aristotle's defence, that this question of just punishment—which nowadays we discuss in terms of deterrence, reform, retribution, and so on—is a modern question, not familiar in Aristotle's day. On the contrary, it figures quite largely in Plato's *Laws*,[20] and Plato himself attributes the deterrence theory to Protagoras, roughly two generations earlier.[21] I think Aristotle has no good excuse for his omission.

But perhaps something can be said in mitigation. While the notion of just punishment certainly is an important case of 'particular justice', perhaps it is not best viewed as falling under the general heading 'justice as fairness'. For fairness is always, as I have said, fairness between two or more people. So adjudicating between two parties, with a view to a possible redistribution of their present assets, is certainly a context in which fairness applies (all the way from a full-blown legal case to a mother attempting to adjudicate between her quarrelling children). But, it may be said, punishment is due not so much for an offence against a person as for an offence against a law (or other edict), and so it is not essentially a case of dealing fairly between two persons. This is clearly the case with victimless crimes (such as exceeding the speed limit, but harming no one thereby), but it is tempting to extend the principle to all cases where it is punishment that is in question, rather than (or as well as) restitution. Indeed, contemporary discussions very often take this line. Punishment, then, may certainly be just or unjust, and is an important case of 'particular justice', but (on this account) it is fair or unfair only in a quite different sense (e.g. if one person is punished more severely than another, for the same offence, and where there are no relevant distinguishing features).

---

[19] Of course practice does not always follow theory very closely. For example, in a case of slander, which is a civil case, the jury will often award 'punitive damages'; and in a case of burglary, tried under criminal law, still the stolen goods, if found, will be restored to the owner. (But, if they are not found, the legal process gives the victim no compensation.) However, it is the theory which explains why the very same offence may be tried twice, once under civil law and once under criminal law.

[20] Books IX-XI, *passim* (with the preamble at IV.719e–721e). There is an explicit distinction between restitution and punishment at 861e–863a.

[21] Plato, *Protagoras* 323c–324c.

4. Justice in exchange (V.5)

At the end of V.2 Aristotle promised us two forms of particular justice, namely justice in distributions and in rectifications. These topics are apparently dealt with in V.3 and V.4. V.5 then opens with a different idea, that justice is what is 'reciprocal' (*antipeponthos*, more literally 'suffered in return'), an idea that is credited to the Pythagoreans. This is apparently the idea familiar to us as the *lex talionis*, 'an eye for an eye, and a tooth for a tooth'; Aristotle cites a version credited to Rhadamanthus:[22] 'If he suffers what he did, straight justice will be done'. But he objects that this is not in fact the correct account either of justice in distribution or of justice in rectification (1132^b21–8).

He is clearly right about distribution; the principle of reciprocity is evidently not relevant to this, and no further argument is needed. He remarks that some people wish the principle to govern rectification, but again he claims that they are mistaken. Now, given his *own* account of rectification, i.e. as concerned always with restitution and never with punishment, this too is a correct point. For the principle claims that what the victim has lost (e.g. an eye, or perhaps his life) the offender should lose in return. So there is an equal *loss* on both sides, but no restoration of the status quo, and that indeed is not restitution. In fact the principle aims to be a principle of just *punishment* (and one in line with the 'retributive' theory), not of recompense. But Aristotle's own objection to it is curious. If I may modernize his example a little, it is that when a policeman strikes a civilian, he should not be struck in return; but when a civilian strikes a policeman he should not only be struck in return but punished in addition (1132^b28–30). How, one asks, does *that* fit with Aristotle's own theory of rectification? The plain answer is that it does not, for here Aristotle himself has slipped into thinking of punishment and not of restitution. Given this, his comment is of course fair, and plain 'reciprocity' needs to be amended to take account of it. A policeman, apprehending a suspect, is entitled to use force if his arrest cannot be achieved without it, whereas the suspect is not entitled to use force in order to resist arrest. That is why the one is not subject to punishment but the other is.[23] But the more significant point is that although Aristotle has in fact shifted his consideration here, from restitution to punishment, he himself fails to notice the fact. At any rate, he writes as if the example is already covered by what he has said in V.4, though plainly it is not.

But let us move on. Reciprocity, says Aristotle, is not the right account either of justice in distribution or of justice in rectification, but it *is* the right account of another kind of justice, namely justice in 'associations for exchange'. We must add, however, that what is needed here is not simple reciprocity but rather reciprocity 'in accordance with a proportion', for that is what holds a state together (1132^b31–4).

---

[22] A mythical son of Zeus and Europa, renowned for his justice. He became on his death a judge in the underworld.

[23] It is also relevant to ask, as Aristotle notes, whether the striking was deliberate or unintended (1132^b30–1).

I think it is clear that Aristotle means to introduce here a further form of particular justice, i.e. of justice as fairness. The basic question is: what constitutes a fair exchange? As his own discussion shows, this has as a special case: what constitutes a fair price? To this *we* naturally add, though Aristotle does not, the closely related questions of a fair wage, a fair profit, and so on, for the general area is, as one might say, justice in economics. It would seem that Aristotle had not anticipated this topic when he announced at the end of V.2 that there are just two forms of particular justice, concerned respectively with distribution and with rectification ($1130^{b}30$–$1131^{a}1$), but it is a natural enough continuation of the topic 'justice as fairness'.

(One *could* suggest[24] that it is intended to be part of the discussion of justice in rectification, namely the part concerned with voluntary transactions, of which the first examples were buying and selling ($1131^{a}2$–3). For there is no explicit discussion of such transactions in V.4, and it might be held that the summarizing sentence is meant to show that V.4 was restricted to involuntary transactions ($1132^{b}18$–20).[25] But in that case Aristotle's thought will have to be that where two people have made an unfair exchange this needs to be rectified, even though each entered into the exchange voluntarily, and each kept his side of the bargain. This seems to me improbable. So far as I am aware, no legal system does set out to 'rectify' such transactions (unless perhaps deception was involved), and besides V.5 begins by saying quite clearly that 'reciprocity' is *not* the right account of rectificatory justice. ($1132^{b}23$–32))

Just what Aristotle means by 'proportionate reciprocity', and by saying that it is produced 'by a combination in accordance with the diagonal' ($1133^{a}5$–7),[26] is obscure. The basic problem is clearly this: if a cobbler and a builder are to make a fair exchange, then we must establish how many shoes are of the same value as one house. But no help on this question seems to come from the kind of proportion that Aristotle cites: 'as the builder is to the cobbler, so the number of shoes must be to the house' ($1133^{a}22$–3; cf. $^{a}31$–3). I shall therefore pay no attention to this issue.[27] A more comprehensible suggestion is that the system is built on need (*chreia*),[28] and that money, which allows the value of different things to be measured by a common measure, may be viewed as a representative of need ($1133^{a}25$–31). Now the notion of need is not of course quite the same as our notion of demand—for example, the need for shoes may remain constant while the demand for them increases, as people become more fashion-conscious—but undoubtedly the two are closely related. So I

[24] This is Burnet's view (1900, *ad loc.*) There is a succinct refutation in Hardie (1968: 193–4).

[25] 'Hence what is just is intermediate between a certain kind of gain and loss, [i.e.?] those that are involuntary' (Irwin's translation simply omits the last four words). But it is easy to supply an alternative interpretation. When a transaction is freely entered into, but then one party fails to keep his side of the bargain, the other's resulting loss is of course 'involuntary'.

[26] In Ross's translation 'by cross-conjunction'. The basic idea is of a proportion that is naturally represented in the form $a_1/b_1 = b_2/a_2$.

[27] An interpretation which makes some sense is offered by Ross (translation, *ad loc.*) and largely followed by Hardie (1968: 196). This has no textual support. A different interpretation is offered by Judson (1997), but it seems to me that this does not in the end make much sense.

[28] Ross translates the word as 'demand', but I regard this as anachronistic.

think one may fairly say that what Aristotle gives us here is a first essay on how the value of things is in practice fixed by what we now call 'supply and demand'. It is of great interest as a pioneering attempt to explain how 'the market' actually works.[29]

If this is on the right lines, then we may add that, from our own contemporary perspective, Aristotle has missed the force of his own question. The market, left to itself, *will* determine the price of things. But is it *just* and *fair* that prices should be determined in this way? Or should we rather say that government should step in, and interfere with the 'natural' working of the market, in order to bring about a 'fairer' set of prices? If so, what kinds of interference are justified? Clearly, these are questions of great contemporary interest and importance, but it would be wholly unreasonable to reproach Aristotle for not seeing them. One must first get a clear view of how the market does work, before one can seriously ask whether interference would be justified.

I remark here that while this seems to be a perfectly good area in which to apply the notion of fairness, what is in question is whether an institution or arrangement is a fair one, and not whether anyone has *acted* unfairly. For example, if one believes that the market, left to itself, delivers an unfair distribution of goods, this is evidently not because any person, or group of people, is acting unfairly. Or again, suppose that we are discussing the Common Agricultural Policy of the European Community, and asking such questions as 'Is it fair to the consumer?' 'Is it fair to the farmer?' 'Is it fair to those outside the Community?' One who holds that the answers (or some of them) are 'no' will not usually conclude that those who devised and effected the policy were 'acting unfairly'. It is not *they* who were unfair, but the policy that they introduced. Similarly with the simple case that Aristotle discusses of an unfair exchange; it is the exchange itself that is fair or unfair, and it does not follow that one of the parties to the exchange was acting unfairly (cf. 1136^{b}9–13, on Glaucus and Diomedes). There are plenty of further cases. For example, a particular law may be stigmatized as unfair (e.g. one that sets the age of consent differently for heterosexual and for homosexual sex), or a particular custom (e.g. that women change their names on marriage whereas men do not), and so on indefinitely.

Suppose that, very charitably, we take Aristotle's discussion of 'reciprocity' to extend to all cases where we speak of the fairness or unfairness of (methods or) outcomes, irrespective of whether any particular agent can be said to have acted fairly or unfairly. Suppose too that we extend his notion of a fair distribution, as I have indicated, and add fair redistributions (to put right a wrong done), these both being cases where there is an agent who distributes or rectifies, either fairly or unfairly. Then one's first reaction might be that this yields a good coverage of the topic 'justice as fairness', and that Aristotle is to be congratulated upon it. Where his coverage mainly slips is in *other* areas of 'particular justice', not specially connected with fairness. I have mentioned two: the notion of just punishment is in effect omitted, and the possible virtue of respect for the law (or, more generally, the

---

[29] Finley (1970) argues that it has no such interest.

rules of any institution) is quite misdescribed in V.1 as the whole of virtue.[30] But we have only to reflect a little upon his own account to see that there is something else, rather important, that has not been given the attention it deserves.

When introducing his discussion of justice in rectification, Aristotle lists a number of actions which lead to situations which need rectifying (e.g. breach of contract, theft, assault, and so on). To act in any such way is to act unjustly, and one who does so habitually is an unjust person. Is this a matter of being unjust in the universal sense or in the particular sense? To judge from what he says about adultery at 1130ª24–8, his view is that it is particular injustice when the motive is greed (*pleonexia*)—for in this case the offender will be aiming for more than his fair share of something—but will conflict with some other virtue when the motive is something else. Let us, for the sake of argument, go along with this, and accept that the cheat, the thief, the mugger, and so on very typically are motivated by greed. But then, should there not be a corresponding form of particular justice (i.e. justice as fairness?), which one manifests simply by *not* being a criminal of this kind? If so, it seems to be a particular virtue which Aristotle fails to mention explicitly. Moreover, it would seem that he should not be allowed to say that the point is covered simply by justice in the universal sense, on the ground that he has equated that with lawfulness, and the actions in question would offend the law. For his idea appears to be that universal justice merely *sums up* all the various particular virtues, so nothing can fall under it which does not also fall under a separate and particular virtue. If this is right, then there is according to his own way of thinking a further form of particular justice, which is the justice of an agent, but is quite different from the virtue shown in justly distributing or rectifying. It is most easily characterized in a purely negative way: it is the virtue of *not* being led (by greed) into any of these various offences. Clearly this is going to create a difficulty for the doctrine that virtue is a mean, which I shall come to shortly, but that is not the point I wish to make here. (For, as we shall see, the virtues of the just distributor and the just rectifier also create a difficulty for this doctrine.) Instead, I mean to draw attention to the oddity of a particular virtue which is only negatively described. Looking for a more positive description, one might not unreasonably suggest that it is the virtue of respecting the rights and interests of other people (whether or not those rights and interests are protected by law). While this may be called a kind of fairness—i.e. fairness between oneself and others—that is perhaps not the most natural way of characterizing it. But in any case it seems to be a further form of particular justice, implicit in Aristotle's account, though not given due and explicit recognition.

Whether there are yet further gaps in his account is a question that I will leave to you.

---

[30] With charity, one might say that V.6–7 goes some way towards redeeming this error. See pp. 71–2.

5. Justice and the mean

At the start of his discussion in V.1 Aristotle had said:

With regard to justice and injustice, we must consider what kinds of actions they are concerned with, what sort of mean justice is, and what are the extremes which it lies between. (1129^{a}3–5)

At the end of his discussion of 'reciprocity' in V.5 he says, 'We have now said what justice and injustice are,' and he at once proceeds to an account of how just action is a mean (1133^{b}29–1134^{a}13). The paragraph then ends with:

Let what we have said be our account of the nature of justice and injustice, and similarly of what is just and unjust universally. (1134^{a}14–16)

Apparently the discussion is concluded. Of course, in our text there are still six more chapters to come, but I think it is a reasonable inference that at one stage Aristotle did intend his discussion to end here. The later chapters are afterthoughts which perhaps he tacked on himself, without properly integrating them with his overall discussion, or perhaps were appended by his editors, when they came to sort through his papers after his death. (Possibly, indeed, some of these chapters are *earlier* thoughts, which Aristotle himself would not have preserved. For example, he might have thought that what is now V.8 would be superseded by III.1.) In any case, the discussion of justice and the mean at the end of V.5 would appear to be Aristotle's final word on that topic, for it is not treated again. But we have had some brief hints earlier.[31]

One can see in advance that he is faced with a problem here, for his general scheme requires that each virtue of character be located between two opposing vices, yet justice seems to be opposed simply to injustice; there is not also *another* vice which is opposed to it 'in the opposite direction'. Now Aristotle has two main ways of exhibiting a virtue as a mean. In one, he finds an associated emotion, so that the virtue in question involves feeling that emotion just to the right amount, whereas there are distinguishable character traits which can be described as having a tendency to feel that emotion 'too much' or 'too little' (i.e. 'too often' or 'too strongly' or both, and 'too seldom' or 'too weakly' or both). In the other, he finds a range of actions which can be summed up in the same way. The first method is what is standardly taken to be his official doctrine, and is pursued with courage, temperance, and good temper; but the second seems to be the method that is actually pursued in some other cases, e.g. with generosity and with honesty.

---

[31] I am sorry if this account is confusing for those using Irwin's translation, for Irwin transposes all of 1133^{b}29–1134^{a}35 — i.e. what is in our text the end of V.5 and beginning of V.6 — to 1135^{a}5, i.e. to near the end of V.7. This does indeed increase the appearance of a single connected discussion, but I am sure that the appearance is misleading. (I mention here one small indication. At 1134^{a}23–4 Aristotle says that the relation of reciprocity to justice has been discussed earlier *eirētai proteron*. On Irwin's reconstruction, this sentence follows *immediately* on the discussion of reciprocity. But that is not how Aristotle uses 'earlier'.)

One strongly suspects that one of his reasons for associating (particular) justice with being greedy (or grasping, or wishing to do better than the next man) was that this was supposed to be the associated emotion. For example, when discussing distribution he says, 'the man who acts unjustly has too much, and the man who is unjustly treated too little, of what is good' (1131^{b}19–20; cf. 1138^{a}28–31). Clearly he is *thinking* of a case where the unjust distributor is himself one of the recipients, and allocates more than he should to himself, and consequently less than he should to the other(s). In such a case, his motive is likely to be greed. But the objections are quite clear. Even where the distributor is one of the recipients, he may allocate the right amount to himself, but still distribute unjustly, giving too much to one of the others and too little to another. (He may, for example, be motivated by favouritism, or by spite.) But very often the distributor is not one of the recipients anyway, and this is yet more obvious in the case of 'rectifications'; the judge in such cases is very unlikely to be one of the parties to the dispute, yet it is still he, and not the parties involved, who will judge justly or unjustly. If he does judge unjustly, then it *may* be that his motive is greed—e.g. if he has been offered a bribe—but clearly it does not have to be; his decision may be due purely to carelessness and inattention.

Even at a later stage, when he has apparently seen (or anyway half-seen) the difficulty with his position, still Aristotle tries to hang on to it. Speaking with distribution in mind he says,

But if he judged unjustly, and did it knowingly, then he himself is greedy, either for gratitude or for revenge. So one who has judged unjustly for these reasons has got too much, exactly as though he were to take a share of the goods unjustly divided. For indeed one who judges about land may receive for it not land but money. (1136^{b}34–1137^{a}4)

But clearly this is very special pleading, and should not be allowed. One cannot reasonably count as 'greed' just *any* desire—for gratitude, for revenge, or even simply the desire not to be bothered—that may lead to an unjust judgement.

The truth is, then, that there is no special association between particular injustice and greed, and even if there were it would not in fact help us to explain how justice is a mean.[32] For perhaps we may accept that there is a disposition which can be described as wanting more than one's fair share, which is a vice. And we may add also that there is a 'mean' disposition of wanting just one's fair share, which might be called virtuous. (Though in fact it would not be the virtue of justice.) But can we add that the disposition to want *less* than one's fair share is another, opposing, vice? It certainly would not seem so; and indeed Aristotle himself appears to imply that taking less than one is entitled to is a mark of decency (1136^{b}20–1), and of generosity

---

[32] I would claim that the same applies to the variety of particular justice which I have described negatively as refraining from (illegal?) acts which harm others, and positively as respecting their rights and interests (p. 66). But for the most part my discussion ignores this kind of particular justice, since for the most part Aristotle's does too.

($1120^a31–^b1$; cf. $^a19–21$), which is a virtue and not a vice.[33] We shall not, then, reach a defensible doctrine of the mean in this way.

This appears to be Aristotle's own opinion too, at least eventually, for in his final discussion at the end of V.5 he makes no mention of greed, or any other emotion, but speaks entirely of just and unjust *actions*. He says:

> It is clear that acting justly is a mean between acting unjustly and being treated unjustly, for the one is to have too much and the other too little. But justice is a kind of mean, not in the way that the other virtues are, but because it aims at the mean, whereas injustice aims at the extremes. ($1133^b30–1134^a1$)

The first sentence here evidently cannot be accepted, for there is no proper opposition between acting unjustly and being treated unjustly. The best that one can extract from this line of thought is that (*a*) being treated justly is a mean between being treated unjustly in one way, i.e. being allocated too much, and being treated unjustly in the other way, i.e. being allocated too little; and similarly (*b*) acting justly towards a given person is a mean between two unjust ways of acting towards *him*, i.e. giving him too much or too little. But since being *treated* either justly or unjustly manifests neither vice nor virtue, it can only be (*b*) and not (*a*) that is relevant here. And (*b*) still does not give us what we want, for the two ways of acting unjustly to a person do not manifest *different* vices. On the contrary, to act in one of these ways to one person (possibly oneself) is automatically and at the same time to act in the other way to another (who again might be oneself).[34] So even in this way one cannot restore Aristotle's standard scheme of one virtue between two opposing vices, each diverging from it in opposing directions, so that *some* people manifest the one while *others* manifest the other.

With some charity, we may say that Aristotle appears to admit this when he goes on to say in the second sentence that justice is not a mean in the way that the other virtues are. It is a mean, i.e. a middle (*meson*), only in the sense that what it aims for is a 'middle' outcome,[35] that is to say an 'equal' outcome (*ison*), that is to say a 'fair' outcome (*ison*, once more). As we have been told, the fair outcome of a distribution is one that treats the recipients in proportion to their worth, and the fair outcome of a rectification is one that treats both parties equally (irrespective of their worth).[36] It is in that sense that justice 'aims for a mean', and so—by a transference

---

[33] One might, of course, suggest a vice of pusillanimity—to be compared, perhaps, to such alleged 'vices' as insensibility (not desiring even normal pleasures of eating and drinking) and self-deprecation (understating one's own merits). But nothing in the text suggests this. (And again the alleged vice would not be a form of injustice, for Aristotle holds that one cannot treat oneself unjustly. See next note.)

[34] In the later chapters Aristotle will take up the question whether assigning to oneself less than one's fair share may be described as 'acting unjustly' towards oneself. (He answers in the negative.)

[35] As we have noted (p. 43), Aristotle says in book II that *every* virtue 'aims for what is middle' (1106^b28), so how does this produce a distinction? I presume because the 'middle' at which justice aims is not that kind of 'moderation' which one could depart from in two *different* ways, characterized as 'too much' and 'too little'.

[36] In the subsequent elucidation ($1134^a1–13$) Aristotle mentions only distribution explicitly, though it is easy enough to apply what he says to rectification. But 'justice in exchange' has apparently dropped out of view once more (as at $1130^b30–1131^a9$), and with some reason. For, as I have noted, when two people agree on an exchange that is in fact unfair, no one need have *acted* unjustly.

which is not too unnatural—can be said to *be* a mean. But there is no disguising the fact that it forms a counter-example to Aristotle's general theory of virtue of character. How serious is this?

6. A general observation on 'virtue of character'

Well, in one way it is quite serious, for it shows that the other counter-example I have already mentioned, i.e. honesty (p. 49), is not an isolated case. Both honesty and justice have it in common that there is no one emotion which is specially associated with them, which can be felt too much or too little, nor any one range of actions which can be said to diverge from the right course either in the direction of too much or in the direction of too little. It is not too difficult, if one goes outside Aristotle's own list, to think of other virtues in a like case. For example, when one is asked to write a character reference, one is often instructed to comment on the candidate's honesty, trustworthiness, and reliability. While these three virtues are no doubt related, they are not the same as one another, but they all fail Aristotle's scheme for similar reasons. Or again, consider prudence, in the sense of taking thought for the future consequences of what one does, and its single opposite, imprudence. But here someone is likely to protest that prudence should be counted, not as what Aristotle calls a virtue of character, but rather as what he calls a virtue of intellect.[37]

This brings to the fore the overall straitjacket that Aristotle imposes: *every* virtue must be either what he calls a virtue 'of character', or what he calls a virtue 'of intellect'. The primary distinction (in I.13) is that the former are virtues of that part of the soul that has desires and emotions, and the latter of that part that has reason ('in itself'). But a further distinction (in II.1–4) is that the former are acquired by training and habituation, whereas the latter are not (but rather 'by instruction', at least for the most part, 1103^{a}15–17). A further distinction, imported simply by Aristotle's nomenclature, is that the former can fairly be called virtues 'of character', whereas the latter are not naturally so viewed, but I would say myself that little weight can be put on this. Now, all the examples that I have just noted as departing from Aristotle's standard scheme—i.e. being fair-minded (in all its forms), honest, trustworthy, reliable, *and* prudent—can very reasonably be said to be acquired by training rather than instruction (and to be traits of character, for what that is worth). Moreover, the reason why training is required, rather than mere instruction, is that one who fails to exhibit these virtues is (usually) led astray by some emotion, and it is fundamentally the emotions that require training, for they will not yield to mere instruction. This is certainly a good reason for associating these virtues with that part of the soul that has emotions. But there is no *one* emotion

---

[37] The word 'prudence' has actually been used as a *translation* of Aristotle's word *phronēsis* (e.g. in Thomson 1955), and *phronēsis* is one of the two intellectual virtues, standardly rendered 'practical wisdom'. I would say that the treatment of *phronēsis* in book VI scarcely ever in fact mentions what we call prudence—though perhaps it should have done. (The exception is 1141^{a}26–8, where Aristotle notes that an *animal* which has prudence, i.e. forethought, is said to be *phronimos*.)

with which each is associated, and consequently no single scale, ranging from 'too much' to 'too little', on which they can be said to occupy a 'middle'.[38] That is why they do not fit Aristotle's general account. We might, then, suggest *three* different kinds of virtues: those associated with some one emotion;[39] those associated in a general way with the emotions, but with no one in particular; and those associated purely with the intellect. But I do not wish to suggest that it will be easy to draw the line in particular cases; we have at best a spectrum, and at worst a mere miscellany.

Appendix. V.6–11

Chapter 6 opens with a brief paragraph which introduces a distinction between being a just or unjust person (which is a disposition) and doing something just or unjust (which may have a different explanation) (1134[a]17–23). This theme appears to be resumed in the last few sentences of chapter 7 (i.e. 1135[a]5–15), and is then treated more fully in chapter 8. So one may sympathize with Irwin's reconstruction, which follows the order just suggested.[40] In the remainder of chapters 6 and 7 we have a continuous discussion of what Aristotle calls 'political justice', introduced by the comment that our enquiry concerns not only what is just without qualification (*haplōs dikaion*) but this topic too.

By 'political justice' Aristotle means the behaviour of the free citizens of a community (*polis*) that is governed by law (1134[a]26–32). Now since in V.1 he had identified lawfulness with that universal justice that is the whole of virtue (in relation to other people), one might suppose that he is here reverting to the topic of universal justice. On the other hand, he apparently associates the corresponding injustice with greed when he says that it is 'taking more for oneself of what is unconditionally good' (1134[a]32–4; cf. 1131[b]19–20), and this of course was associated in V.1–2 with particular justice. The best reconciliation seems to be that he is in effect here recognizing something like the point I was urging on pp. 57–8, that respect for the law is a particular 'part' of justice, though not one of the 'parts' analysed in V.3–5. This is reinforced by the fact that it is only the behaviour of free and equal citizens towards one another that he regards as governed by the law—'equal either proportionately or arithmetically' (1134[a]27–8)—and he distinguishes this from the unequal relations that obtain between a master and his slaves, or a father and his children, and, to a lesser extent, between a man and his wife (1134[b]8–18). In these cases there can only be something 'analogous' to justice (i.e., one presumes, to 'political' justice).

This leads him in chapter 7 to an interesting but rather confusing distinction between political justice which is 'natural', and that which is merely 'legal' (*nomikon*), by which he evidently means 'conventional' (*sunthēkēi*, 1134[b]32). The distinction is confusing because he apparently begins with the idea that what is natural 'has the same force everywhere alike, and does not depend on what anyone may think' (1134[b]18–20), and this, he says, leads some

---

[38] If there is in some cases a suitable scale of *actions* (as with generosity), that is a mere accident.

[39] Or perhaps: some *few* emotions. (Recall that Aristotle himself associates courage with *two*, i.e. both fear and confidence.)

[40] It seems probable that the editors placed the opening of ch. 6 where it is because the remainder of ch. 6 appears to contain a parenthetical reference to its thought at 1134[a]32–3. But I observe (*a*) that the first eight words of our present ch. 8 mark it as beginning a new topic (i.e. 'Since what is just and unjust is as has been said'), and (*b*) that the final words of our present ch. 7 promise a fuller discussion 'later', and yet there appears to be no suitable discussion 'later', unless it be ch. 8 itself. So my own guess would be that ch. 8 was written to *supersede* the initial treatment in 1134[a]17–23 and 1135[a]5–15.

people to suppose that all law is merely conventional (1134^{b}20–7). But, he responds, 'this is not how things are, though it is in a way' (1134^{b}27–8). His point seems to be this. Some laws are *purely* conventional, i.e. in cases where some law or other is required, but it does not matter exactly what this law is. To give a modern example, one must have a law saying which side of the road to drive on, but it does not matter which side is chosen. To cite one of his own examples, 'in this case one sacrifices a goat, rather than two sheep' (1134^{b}22). He compares a system of weights and measures, where again the point evidently is that it does not matter which system is adopted, but once a particular system is adopted it must be adhered to (1134^{b}35–1135^{a}3). Some laws, then, are purely conventional in this sense, and these of course differ from one society to another. But he also wishes to maintain that some are not, and these others—he gives no examples—may in contrast be called 'natural', even though they *too* differ from one society to another (1134^{b}29–30). This is the admission that his opponents are right 'in a way'. But where is the argument to show that they are also in a way wrong? All that we are offered seems to be (i) the appeal to us to recognize that not all laws are purely conventional, in the way illustrated (1134^{b}30–4), (ii) the analogy that while being right-handed is 'natural', still it is not universal (1134^{b}34–5), and (iii) the claim that only one law is by nature *best*, for *all* societies (1135^{a}3–5). While we may perhaps accept that these points help to clarify Aristotle's own position, still they would hardly convince his opponent.

The remaining chapters are of less interest, and I comment on them only briefly. Chapters 8, 9, and 11 may be viewed as forming a continuous discussion. The main interest in chapter 8 is what it has to say about 'voluntary' action, and I shall make some comments on this in the appendix to my Chapter V. This chapter leads quite naturally into a discussion of various problems about justice and injustice in chapter 9, in particular the question whether one can act unjustly towards oneself. This topic is further resumed (and interwoven with some others) in chapter 11. But I think it is not unfair to say that both of these chapters, considered either individually or together, give one the impression of a somewhat scrappy and ill-organized treatment. They are interrupted by chapter 10, which deals with a different issue altogether, namely the relation between law and 'equity', and is very nicely organized. (I have already touched on this topic on pp. 56–7.) It might be natural to view this chapter as continuing the treatment of the law, which was the main focus of (most of) chapters 6 and 7, and to consider reorganizing the text accordingly. But it would be misleading to attempt to give the impression that these chapters 6–11 form a discussion that is both continuous within itself and continuous with what has preceded them. They are, as I said at the outset, miscellaneous further essays or notes on justice, not properly integrated either with one another or with the scheme announced in V.1–2.

Further reading

Hardie (1968, ch. 10) gives a useful account of all of book V, paying more attention than I do to the second half of the book.

Essential reading on this topic is Williams (1980), who brings out more strongly than I do how justice differs from other more typical Aristotelian virtues. There is a response to Williams in O'Connor (1988).

There has been some discussion of Aristotle's account of distributive justice; a helpful treatment, which also contains much information on Aristotle's *Politics*, is Irwin (1988a, ch. 20); a treatment devoted almost entirely to the *Politics* is Keyt (1991). It is useful to compare here some modern accounts which are not specially concerned with Aristotle; I suggest Rawls (1958), which puts in a brief way some of the main points of his lengthy (1971), and for a different approach Nozick (1974, esp. ch. 7). I am not aware of anything that I would wish to recommend on Aristotle's account of justice in rectification. His puzzling account of 'justice in exchange' is interpreted in one way by Ross (in a footnote to his translation; this view is elaborated by Hardie 1968: 196–201), and in a different way by Judson (1997); I confess that neither of these interpretations seems to me very convincing. Those interested in economic theory will be provoked by Finley (1970), who argues that Aristotle makes *no* contribution to 'economic analysis', properly so called.

For some discussion of 'political justice' (to which I pay little attention), see e.g. Yack (1993, esp. ch. 5).

Chapter IV

Virtues of intellect (Book VI)

1. The introduction (1138ᵇ18–34)

VIRTUE of character was defined in II.6 as a middle state 'determined by reason, and in the way that the man of practical wisdom would determine it'. That definition is recalled at the start of VI.1, in the form 'the middle is as correct reason says',[1] and we are apparently promised an account of what correct reason does say. For, if we do not know this, we will be none the wiser (1138ᵇ29–30). But this short introductory section ends on a rather different note, as Aristotle concludes that 'we must determine what correct reason is, and what its definition is' (1138ᵇ33–4). There is surely a difference between knowing what correct reason says, and knowing what it is, and apparently either knowledge might be obtained without the other.

In fact it is the second description of the task that fits the discussion to come, and not the first. When we reach the end of book VI we really are 'none the wiser' on what correct reason *says*, i.e. on its conclusions about the right dispositions to have and the right way to act. We are even left in some considerable doubt on just how it is supposed to reach these conclusions—i.e. what premises it starts from, and how they are obtained. These are issues which I shall take up towards the end of this chapter. For the most part Aristotle concentrates upon saying what it is, and this only in the most general terms. Basically, his approach is to distinguish the relevant form of reasoning from other forms which do not contribute to virtue of character.

To outline his plot very briefly: he begins with a distinction between theoretical reasoning and practical reasoning, and hence between theoretical excellence and practical excellence. These are the two main excellences (i.e. virtues) of the reasoning part of the soul, and Aristotle certainly presents them as if they are quite independent of one another, so that one can in principle have either without the other.[2] The distinction is introduced in chapters 1–2, and then further clarified in chapters 3 and 5–7, as the nature of each is expounded more fully. Evidently it is practical reasoning, and not theoretical reasoning, that is relevant to the determination of the mean in conduct. But in chapter 4 he turns aside to introduce a second distinction, between practical reasoning properly so called and what may be called 'technical' reasoning, i.e. the reasoning employed in a craft or skill (*technē*). In a broad sense, each is a kind of reasoning concerned with what to do, and is thereby distinguished from theoretical reasoning; nevertheless they need separating from

---

[1] Ross as usual translates *orthos logos* not as 'correct reason' but as 'the right rule', and (at the end of the section) takes its *horos* to be not its definition but rather 'the standard that fixes it'. Recall my Ch. I. n. 29.

[2] In this he conspicuously disagrees with Plato.

one another. In our vocabulary, it is very natural to say that the distinction he is aiming at here is that between moral excellence and technical excellence, but one must reiterate that 'moral' is our word and not Aristotle's; his own word is 'excellence of character'. These, then, are the main distinctions that he draws between different kinds of reasoning, and hence different excellences of the reasoning part. But there are also some subdivisions, as we shall see. In fact chapters 8–11 may well be regarded as going on to introduce subdivisions within practical thinking, and it is not until the end of chapter 11 that Aristotle tells us that he has now said what excellence (virtue) in practical and in theoretical thinking is.

The final two chapters are introduced as stating and then resolving various puzzles about these virtues, but their primary interest is in what they have to say about the relation between practical wisdom on the one hand, and virtue of character on the other. This, as we shall see, is very puzzling.

2. The theoretical and the practical (VI.1–3, 6)

Aristotle begins his discussion (at 1138^b35) by reminding us that in I.13 the soul had been divided into a part that has emotions and a part that has reason. Virtues of character concern the former, and virtues of intellect the latter. But he now proceeds at once to subdivide the part that has reason, and the principle of his distinction is clear: one part deals with facts that are necessary, and could not be otherwise, the other with what is contingent and could be otherwise. This distinction is relevant because theoretical knowledge, and theoretical reasoning, concern only the former, whereas practical reasoning must concern the latter, since it must take account of the particular situation in which one is placed, and that is always something that could have been otherwise. Let us first elaborate Aristotle's conception of theoretical knowledge.

He thinks of it as involving two main abilities, first the ability to deduce conclusions from premisses (chapter 3), and second the ability to grasp those ultimate premisses from which deduction must start (chapter 6). So his ideal is that of a deductively organized body of knowledge, such as (in his day) was exemplified by geometry, but in fact by little else. It starts from axioms (or 'first principles') which are necessary truths, and then proceeds purely by deduction from these axioms, so that the conclusions reached are also necessary. This much may fairly be inferred from what he tells us here (in chapters 1, 3, and 6), but is also amply confirmed by his fuller account in the *Analytics*, to which he twice refers us (1139^b27, $^b32$). But I should add two further elucidations. First, Aristotle does not himself think that such theoretical knowledge is available *only* in the case of geometry (or, more generally, mathematics). On the contrary, his usual list[1] of the main areas of theoretical knowledge is (i) theology (the study of god), (ii) mathematics, and (iii) 'physics' (i.e. the study of nature, *phusis*), which includes all that we would call the 'natural' sciences, for example biology and psychology, as well as what we call

---

[1] See e.g. *Metaphysics* E.1.

'physics'. No doubt theology and physics were not in fact presented as axiomatized bodies of knowledge in his day, but he clearly felt that in principle they could be, and that this was the ideal to work towards. Second, we must note that his claim that theoretical knowledge deals only with *necessary* truths should not be mis-understood. He did not distinguish as we do between various kinds of necessity—e.g. analytic necessity, metaphysical necessity, and physical necessity—and I am sure that he would have welcomed Newton's *Principia Mathematica* as a statement of the *necessary* truths that lie at the basis of physics (i.e. Newton's three laws of motion, and his law of gravitation). It is not the least bit easy to give a suitable criterion to fit Aristotle's conception of necessity, and I shall not attempt it. But at least this much can be said: he takes it as obvious that the truths in question must be *universal*—that is to say, they hold for all places and for all times[4]—and that this is a consequence of their being (in the appropriate sense)[5] necessary (cf. 1139^{b}22–4, 1140^{b}31–2). That is why he says that theoretical knowledge concerns universals only, and not particulars, and this is his main contrast between theoretical and practical knowledge. Presumably it is universals and particulars that are distinguished at 1139^{a}10 as the two different kinds of objects of thought.[6]

In what I have said so far I have been deliberately avoiding Aristotle's own terminology for the various faculties and excellences of the soul involved. This is because his terminology is confusing, and especially so when the book is read in translation, the root cause of the difficulty being that Aristotle is in fact employing perfectly ordinary Greek words, but twisting them into his own special and technical meaning. So, when translating, there is a strong temptation to find English words which express Aristotle's own technical use, rather than the ordinary use of these words. I hope it will be useful if I here pause for a brief glossary on some of the more crucial words.

Epistēmē. Elsewhere this word is standardly translated 'knowledge', and it covers knowledge of any kind. But Aristotle (influenced by Plato) is almost always inclined to think that knowledge *proper* is always of universals.[7] Here he restricts the word still further, so that it is explained in VI.3 as the knowledge of conclusions deduced from universal and necessary premisses. Hence both Ross and Irwin translate the word as 'scientific knowledge', or sometimes as 'science'. (For example, the part of the soul concerned with theoretical reasoning is called by Aristotle 'the *epistēmonic* part'; both Ross and Irwin translate this as 'the scientific part'.)

Nous. In common usage this is a very general word, and might be translated simply as 'thought' or 'understanding' or 'intelligence'. In VI.6 Aristotle assigns to

---

[4] Elsewhere (e.g. *Posterior Analytics* 96^{a}17–19) he admits that some basic truths of physics may hold only 'for the most part'. Since his discussion in the *Ethics* omits this qualification, I shall say nothing about it. (For some discussion, see e.g. Reeve 1992, sect. 2.)

[5] In *another* sense what has already happened is necessary (i.e. cannot be otherwise, 1139^{b}8–9). (We might wish to distinguish here between what *cannot now be* otherwise, and what *could not have been* otherwise, but Aristotle makes no such distinction.)

[6] The passage also alludes to Aristotle's strange doctrine that when the soul thinks of an object it becomes similar to that object (*De Anima* III.4).

[7] A well-known (and problematic) exception is *Metaphysics* M.10.

nous the task of reaching the 'first principles' from which theoretical knowledge starts (and he tells us that it works 'by induction' from particular cases, 1139^{b}26–31). In VI.11 (1143^{a}35–^{b}5) *nous* is assigned what seems to be a different task in practical reasoning. With a view to the tasks assigned (and with a view, too, to his own conception of what Aristotle means by 'induction'), Ross translates the word as 'intuitive reason', whereas Irwin looks more to the common usage and translates it as 'understanding'. In this case, where the word is being used technically I shall leave it untranslated, but quite often it is also used in its ordinary and general sense (e.g. in VI.2 at 1139^{a}18, ^{a}33, ^{b}4).

Sophia. A traditional translation is 'wisdom' (as in '*philosophia*' = 'love of wisdom'), and Irwin translates accordingly. But Aristotle restricts this word to the virtue of the theoretical part of the soul, which is not at all a natural restriction, either of the Greek *sophia* or of the English 'wisdom'. Hence Ross prefers the artificial phrase 'philosophic wisdom', and I shall similarly use 'theoretical wisdom', in order to bring out Aristotle's intended contrast with what is practical.[8]

Phronēsis. The traditional translation, *when translating Aristotle*, is 'practical wisdom', which Ross uses and which I shall follow. This is because for Aristotle *sophia* and *phronēsis* are respectively the excellences of the theoretical and the practical parts of the reasoning soul. But one has to add that, outside Aristotle, one would never think of using such a translation. The root meaning of the word is contained in the verb *phronein*, which simply means 'to think', and so one might translate it as 'being thoughtful' or as 'being sensible' (or perhaps 'prudent'?), or as 'being intelligent'. Irwin prefers the last. While this is a perfectly reasonable attempt to capture the ordinary use of the word, it gives no hint of the very special meaning which Aristotle assigns to it in the *Ethics*.

I now leave this awkward question of how best to translate Aristotle's peculiar terminology, and come back to his thought, which is more straightforward. The main thought so far is that we should distinguish between a theoretical and a practical part of the soul, because each deals with a different object—the former with necessary and universal truths only, and the latter with truths that are particular and contingent, since it is concerned with action, and action takes place in particular circumstances that could have been otherwise. Is this a good reason for distinguishing two 'parts' of the reasoning soul? Well, I think the basic answer is 'no'.

Let us start with a simple point which accepts Aristotle's own view of what 'theory' is, and uses the most promising example of it, namely geometry. We may accept that geometrical theory is concerned wholly with universals from beginning to end, but nevertheless it is quite clear that this theory also has *practical applications* in which it is applied to particular situations in working out how to act. Indeed, the very etymology of the word 'geometry' shows this, for it means literally 'the measurement of land', and it is clear that the ancient Egyptians did apply it for

---

[8] In ordinary Greek *sophia* may mean just skill or expertise *at* a particular task. Thus, as Aristotle notes at 1141^{a}9–12, one may say of someone that he is a *sophos* sculptor, meaning by this that he is an expert sculptor. But, as he makes clear, this is not his own usage of the word.

this purpose. Or, to take another example that was surely familiar to Aristotle, Thales is credited with introducing a way of applying the geometrical theory of similar triangles in order to determine the distance from land of a ship at sea. If we think of things from our own perspective, the scope of what we call 'applied science' is of course enormous, and yet Aristotle appears to be denying its existence. For science proper, he says, belongs to one part of the soul, the 'theoretical' part, whereas the 'practical' part is concerned with a different type of proposition altogether. It follows that theory is never relevant to practice, which is an evidently absurd result.

This shows that, even on Aristotle's own conception of what theoretical reasoning is, it is quite possible for a piece of theoretical reasoning to form *part of* a piece of practical reasoning. The converse cannot be clearly shown, at least so long as we accept his view that theoretical reasoning concerns universals throughout.[9] But we should not accept it, and nor indeed should he. For he himself accepts that *nous*, which provides the axioms of a deductive science, both is a part of theoretical reasoning and works by induction from particulars. Nowadays we might wish to put it rather differently: the axioms of a deductive science are confirmed or disconfirmed by whether or not their consequences fit with the particular facts of observation and experiment. But in either case it is being admitted that the evaluation of particulars, i.e. particular observations and experiments, is a part of 'theoretical reasoning'. There is therefore no strong contrast with 'practical reasoning' on this point. But, in order to put my objection more strongly, I need to shift from Aristotle's own conception of a theoretical enquiry into a more general notion.

Aristotle's dominant conception of practical thinking is that it is thinking with a view to action. The natural contrast to this is just thinking undertaken with any other purpose in view, but perhaps we may be allowed to specialize this a little (in Aristotle's direction) by saying that our contrast is with thought which aims simply at knowledge, i.e. 'knowledge for its own sake'.[10] Then the fundamental criticism is this: exactly the *same* thinking may be undertaken for either purpose, but it is surely ridiculous to credit it to one part of the soul in the one case, and to a different part in the other. We may illustrate this with an example of Aristotle's own: at 1112ᵃ28–9 he tells us that 'no Spartan deliberates about how the Scythians would best be governed'. His point is that what he calls 'deliberation' is assigned a role only in practical thinking, i.e. thinking with a view to action, and no Spartan will intend to *act* on any such thoughts. For the Scythians are too far away, are notoriously ungovernable anyway, and clearly would pay no attention to what a Spartan might think. But surely, a Spartan might *think* about the problem anyway, even if Aristotle will not allow this thinking to be called 'deliberation'? Let me update the example. An undergraduate studying politics may well be asked to think about the merits

---

[9] Kenny (1978: 171) is I think alone in supposing that Aristotle believes that particulars have a part to play in theoretical reasoning.

[10] This therefore sets aside several mental activities which have neither of these aims, e.g. from mere daydreaming on the one hand to, on the other, all the concentration that goes into an attempt to appreciate some 'difficult' work of art.

and demerits of the American constitution, or of the British electoral system, and so on. Again, an undergraduate studying economics may well be asked why the Russian economy is so fragile, and what would best be done about it. So far as Aristotle is concerned, these thoughts would not be *practical*, since they are not undertaken with a view to action; the undergraduate knows perfectly well that in his present position he cannot do anything about it, and very probably he will never even wish to. So from his point of view the thinking is in my broader sense 'theoretical'; it is undertaken purely with a view to increasing his general understanding of politics or economics, and is a case of pursuing knowledge 'for its own sake'. If this is admitted, then the objection to Aristotle is clear. For the very same thinking may perfectly well be undertaken by someone who *is* in a position to do something about it, who intends to do something about it, and who clearly is thinking 'with a view to action'. I do not, of course, claim that there are *no* differences between the two cases, and I take up this topic in a moment, but the *reasoning* involved is not different. It is therefore perverse of Aristotle to posit two different parts of the reasoning soul, one for the first case and one for the second.

Admittedly I obtain this criticism only by abandoning Aristotle's own, very strict, view of what theoretical thinking is. For clearly in my example the thinking will not take place within a deductively organized body of knowledge, confined to universals, and unconcerned with particular facts. (Evidently, particular facts will have a very large part to play in such an enquiry.) But one cannot rely on this point to rescue his overall position. For, if one does, then it will turn out that a great deal of our reasoning is *neither* practical *nor* theoretical, as he explains these terms. (For example, all historical reasoning will fall between the two types acknowledged.) But then this reasoning is not attributed to either of his two parts of the reasoning soul, which is surely an absurd result. So should we multiply the parts yet further? It is surely better to conclude that no such division into parts was required in the first place.

If the reasoning involved can be the same, whether the enquiry is practical or theoretical, what is it that makes the difference? Aristotle is clearly right in the main claim that he goes on to make in VI.2: in the theoretical case the aim is simply truth, but in the practical case it is 'truth in agreement with correct desire'; what makes the difference is the presence of desire[11] (1139^a21-31). As he goes on to say, this is because action follows from choice,[12] and choice is a kind of combination of reason and desire ($1139^a31-{}^b5$). It was indeed defined in this way in the discussion in III.2–3. The reasoning involved in choice is what Aristotle calls deliberation. As this is described in III.3, two things must be *given* for deliberation, on the one hand an end or goal, and on the other hand the circumstances one is presently in, which make certain courses of action possible and others not. The deliberation is then an enquiry—a kind of seeking—into what action both is possible in the circumstances and will lead to the goal in question. (If more than one will, then the enquiry is into which will do so in the best way.) So desire comes in in the first place because to say

---

[11] The word is *orexis*, i.e. Aristotle's general word for any kind of 'appetition'.
[12] Irwin (as always) translates 'decision'.

that something is an end or goal is to say that it is desired. But also, when the deliberation is completed, and points to a particular course of action, then this also is desired, because it is now seen as the only, or the best, way of achieving what was originally desired. That is how a choice—i.e. a 'choice-beforehand' in the full sense, a deliberated choice—combines both reason and desire. And it will then lead directly to action (at least in the usual case).[13] I think this is a very nice general description of what is special about *practical* reasoning: it is reasoning that is involved in choice, which is a combination of both reason and desire. To have a choice at all, the reasoning must 'affirm' and the desire must 'pursue' the same thing; to have a *good* choice the reasoning must be 'true', in the sense of making no mistakes, and the desire must be 'correct', in the sense that the end or goal desired must be one which it is right to desire (1139ᵃ23–6, slightly corrected).[14]

Several commentators have claimed that Aristotle's conception of choice is more restricted than my discussion has suggested. Anscombe (1965) introduces this view by drawing attention to two things that Aristotle says, which I myself am inclined to set down merely as somewhat careless generalizations on his part, but which she takes seriously. Her first point is that Aristotle describes a 'weak-willed' person (i.e. one who is akratic because weak, 1150ᵇ19–22) as initially making a choice, but then not sticking to it, so that what he actually does is not what he has chosen to do. Indeed, he frequently says that the akratic man does not choose to do what he does. But he also admits that what such a man does may be the result of clever calculation, and so (in one sense) good deliberation (1142ᵇ17–20).[15] (To paraphrase Anscombe's example: I may be well aware that adultery is wrong, and yet be sorely tempted by my neighbour's wife. In the end I give in to temptation, and plot how to achieve her seduction.) When the akratic man acts on his calculation, then he is acting on a combination of desire (*orexis*) and reasoning, and yet if we take Aristotle strictly he is not acting on a choice. From this she infers that Aristotle will count something as a choice only when the desire (*orexis*) that is acted on is what he calls a wish (*boulēsis*) and not when it is a bodily desire (*epithumia*) or other emotion (*pathos*).

Her second point depends upon an unexpected sentence in our passage VI.2. For the most part Aristotle simply says in this passage that choice combines thought and *orexis*, but at 1139ᵃ33–4 he implies rather that it combines thought and a state of character. This apparently limits still further what is to count as a 'choice', for presumably not all wishes (*boulēseis*) express one's character, but only those that reflect one's conception of the good life in general. (For example, I may have a wish to sail a yacht one day from England to France, and this may explain several of my

---

[13] Of course a deliberation which takes place at one time may lead to a decision to do something at a *later* time. Another apparent exception is a case of 'weakness of will' (*akrasia*, to be discussed in Ch. VI).

[14] It was noted by Greenwood (1909: 175) that the condition that the reasoning should affirm what the desire pursues is a condition on *all* choice, and not just (as Aristotle seems to say) on a *good* choice.

[15] Ross emends the received text, surely rightly, to read *ei deinos* for *idein*. He translates accordingly. But, strangely, Irwin does not accept this emendation, and translates the text as given (but as obelized in Bywater's OCT).

actions—e.g. my taking a course in navigation—but one would hardly say that it expresses my character, or my views on what a good life would be. I may surely say that it is something that I would like to do one day, without thinking it at all important as an ingredient in the good life.)

Anscombe's ideas have had some support,[16] but I doubt whether we should take them seriously when we are considering what to count as a choice, and hence what to count as practical thinking. No doubt it is fair to distinguish, as Aristotle does, between various different forms of desire (*orexis*), but any of them can combine with reasoning to lead to action, and that is all that is needed to mark off practical thinking. If he does seem sometimes to inflate the notion of choice to something grander than this, it is very probably because (as here) he is specially concerned with *good* practical thinking, and in particular that special excellence in practical thinking that characterizes the practically wise man. But before I come to a direct consideration of this, I pause for a brief word on Aristotle's other distinction between different types of reasoning, which is a distinction within the broad category of reasoning concerning what to do. For this may be either practical reasoning in the proper sense, or merely 'technical' reasoning.

3. The technical (VI.4)

Aristotle has said that the aim of the reasoning part of the soul is always truth. In chapter 3 he makes a fresh start and enumerates five things by which we achieve truth, namely expertise (*technē*), (scientific) knowledge (*epistēme*), practical wisdom (*phronēsis*), theoretical wisdom (*sophia*), and finally *nous*. As it then turns out, both (scientific) knowledge and *nous* are parts of theoretical wisdom, and I shall say no more on this topic. Let us turn, then, to the difference between expertise and practical wisdom, which is expounded in chapter 4. Each is said to 'achieve truth', which presumably means that each enables us to get something *right*, but Aristotle tells us that it is a different thing in each case, in one case a making (*poiēsis*),[17] and in the other an action (*praxis*). This, I think, points very much in the right direction, but it needs a small correction. For why should one not say that making something *is* one kind of action?

The notion of a craft, or skill, or particular branch of expertise is very broad. It includes, for example, making shoes and painting pictures, but also playing the violin and curing the sick. In each case one can say that there is a thing that is made, but sometimes it is a straightforwardly concrete object (a shoe, a picture) and sometimes something rather less concrete (music, health). At any rate in the former case, it is often tempting to say that a 'technical' appraisal will simply be an appraisal of the object produced (is it a good shoe? for example, is it comfortable,

---

[16] e.g. Cooper (1975: 47 n.), Wiggins (1978–9: 252–4), Kenny (1979: 96–100), Irwin (1980: 128–9; 1988a: 336–8, 598 n.), Reeve (1992: 87–8). Both Irwin and Reeve attempt to bolster their case by reading *boulēsin* for *bouleusin* at 1113ᵃ12. On the other side, see Charles (1984: 151–5).

[17] Irwin not unreasonably prefers 'production'.

smart, hard-wearing, and so on?). As Aristotle has noted earlier, and in a different context, 'what is produced by a skill has its good in itself' (1105^a27–8). But a little reflection shows that even in this case that approach is over-simple, for it may be that there is nothing specially remarkable about the product, but nevertheless conspicuous skill was shown in its making, e.g. if the maker was working with unusual and recalcitrant materials, or with scarcely any of the usual tools. In other cases— for example, making someone healthy—it is even more obvious that the skill lies much more in the ability to treat difficult cases than in the product considered on its own. But in any case we may surely accept that there is such a thing as skill in making, and it may reasonably be counted as one kind of intellectual excellence.

Now a case of making something may perfectly well be considered as a case of acting in a certain way, and to this extent it seems to me that Aristotle's way of trying to draw the distinction he is after is mistaken. But we can easily reinstate it in this way: to appraise a series of bodily movements *as* a piece of making (a production) is not the same as to appraise it *as* an action. For it may perfectly well be good in the one way while at the same time being bad in the other. For example, one may well say: 'That is an excellent painting, but you should not have done it' (perhaps because it will be offensive, perhaps because you should have been doing something else at the time, or for all kinds of other reasons). In some cases, the judgement on the making may be a ground for the judgement on the action ('That is a rotten painting. You shouldn't waste your time on painting, for you're no good at it. Stick to poetry'), but the judgement on the action can never be a ground for the judgement on the making. This is because the judgement on the action is an *overall* judgement, taking into account *all* considerations for and against acting in that way, at that time, in those circumstances, and so on; whereas the judgement on the making is made from a limited point of view, taking into account *only* the goodness of the product and the skill that went into its production. Thus if, as legend has it, Nero fiddled while Rome burnt, then his playing may have been, from the purely technical point of view, excellent; but at the same time, and taking everything into account, it was a manifestation of his utter depravity. And the goodness of the one does not in this case do anything to redeem the badness of the other.

So much, then, for merely technical appraisal. Let us come back to the appraisal of actions, and to Aristotle's conception of the virtue of practical wisdom, which ensures that one's actions are always good actions.

4. What practical wisdom is

Sarah Broadie describes one conception of practical wisdom in this way:

Practical wisdom . . . is like a craft such as medicine: it seeks to realise, not health, but the human good without restriction; and in this it takes its cue from an explicit, comprehensive, substantial vision of that good, a vision invested with a content different from what would be aimed at by morally inferior natures. This blueprint of the good guides its possessor in all his deliberations, and in terms of it his rational choices can be explained and justified. A choice shows practical wisdom only if two conditions are satisfied: (1) given the facts as seen by the agent, enacting the choice would lead to the realisation of his grand picture; (2) his grand picture is a true or acceptable account of the good. (Broadie, 1991: 198)

She calls this the 'Grand End' theory, claims that it is often adopted by interpreters of Aristotle,[18] but argues that it is in fact quite a wrong account of Aristotle's own views. Essentially she offers three objections to it: (i) that if this were Aristotle's view it would be an 'astonishing lacuna' in his account of practical wisdom in book VI that he fails to spell out the conception of the 'Grand End' that characterizes the practically wise man (p. 193); (ii) that in practice no one does have any such 'Grand Conception' of the end, which figures in all his deliberations, and yet there are people whom we can recognize as practically wise (pp. 200–1); (iii) that what Aristotle says about how we do attain knowledge of the end could not explain how one might reach any such 'Grand Conception' (p. 243). I shall take these three points in turn.

(a) The 'astonishing lacuna'. Broadly speaking, one must accept that there is a gap in book VI: Aristotle does not make any real attempt to spell out just how the practically wise man conceives of the ultimate end, *eudaimonia*. He seems much more interested in a different point, which gives a clear contrast between practical and theoretical wisdom, namely that the former deals with contingent and perceptible particulars, whereas the latter (on his account) does not. This is the first thing he says about practical thinking (i.e. that it deals with what can be otherwise, 1139^a8), and it dominates much of the ensuing discussion (1140^a30-^b4, 1141^b8-23, 1142^a11-30, 1143^a25-^b14). Admittedly it is occasionally mentioned in these passages that practical thinking requires a universal premiss too, as well as knowledge of particulars (1141^b14-15, 1142^a14, $^a21$). But no stress is put on this, and certainly there is little effort to explain it. The emphasis is all on the particular.

To this the proponent of the 'Grand End' theory might make three responses. There is no doubt that Aristotle does think of practical reasoning as requiring a conception of the end to be achieved, as I shall demonstrate shortly, and this is presumably what is referred to as the 'universal' premiss. But if, as seems to be the case, he makes no effort here to spell out just what this premiss is, that may be (*a*) because he takes it to be obvious, or (*b*) because he does not claim to know (for he does not claim to be 'practically wise' himself), or (*c*) because his account is elsewhere, and does not need to be repeated here.

Response (*a*) is unconvincing, for it has been pointed out early in book I that different people have widely different views on what *eudaimonia* is (1095^a17-30). But this at once points to response (*c*), since Aristotle does try to tell us what

---

[18] She cites as clear adherents Cooper (1975: 96–8), Kenny (1979: 150–1), MacIntyre (1988: 131–3), adding that many others could be included. I fill this in a little: one might add Anscombe (1965: 155), Sorabji (1973–4: 206–7), Irwin (1975, *passim*), Wiggins (1978–9: 253–4), Annas (1993: 66–84). On the other side one might cite Fortenbaugh (1975), Urmson (1988, ch. 6), and those who argue that practical wisdom is basically a matter of correctly perceiving particular situations, e.g. Woods (1986) and McDowell (most recently 1998, but also as far back as 1979). There is an explicit response to Broadie in Kraut (1993). (Her title 'the Grand End theory' is taken from Cooper 1975: 59.)

eudaimonia really is, in I.7 and in X.6–8. So the gap that Broadie draws attention to is filled elsewhere, and that is a perfectly good explanation of why it is left as a gap in book VI. There is surely some truth in this response, but perhaps not enough. For, as I shall argue later (in Section 5), what is said in I.7 and in X.6–8 does not *really* satisfy the urge that we feel for something more. So I am tempted to invoke response (*b*) in addition: if there are important things that are missing in Aristotle's account, that may be because he did not himself know how to fill them in. He has a theory which demands that the wise man be able to provide certain explanations, but he might admit—if directly tackled—that he cannot provide them himself, and that is why he preserves a discreet silence. (We may compare Plato, *Republic* VI, 505a–506e, where it is quite natural to see Plato as admitting that the 'knowledge of the good' of which he speaks is not something that he has attained himself. But his theory demands that it be possible.) In my own view, this last response is essentially right, so I do admit that there is the 'lacuna' that Broadie points to, though I do not admit that it is 'astonishing'. Aristotle may have concealed it from himself by the thought that he does elsewhere discuss what *eudaimonia* is, and it *may* be that he would have admitted—if pushed—that there are gaps which he has not filled and does not know how to fill. For he does not claim to possess himself all the wisdom that his ideal practically wise man is supposed to have.

Broadie, of course, thinks that there is not actually any such lacuna, for she claims that Aristotle anyway does not attribute to the practically wise man any 'Grand Conception' of what *eudaimonia* is. Let us move on to evidence against this claim.

(b) No one has such a 'Grand Conception'. In so far as Broadie is claiming that no one does, in practice, have a fully worked-out picture of what *eudaimonia* consists in, and that no one does, in practice, refer to such a picture in all their deliberations, I agree with her entirely. But the question is not whether I agree but whether Aristotle agrees, and surely the signs are that he did not. I begin by recalling the brief passage from the *Eudemian Ethics* already cited (p. 15): 'not to have ordered one's life in relation to some end—an aim that one will have in view in all of one's actions—is a mark of extreme folly' (1214^{b}6–11). Here it is taken to be just obvious that *every* sensible person has exactly such a 'Grand Conception' as Broadie complains of, and not only those who are practically wise. But let us set this aside, and look now at some passages from our book VI which—as it seems to me—do clearly attribute to the practically wise man just such a conception of *eudaimonia*.

We may begin with chapter 5, where practical wisdom is first introduced. Aristotle says:

It seems to be characteristic of the practically wise man to be able to deliberate well about what is good and advantageous for himself, not in some particular respect—e.g. what conduces to health or to strength—but about what conduces to living well in general. (1140^{a}25–8)

Here the phrase 'living well in general' evidently refers to the supreme end, *eudaimonia*, and clearly the wise man must have some conception of what this is if he is to work out what conduces to it. A very similar point is made again in chapter 7, where Aristotle is contrasting theoretical with practical wisdom, the main point of the contrast being that theoretical wisdom is not concerned with human beings, but with other objects, more divine (1141^{b}1) and more to be valued (^{b}3).[19] On the other hand practical wisdom is directed at human concerns, which one can deliberate about:

For we say that good deliberation is especially the task of the practically wise man . . . and the one who is without qualification good at deliberation is the one who aims, by reasoning, at what is best for man of all things attainable in action. (1141^{b}9–14)

Here again 'what is best for man of all things attainable in action' is surely a way of referring to the supreme end, which is *eudaimonia*. A version which is even more explicit appears in chapter 9. Until this point, it has appeared that practical reasoning (and hence practical wisdom) simply *is* good deliberation, but in chapter 9 we appear to be offered a distinction:

One may deliberate well unconditionally, or with reference to a particular end. Unconditionally good deliberation is deliberation which is correct about what is unconditionally the end, whereas a particular kind of deliberation is deliberation which is directed to a particular end. So if good deliberation is characteristic of the practically wise man, it will be a correctness concerning what is expedient for that end of which practical wisdom is the true apprehension. (1142^{b}28–33)

Once more the phrase 'what is unconditionally the end' (*telos haplōs*) is naturally taken to be a reference to the supreme end, *eudaimonia*,[20] and the last sentence explicitly tells us that the practically wise man does have a 'true apprehension' (*alēthēs hupolēpsis*) of this.[21] But this true apprehension of the end is here explicitly distinguished from his deliberative ability. Indeed, on the face of it, this true apprehension is *identified* with practical wisdom, and deliberative ability is regarded as something different. But obviously Aristotle cannot have quite meant this; good deliberation is certainly a *part* of practical wisdom, even if—as this passage suggests—it is not the whole of it. For we must also allow for the true apprehension of the end itself.

Taken together, these three passages clearly do attribute to the practically wise man a true conception of the supreme end, *eudaimonia*. But they say nothing of what that conception is, so perhaps Broadie can still maintain that it is not the 'grand' conception that she objects to. On this point we obtain more insight from an initially surprising passage in chapter 13. Since there are several things about that

---

[19] Aristotle is thinking, apparently, of 'what composes the earth and the heavens (the *cosmos*)', i.e. of what we might call 'fundamental physics'.

[20] Cf. 1097^{a}33–^{b}6: only *eudaimonia* satisfies the conditions for being 'unconditionally end-like' (*haplōs teleion*).

[21] An interpretation due to Burnet (1900) construes this sentence as saying that practical wisdom is the true apprehension of what is expedient for the end, and not of the end itself. Grammar would permit this, but it is not the natural reading, nor that which is expected in the context. (See e.g. Kenny 1979: 106–7.)

chapter that are initially surprising, I shall first digress a little to locate the relevant passage in its overall context.

At the end of chapter 11 Aristotle says that he has now concluded his account of what practical wisdom is (1142^{b}14–17). Chapter 12 then opens with some problems that arise out of that account, and the chief among them is the relation between practical wisdom and virtue of character. I shall discuss shortly what chapter 12 has to say about this, but for the moment I set it aside, only remarking that it is this general topic that continues into chapter 13.

The chapter opens with a distinction between 'natural virtue' and what I shall call 'full virtue' (*aretē kuria*),[22] the former being virtue that one is born with, and the latter the improved state that results from this when thought[23] is added. Full virtue, then, involves practical wisdom, though natural virtue does not (1144^{b}1–17). The distinction is surprising, for II.1 has roundly insisted that *no* virtue is in this sense 'natural', and it has said that virtue is acquired not by thought but by training and habituation.[24] To harmonize Aristotle's claims in these two places, we should perhaps think not of two levels of virtue, but of three. The lowest level is natural virtue, accepting for the sake of argument that some people are born just, or temperate, or courageous (1144^{b}5–6); the second level one might call 'trained virtue', as that is described in book II. On my account, successful training will lead one to act (by and large) as the man of practical wisdom *would*, but one can be trained to act in this way without yet acquiring practical wisdom oneself. (This is the situation envisaged in the problem developed at 1143^{b}28–33. To put it in the terminology of chapter 13, one who is so trained will act '*in accordance with* correct reason', but not yet '*with* correct reason',[25] 1144^{b}24–8.) Finally, there is the level of 'full virtue', which is the state of one who himself has practical wisdom. Anyway, setting this issue to one side, as not central to our main problem, let us see what Aristotle does have to tell us here about 'full' virtue.

It leads him at once into a discussion of Socrates' position. Socrates thought that all the virtues were 'practical wisdoms'—this is Aristotle adapting to his own terminology the Socratic thesis that virtue is knowledge—which Aristotle takes to be a mistake; but it is correct to say that all ('full') virtues involve practical wisdom (1144^{b}17–21). Moreover, Socrates inferred that all the virtues are 'one', at least in the sense that they are inseparable: you cannot have one but lack another. Aristotle, to one's surprise, *agrees*. At the level of natural virtue (and, one might add, at the level of 'trained virtue') such separation is entirely possible, but at the level of full virtue

---

[22] Ross's rendering, 'virtue in the strict sense', captures the sense of the Greek very much better. But Irwin's rendering is not seriously misleading, and is more convenient, so I adopt it.

[23] The word is *nous* (1144^{b}9, ^{b}12), but presumably it is not here being used in its technical sense.

[24] The contradiction is not removed by thinking of *EN* VI as part of the *EE* rather than the *EN*. For *EE* 1220^{a}38–^{b}7 presents the same view as *EN* II.1 (though more briefly). But *EE* III does look forward to a distinction between 'natural' virtue and that which involves practical wisdom (1234^{a}28–30). In addition, *EN* VII refers in a casual way to virtue that is either 'natural' or 'habituated' (*ethistē*, 1151^{a}15–19; I discuss this passage below.) It is not clear from these texts whether Aristotle means to identify 'full' and 'habituated' virtue.

[25] Irwin's constant rendering of *kata* not as 'in accordance with' but as 'expressing' is here very misleading.

it is not: 'For as soon as this one thing, practical wisdom, is present, all the virtues will be present too' (1144^{b}32–1145^{a}2). But, one asks, why on earth should Aristotle accept that?

Socrates' reasons are clear.[26] He believed that virtue is knowledge, and the knowledge in question simply is knowledge of what is good and what is bad. (To counter the objection that one may know what is good but still not desire to do it, Socrates replies that this is not in human nature, for all men just do desire what is good.) But it is a consequence of this that all the various different virtues, which men think of as quite distinct from one another, are actually the *same* knowledge. So in fact one cannot have any one of them without having all the others too. Is Aristotle in effect accepting the same position? If so, then he must suppose that practical wisdom involves just that 'Grand Conception' of the end that Broadie denies. It certainly cannot be nothing more than the ability to deliberate well on how to achieve the various goals set by training and habituation, for clearly one could have this ability while still some of one's goals were correct and some incorrect. It must be something 'higher', and must in fact involve what Socrates calls 'knowledge of the good', and what is in Aristotle's vocabulary a 'true apprehension' of the ultimate end, *eudaimonia*. To make sense of what he says here, we must take it that this includes a clear view of *all* the different things that go together to make up *eudaimonia*, and of how they fit together. And this is the *same* apprehension, whatever virtue is concerned. Unless Aristotle does share Socrates' picture, at least to this extent, one cannot at all understand why he should endorse the Socratic thesis that all the virtues are 'one', in the way that he here does. No doubt, there are still questions to be asked, and in particular: why is Aristotle not vulnerable to the same objection as Socrates is, namely that one may know what is good but still not do it? I shall say *something* on this later. Meanwhile I offer a different route to essentially the same conclusion.

One might argue for the unity of the virtues in this way. As I remarked on p. 52, it would seem at first sight that the virtues may conflict with one another. For example, honesty may demand that one tell the truth, whereas compassion may advise that one should not say what will hurt. This situation seems inevitable while the virtues remain at my 'second level', instilled purely by training. No doubt, children can be trained to tell the truth without much intellectual appreciation of why this is a good thing. As they grow older, they can be brought to assimilate various exceptions to this general rule. But, it may be suggested, the exceptions come only where some *other* virtue demands something else. Consequently, they cannot recognize the exceptions unless they do have the other virtue too.[27]

---

[26] Aristotle's knowledge of what Socrates thought is no doubt based upon the same source as ours is, namely Plato's early dialogues. I am here simply assuming that he and we are pretty much in agreement on this topic.

[27] It is noted by Kraut (1988) and more fully by Telfer (1989–90) that this step is fallacious. One must *know* what the other virtue demands, but this is compatible with not yet *having* that virtue, as would be the case with the self-controlled man (*enkratēs*). (I add that a similar point may be made concerning the problem brought by Irwin (1988*b*): I may know what a 'large-scale' virtue such as magnificence requires, without myself having the resources to practise that virtue; whether I count as 'having' the virtue in such a case may be given one answer in one context and another in another.)

Generalizing this line of thought, one cannot fully possess any one of the virtues unless one does have all the others too, for this is required in order to be in a position to recognize where an exception may be appropriate. But the fully virtuous man, of course, will not only *recognize* these cases of apparent conflict between the virtues, he will also *resolve* them in the correct way, and how is he to do that? Well, the only answer that evidently suggests itself is that he must have an overall conception of just how each of the various virtues do fit together with one another to make up the final end of life, *eudaimonia,* for this is what enables him to see which should be followed in this or that particular situation. Here, then, is a reason for saying that *full* virtue requires practical wisdom, but the feature of practical wisdom that it relies on is again the true apprehension of the supreme goal, in all its articulated detail.[28] If this is Aristotle's thought, then once more it is what Broadie calls the 'Grand Conception' that is required.

I add here that there is nothing in Aristotle's account of practical wisdom which leads one to suppose that he must have thought it a common attainment, and this last point surely speaks in the opposite direction. For it must be rare for someone to be in an intellectual state which by itself ensures that he has *all* the virtues. Hence 'full' virtue must be equally rare, for it is said to require this practical wisdom. No doubt ordinary people will approach to practical wisdom, and to 'full' virtue, to various degrees, but the ideal that Aristotle has in mind is surely not commonly achieved. We need not agree, then, with Broadie's thought that the 'Grand End' theory makes practical wisdom, and hence full virtue, too difficult to achieve and too difficult to recognize (pp. 200–1). For it may well be that Aristotle *meant* it to be difficult. But this brings me on to her third objection, which concerns what Aristotle says on how this wisdom is acquired.

(c) How practical wisdom is acquired. As I have argued, the practically wise man must have a true conception of what *eudaimonia* is. But how does he reach it? In several passages in book VI, and particularly in chapter 12, Aristotle seems to credit the attainment to one's training in virtue. But then Broadie can fairly argue that this cannot actually provide the 'Grand Conception' that her opponents require, and so it cannot be what Aristotle intended. Let us turn, then, to what he does say in chapter 12, and to why he says it.

The account of what the two intellectual virtues are, i.e. practical wisdom and theoretical wisdom, is brought to an end at the close of chapter 11 ($1143^{b}14$–17). Chapter 12 then takes up various problems that arise out of this account, and the first is: what *use* are they? What most concerns us, and most concerns Aristotle too, is the question 'What use is practical wisdom?'[29] This question is asked both from

---

[28] For the line of argument, compare Annas (1993: 66–84).

[29] The question 'What use is theoretical wisdom?' is treated only briefly. It is claimed that, since both are virtues, both are anyway worthy of pursuit, even if they are of no use ($1144^{a}1$–3), and that in fact both are of use. Theoretical wisdom is so because its exercise is at least a part (if not the whole) of *eudaimonia* ($^{a}3$–6), and practical wisdom because it, together with virtue, promotes theoretical wisdom ($^{a}6$–7; cf. $1145^{a}6$–11). I take up this latter thought in Ch. IX. But for the present I am concerned with how practical wisdom and virtue co-operate with one another.

the point of view of one who is already a good person, and so possesses the various virtues of character (1143^{b}20–8), and of one who is still in the process of acquiring them (1143^{b}28–33). In the first case, it is not clear how the added knowledge would improve the disposition he already has; and in the second case all that is needed is a knowledgeable instructor, which is not to say that one needs to acquire this knowledge oneself. ('Though we [all] wish to be healthy, still we do not [all] learn medicine', 1143^{b}32–3.) The general thrust of Aristotle's answer is to describe (full) virtue as resulting from a kind of *co-operation* between practical wisdom on the one hand and what he still calls 'virtue' on the other. The overall theme is that neither can exist without the other. But in the course of working out this theme he certainly seems to be assigning different tasks to each, in such a way that it is virtue which provides the appropriate conception of the end, and practical wisdom which works out (by deliberation) how to put this into practice.

His first move in this direction is to claim:

The function [*sc.* of man?] is achieved in accordance with both practical wisdom and virtue of character; for virtue makes right the target aimed at, but practical wisdom makes right what promotes this. (1144^{a}6–9)

This thought is apparently repeated a little later:

Virtue makes the choice correct, but what needs to be done for its sake is provided not by virtue but by another capacity [which Aristotle goes on to call 'cleverness' (*deinotēs*). (1144^{a}20–2)

Here it seems clear that when Aristotle refers to 'the choice' he means 'the end aimed at'; this is what virtue provides, whereas 'cleverness' is clearly provided by practical wisdom, and seems in fact to be simply a new name for what has earlier been called deliberation.[10] (The difference is just that 'cleverness' is an ethically neutral term, which applies to anyone who is good at calculating how to achieve his goals, whatever they may be. So practically wise men are clever, but so also are those who are quite unscrupulous.[11] By contrast, Aristotle confines 'good deliberation' to those whose goals are good.) Practical wisdom, then, involves cleverness, but is more than mere cleverness, just because it requires virtue in addition (1144^{a}22–31), and this in turn is because

Practical inferences have as their starting-point (*archē*) 'since the end and the best is such-and-such' (whatever it may be—for the sake of argument let it be whatever you like), and this is not apparent to one who is not good. (1144^{a}31–4)

Thus cleverness (or good deliberation) is only called 'wisdom' when the end aimed at is a good one, and such an end 'is apparent' only to one who is virtuous.[12] So, as

---

[10] Urmson (1988: 82) wishes to distinguish thus: deliberation is what *plans* how a goal is to be achieved, whereas cleverness determines how to *execute* that plan. But I see no good reason to credit to Aristotle this tenuous distinction.

[11] The OCT text at 1144^{a}28 should evidently be corrected to *kai tous panourgous* (Allan 1953: 77). Irwin translates the corrected text, but Ross tries to make sense of the uncorrected one.

[12] Again, there is the same objection as in n. 27. The true end must also be 'apparent' to one who is not virtuous but merely self-controlled (*enkratēs*).

he sums up towards the end of chapter 13,

A choice will not be correct without practical wisdom, nor without virtue. For the one makes us achieve the end, and the other what promotes it. (1145^{a}4–6)[33]

On this account, while the practically wise man must *have* a true conception of the end (as 1142^{b}33 has explicitly said), it is not practical wisdom itself which *provides* this conception. That is due to what Aristotle calls virtue, and the contribution of practical wisdom is confined to the deliberation which shows how this end may be achieved.

It is difficult to believe that this is what he really means, and if these passages in chapter 12 stood alone one might offer to 'explain them away' as simply a defensive move on Aristotle's part. He is envisaging an opponent who supposes that one can be perfectly virtuous without practical wisdom, and so asks what use it is (1143 ^{b}20–8). To this he can reasonably reply that, even if the goals are set by virtue (as his opponent presumes), still deliberation is required to work out how those goals are best achieved, and this is uncontentiously the province of wisdom rather than virtue. Yet it does not seem that Aristotle is here conceding merely for the sake of argument that it may be virtue that provides the goals; this seems, rather, to be his own opinion. Moreover, these passages in VI.12 do not stand alone, for Aristotle also says much the same thing elsewhere, where there is no question of defending practical wisdom against an attack.

In the last chapter of book II of the *Eudemian Ethics* he raises the question whether it is virtue or reason (*logos*) that makes one's end correct, and very clearly answers that it is virtue (1227^{b}12–1228^{a}2).[34] The thought of VI.12, that vice destroys one's perception of the end (1144^{a}34–6), is also found earlier, at VI.5, 1140^{b}11–20. But perhaps the most forthright passage outside VI.12 is in VII.8:

Virtue preserves the first principle (*archē*) and vice destroys it, and in actions it is the goal that is the first principle, as in mathematics it is the hypotheses [i.e. axioms]. In neither case is it reasoning (*logos*) that teaches us the first principles, but [in the case of actions] it is virtue, either natural or habituated (*ethistē*), that teaches us right opinion about the first principle. (1151^{a}15–19)

This is surely unequivocal. The goal that Aristotle is thinking of must presumably be an ultimate goal, a genuine *first* principle (and not a mere starting-point for a

---

[33] As Kenny remarks (1979: 105), this sentence *could* be understood as saying that it is practical wisdom which makes us [have a true conception of] the end, and virtue which makes us do what promotes it, since it is virtue that provides not the conception but the desire. But, in view of the preceding passages just cited, it is much more probable that 'the one' is virtue and 'the other' is practical wisdom, and that is what I shall assume.

[34] Curiously this passage in *EE* II begins with the idea that those who say that *logos* makes the choice right are really thinking of self-control (*enkrateia*). Kenny (1979: 82–4) does his best to make sense of this idea. But at any rate the discussion clearly concludes with the view that it is virtue that provides the right end, but 'another capacity' that yields the means to it (1227^{b}39–40). This 'other capacity' is presumably the capacity that *EN* VI explains as 'cleverness' (1144^{a}20–9).

particular deliberation),[35] since it is compared to the axioms of mathematics. It is his standard doctrine that first principles cannot be deduced, for otherwise they would not be 'first'; thus it cannot be reasoning or argument (*logos*) that provides them. So one asks: what *else* could provide them? In the case of theoretical first principles, we have seen that Aristotle's answer is that it is *nous* (which he describes as working by induction from particulars); but here he undeniably says that it is virtue which provides, i.e. *teaches*, the corresponding first principle for action, which is the goal that ultimately is aimed for.

It is true that there is a rather perplexing passage in VI.11 which *can* be interpreted as yielding a rather different moral. In general, chapters 8–11 are devoted to further elaboration of the notion of practical wisdom that was introduced in chapter 5. In chapter 8 the main focus is on the distinction between reasoning about what is good for oneself (which is how practical wisdom was first introduced at 1140ª25–8), and about what is good for the community—either the state as a whole (politics), or just one's own household. (Aristotle argues that in fact there is no firm distinction.) In chapters 9–11 he considers various qualities related to practical wisdom, first good deliberation (chapter 9), and then understanding (*sunesis*,[36] chapter 10), and finally considerateness ((*sun*)*gnōmē*),[37] chapter 11, 1143ª19–24). Then at 1143ª25 he appears to move towards a summary, which is introduced as concerning the four qualities considerateness, understanding, practical wisdom, and *nous*. It is not clear why good deliberation is apparently dropped out at this point, unless perhaps practical wisdom is here representing it, and still less clear why *nous* is suddenly included. In any case, the main drift of the summary is that these four naturally go together, because they are all concerned with particulars, and actions are particulars (1143ª25–35). This, indeed, has been a constant theme in all of Aristotle's discussion. While he does incidentally admit that practical wisdom is concerned with universals too (1141ᵇ14–15), it is the concern with particular things—also called 'the final things' (*eschata*)—that has been stressed throughout.

In the summary, then, Aristotle first applies this thought to practical wisdom, to understanding, and to considerateness, and then turns to *nous*. Since the passage is obscure, I give a rather literal translation:

Nous is of the final things (*eschata*) in both directions. For indeed there is *nous*, but not reason (*logos*), both of the first terms (*horoi*) and of the final ones—the one sort, where proofs are involved, is of the first and changeless terms, and the other sort, in what is practical, is of what is final and contingent and of the other premiss. For these are the starting-points (*archai*) of that for the sake of which. For universals are from particulars. So one needs perception of these, and this perception is *nous*. (1143ª35–ᵇ5)

What I think may fairly be called the *usual* interpretation of the first part of this passage is as follows. *Nous* is of what comes last, in both directions, on the scale of

---

[35] There is an awkward ambiguity about the word *archē* which can be used to refer to *any* kind of 'starting-point'. I discuss this in Ch. IX, Sect. 2. But here the starting-points in question must be first principles.

[36] Irwin (who has used 'understanding' for *nous*) prefers 'comprehension'.

[37] Ross prefers '(sympathetic) judgement'.

more universal versus more particular; that is, it (but not reason) grasps both what is most universal and what is most particular. In theoretical reasoning, it is *nous* that grasps those most universal premisses (first principles), from which proofs proceed;[38] in practical reasoning it is *nous* that grasps the particular and contingent premisses which describe the present situation. Given this much, the next sentence, 'For these are the starting-points of that for the sake of which', should presumably be interpreted, as Cooper suggests,[39] in this way: these particular premisses are what we start from [*when reasoning about*] the goal, [i.e. about how to attain it]. But in that case the next sentence, claiming that 'universals are from particulars', appears to change the subject, since this is Aristotle's usual way of putting his doctrine that we come to *apprehend* universals by induction from particulars. Hence Ross supposes that both sentences must be concerned with apprehension, and construes the first as saying that these particular facts, which are the contingent premisses of deliberation, are the 'starting-points' for our apprehension of the goal. This is not a familiar thought, and (on reflection) not easy to understand; moreover it fudges the Greek at an important point,[40] and leaves us wondering why the sentence begins with 'For'. So Cooper supposes that both sentences concern reasoning which *attains* something and takes 'for universals are from particulars' as a way of saying that we achieve our universal ends by achieving particular objectives. This is not impossible, though certainly it is not the usual Aristotelian usage of this phrase. My own view (proposed rather tentatively) is that Cooper is right about 'For these are the starting-points of [the goal]', and that Ross is right about 'For universals are from particulars', and that the apparent change of subject is to be explained by taking the 'For' in 'For universals are from particulars' as looking back not to the immediately preceding sentence but to the one before it. Aristotle is explaining his claim (which apparently contradicts 1142ª25–30 earlier) that we can reasonably credit *nous* with perception of particulars. His thought is that it is already admitted that (in the theoretical case) *nous* provides universals, but our grasp of universals is derived from our grasp of particulars, so the perception of particulars too must fall under *nous*. That is a line of interpretation which fits well enough with the general drift of the discussion that leads up to it.

But a quite different interpretation is proposed by Kenny,[41] which one might introduce in this way: how are we to understand the phrase 'the *other* premiss'? What is it other than? Well, on this account, the 'other' premiss in practical reason-

---

[38] The word *horos* (literally 'boundary') is used by Aristotle in his logical works to refer to the *terms* of a proposition. (In a proposition such as 'All A are B' it is 'A' and 'B' that are the terms.) But (i) the word is also used (loosely) to refer not to the term itself but to the proposition that contains it, and anyway (ii) it *may* be that Aristotle is here using it with the different meaning 'definition' (cf. *horismos*). For it is a very Aristotelian thought that proofs, in the theoretical sciences, begin from definitions. (Cf. Reeve 1992: 57 n.)

[39] Cooper (1975: 42 n.).

[40] The word 'these' in 'For these are the starting-points . . .' is feminine plural (*hautai*). The noun to be supplied must therefore be 'premisses' (*protaseis*), and Ross's 'these facts' is a fudge, while Irwin's 'these terms' is simply wrong. When Aristotle says that our grasp of universals comes by induction from particulars, he does not call these particulars 'premisses'.

[41] Kenny (1978: 170–2).

ing is the one which is 'other than' the premiss which states the goal. So it is admitted that there is *also* a premiss which states the goal, though nothing is here said about how *it* is grasped, and it is apparently not counted as a 'starting-point' for reasoning about how to attain the goal. Surely this is unreasonable, and moreover contradicted by 1144ª31–3, which does describe this premiss as a 'starting-point'.[42] Kenny therefore proposes that 'the other premiss' should be understood to be the premiss *other than* 'what is final and contingent', i.e. to *be* the premiss that states the goal, as opposed to that which states the particular situation. This has the consequence that *nous* is credited with supplying both the starting-points for practical reasoning, i.e. both the particular premiss and the goal. (Hence the plural in '*these* are the starting-points . . .', for two kinds of premisses are referred to.) The overall thought is then that it is *nous* that supplies *all* the premisses from which reasoning starts; in the theoretical case there is just one such kind of premiss, i.e. a wholly universal premiss, and in the practical case there are two, but all are due to *nous*. One cannot deny that this interpretation has its attractions.[43] On the other hand I do think that the passages that I have cited earlier, from chapter 12 and elsewhere, are strongly against it, for they all cite virtue, and not *nous*, as the source of our grasp of the first principle that is the goal of practical thinking, namely *eudaimonia*. Besides, one must grant that chapters 2–11 as a whole clearly focus upon the point that practical thinking requires a grasp of particulars, so one would expect the same emphasis here too. It may be mildly odd that Aristotle should on occasion say that this grasp of particulars is a kind of *nous*, but it would be odder still if, in this one little sentence in chapter 11, he meant to imply that the other premiss is also due to *nous*. For he shows virtually no interest in that other premiss until he comes to chapter 12, and there he says something quite different.[44]

How should we conclude? It would seem that Aristotle both requires the practically wise man to have a suitably 'grand' conception of his end, and yet denies him the means to attain it.

One view is that Aristotle in book VI has given us a misleading account, misleading because one-sided, but (*a*) this can be corrected from what he says elsewhere, and (*b*) one can also explain why he should have made this mistake. The leading idea can be put in this way: deliberation issues in choice, and Aristotle rightly

---

[42] And we may note, as a curiosity, that 1139ª17–20 denies that perception is a 'starting-point' of action.

[43] Kenny himself goes further, for he takes it that there are both universal and particular premisses in *both* practical *and* theoretical reasoning, and that *nous* provides all four of these types of premisses. I have ignored his claim that there are particular premisses in theoretical reasoning, which seems to me quite unfounded. Kenny bases it on another perplexing passage in book VI, namely 1142ª25–30. Since this passage is peripheral to my main problem, I relegate discussion of it to a note at the end of the chapter.

[44] This vexing passage 1143ª35–ᵇ5 on '*nous* in what is practical' is a well-known crux, and has received many different interpretations. A very thorough discussion (which reaches Kenny's conclusion, but by a quite different route) is in Dahl (1984: 41–5, 227–30). (Dahl denies that *protasis* means 'premiss'; on this see pp. 131–2.) Other discussions may be found in Sorabji (1973–4: 214–16), Wiggins (1975–6: 236–7), Woods (1986: 153–61), Broadie (1991: 254–6).

explains choice as resulting from *both* thought *and* desire. The same combination applies to the goal which is (in one way) the starting-point for deliberation: one must have both an intellectual understanding of what this goal is and a desire for it. Broadly speaking, it is virtue which provides the desire, but it must be something else, of a more intellectual kind, that provides the understanding.[45]

This general idea is quite well supported from something that Aristotle says elsewhere, in *Metaphysics* Λ.7. His overall topic there is the unmoved mover of the universe, and he is trying to explain how something can be a cause of motion without being moved itself. He says:

There is something which makes things move without being moved itself . . . What is desired (*orexis*) and what is thought (*noēsis*) make things move in this way, not by being moved themselves. And the primary instances of these are the same. For what appears noble is wanted (*epithumia*), but what is actually noble is what is primarily wished for (*boulēsis*). But we desire (*orexis*) because of what we believe (*doxa*), rather than believing because we desire. For it is thought (*noēsis*) that is the starting-point. (1072ᵃ25–30)

Although the overall topic is a highly metaphysical issue about the universe as a whole, still it seems clear that this illustration concerns ordinary human actions, and what causes them. And it undoubtedly says that desire follows thought, rather than vice versa: what comes first is the thought that something is good (or noble), and that then leads to the desire for it, which in turn leads to action. Admittedly, the passage would still allow us to suppose that it is mere habituation which provides the thought, as well as the corresponding desire, but the clear separation which it makes between the two does rather suggest otherwise.

There are various ways of trying to work out the idea that it is something of an intellectual kind that provides the thought, i.e. the 'true conception' of what *eudaimonia* is. A pleasantly simple suggestion is that the task of working out what *eudaimonia* is should be assigned to deliberation.[46] It is true that Aristotle says that we do not deliberate about ends, but it is also true that his illustrations of this point (in III.3) are evidently examples of ends that one *could* deliberate about, even if at the time they are being taken for granted. (For example, a doctor may surely deliberate about whether, in this particular case, he should try to heal, 1112ᵇ11–16.) So Aristotle's position must be that deliberation will always take *some* end for granted, though on another occasion that end itself may be questioned, in the light of some higher end that it is supposed to serve. There is a hierarchy of ends (as explained early in book I), and it is only the highest end of all, namely *eudaimonia*, that is never subject to deliberation; in this case we can only deliberate about what conduces to it. But it is nowadays almost universally conceded[47] that Aristotle's phrase 'what conduces to the end' (*ta pros to telos*) includes not only what we might naturally call 'means' to that end but also what may be called the 'parts' or 'constituents' or 'ingredients' of that end. So we have as a fit subject for deliberation

---

[45] Allan (1953) gives a good exposition of this view.

[46] e.g. Wiggins (1973–4: 226–7); Irwin (1975: 568–70); Broadie (1987).

[47] Kraut (1989: 197–237) is an exception. (So far as I am aware, the point was first made by Greenwood (1909: 46–8.)

'what conduces to *eudaimonia*', and this includes the question 'What *is eudaimonia*?', i.e. what does it consist of? It is therefore the part of practical wisdom, and not mere training and habituation, to find an answer to this question.

I am attracted to this view, partly because it seems so simple that Aristotle *could* have thought that it goes without saying (and hence he never explicitly says it). But I confess that it is difficult to square with the texts that we have, and in particular the passage cited on p. 90 from book VII, chapter 8. That says that it is not reasoning (*logos*) that teaches us the first principle of actions, but virtue, either natural or habituated. This cannot simply mean that it is not reasoning but virtue that teaches us to *desire eudaimonia*, for (on Aristotle's account) we *all* desire *eudaimonia* anyway, whether virtuous or not. So it must be concerned with some more or less detailed conception of what *eudaimonia* is, and the claim that this is not due to *reasoning* would certainly seem to rule out the idea that it is due to good deliberation.

But at the same time the passage suggests a different thought. Aristotle denies that first principles are reached by *logos* because he insists that they are not *deduced*. On the other hand he does think that, at least in the theoretical case, they are reached by what is broadly speaking an intellectual process. It involves what he calls 'dialectic', and 'induction', and *nous*. But if this holds for the first principles of other subjects, might we not expect it to hold for the first principles of ethics too? Moreover, when we look at his own accounts of what *eudaimonia* is (in I.7 and X.6–8), can we not say that it is this kind of 'reasoning' that he is himself employing—not, indeed 'deduction' (as he understands that notion), but still recognizably an intellectual process as opposed to a mere reliance on virtuous upbringing? I postpone until my Chapter X a more detailed account of just how Aristotle thinks that he has reached his own view of what *eudaimonia* is, but it is sufficient to say here that many commentators have supposed that he sees no real difference between situation in ethics and in the more theoretical sciences.[48] But we still have to 'explain away' the apparent claim that in ethics it is virtue which provides our true apprehension of the end.

Again, we start with the point that *two* things are needed, both an intellectual grasp of what that end is and a desire for it. Now we all start by desiring *eudaimonia*, but this desire may fade as intellect elaborates what *eudaimonia* really is. Thus, if I come to believe that *eudaimonia* requires a more healthy diet than that which I currently enjoy, or (more seriously) that *eudaimonia* requires me to devote time to theoretical activity, which I am currently not at all disposed to, then desire may rebel: if *that* is what *eudaimonia* is then I do not want it, or anyway do not want it enough, and so I am led into temptation. I become prey to 'weakness of will' (*akrasia*), and while I acknowledge that such-and-such is the better course, still I fail to do it, for I also have contrary desires which retain their strength. Here is really where virtue, i.e. training and habituation in virtue, comes in. For virtue requires that intellect and desire are in harmony with one another, and so long as this holds then I will not be akratic. Moreover, we can explain why Aristotle puts all his

---

[48] An able expositor of this view is Reeve (1992, ch. 1). See also Cooper (1975, sect. 4); Sorabji (1973–4); Dahl (1984, pt. 1).

emphasis on virtue: it is because he thinks that in vice or in *akrasia* the non-virtuous desire obscures one's intellectual grasp of what *eudaimonia* requires, i.e. it prevents this end from being 'apparent' (1144^a31-4). As I shall argue in Chapter VI, there is no good reason to agree with this view, but undoubtedly it is what Aristotle thinks. So although in truth the intellectual grasp of what the end is, and the desire to pursue it, are different, and one can have the first without the second, still Aristotle thinks otherwise. That is why he here puts all his emphasis on the second.

These proposals attempt to retain the view that (in Aristotle's own opinion) there is an intellectual basis to the practically wise man's grasp of what *eudaimonia* is, despite his saying that it is simply virtue which provides the end. But why can we not take his claim entirely at face value (as Broadie would wish)? Early training in virtue teaches one that this, that, and the other ways of acting count as virtuous, and so praiseworthy. It may add that such actions—or, anyway, some of them—count as 'noble' (*kalon*), and so particularly praiseworthy. It may of course add that it is through such actions that *eudaimonia* is achieved. All this the child easily absorbs from his elders when growing up, and this forms his conception of what *eudaimonia* is. The position is not essentially altered when matters become more difficult, and one has to recognize that this is a situation in which honesty would not be appropriate, or in which mercy should override justice, or honour should take second place, and so on. For it is still true that education and training can provide this information, and so refine one's ideas of how best to live. By absorbing this education (and adjusting one's desires appropriately), and by reflecting on it inductively, one may then become wise oneself. On this kind of account, the induction need not be explicit; one may not formulate to oneself the general principles involved, but simply acquire a sort of knack or instinct which ensures that they will be correctly applied to new cases. For this, experience is required, but not what Aristotle calls proofs or demonstrations (1143^b6-14). Nevertheless, the result is practical wisdom. It provides in a way a true conception of the supreme end, *eudaimonia*, just because it enables one to say whether or not any proposed course of action would conduce to it. It shows itself in the ability simply to *perceive* the morally salient features of particular situations, and consequently to say what action is here and now required, and that is why Aristotle so stresses the importance of perception for practical wisdom.[49] The source of such a knack is simply education and training, and it need not include an ability even to *state* the goal aimed at in any informative way, let alone the *reasons* why this or that course would conduce to it.[50]

This conception of practical wisdom is, to us, profoundly unsatisfying. But whether Aristotle would share our dissatisfaction is a moot question.

---

[49] Woods (1986) gives a good exposition of this view.

[50] Compare Plato, *Meno* 97d–100b. The only way to make sense of Plato's argument there is to suppose that by 'belief' (*doxa*) he means a kind of knack or instinct which one cannot formulate, and so cannot pass on to others by 'teaching'. Of course, Plato does not elsewhere accept the view that true virtue is just a matter of 'instinct', and I am sure that he was not himself accepting it when he wrote the *Meno*. But perhaps Aristotle does accept it? (However, to judge from the confused discussion of *EE* VIII.2, he would rather describe such people as 'by nature fortunate', and not practically wise. Cf. $1246^b37-1247^a13$, 1247^b18-28.)

5. Comment

Virtue of character was defined in II.6 as 'a disposition concerned with choice, in a middle relative to us, determined by reason (*logos*) and in the way that the practically wise man would determine it'. As I pointed out when discussing this definition, the last part is important, for not all 'middling' dispositions are virtues, as Aristotle himself observes. So when we ask just *which* dispositions are to count as virtues, the answer that is promised is that this is determined by the reasoning that the practically wise man can supply. Then, when we go on to ask what kind of reasoning this is, it seems clear from book VI that it must be reasoning which shows which dispositions conduce to *eudaimonia* and which do not. But that in turn must clearly depend upon how the wise man conceives of *eudaimonia*. Yet now, when we ask what this conception is, and how it is obtained, we find Aristotle replying that the wise man's conception of *eudaimonia* is given by (his training in) virtue. Clearly, we have gone round in a circle, and the result is that there is *no* way of determining which dispositions are virtues. This is why I say that the position just outlined is profoundly unsatisfying.

Moreover, the circle matters in practice, for there is no general agreement on what is to count as a virtue. Let us take a modern example, where opinions genuinely do differ, say chastity: is this a virtue or not? (For definiteness, let us take as a central element in chastity the slogan 'no sex outside marriage'.) It may well be that I think that chastity is a virtue, and that this is mainly because I was brought up many years ago, when this opinion was common. But you may think that it is not, for you are younger, and have been exposed to rather different influences. Is there any way of settling our disagreement? According to Aristotle's proposal in II.6 there should be, for reasoning can be applied. So we can call in the aid of our practically wise friends, and we can all sit down to engage in an investigation of whether chastity does or does not conduce to *eudaimonia*. But, if we take seriously what Aristotle says in VI.12, this investigation must fail. For we will discover that we all have different conceptions of what *eudaimonia* is, and that these are given simply by our different educations, so that there is simply no more to be said. There is no way of deciding whether one education was better than another, save by relying on the opinions of those who were thus educated. That illustrates why I say that this circle is important.

Must we, nevertheless, accept it? In one place Aristotle seems himself to accept it explicitly. In X.8 he says, 'Practical wisdom is yoked together with virtue of character, and this with practical wisdom, since the starting-points (*archai*) of practical wisdom are in accordance with the virtues of character, and what is correct in character is in accordance with practical wisdom' (1178^{a}16–19). Perhaps we should not read too much into this passage, for the mere statement that two things go together, and that each accords with the other, does not necessarily imply that each determines the other in such a way that reasoning simply cannot be brought to bear. But what alternative is there? In non-Aristotelian theories we may find some; for example, the utilitarian will say that the crucial question is what most conduces to the general happiness, and he will be prepared to criticize particular educations in the light of this criterion. Aristotle, however, takes the goal to be one's own *eudaimonia*, and apparently offers no criterion by which different conceptions of this goal may be criticized. Or, if he does, it can only be in one of the two ways which I outlined earlier, when trying to find some 'intellectual' basis for the true conception of what *eudaimonia* is.

Yet neither of these is particularly convincing. We may claim that it is the task of good deliberation, and hence practical wisdom, to discover what *eudaimonia* is, on the ground that this is a matter of discovering what 'ingredients' it has, and in what proportions, and this is covered by the description 'what conduces to it' (*ta pros to telos*). But we also have to admit that Aristotle himself has nothing to say on how this supposed 'deliberation' could be conducted. Or we may claim that, quite generally, the task of discovering the first principles of any subject is in a broad sense a task for the intellect, even if Aristotle is not willing to call it either 'deliberation' or 'reasoning' (*logos*), as he uses that word. (For he does insist that it is not deduction.) We may add that he himself is using what *we* would call 'reasoning' when (in I.7 and X.6–8) he tries to give his own account of what *eudaimonia* is. Certainly, he does not admit that he is simply relying on how he himself has been brought up, and it would be quite unfair to say that the arguments which he offers there do in the end reduce just to this. For, after all, he does reach a conclusion about the value of theoretical activity which would have surprised almost all the Greeks of his day. Yet, even so, the arguments that he offers in these passages simply do not address the problem that I raise here: how are we to tell which dispositions *of character* are to count as virtues, and which do not? He has nothing to say on how reason, even in a broad sense of that word, can be brought to bear upon this question.

This is connected with another gap in his discussion, namely that he gives us no analysis of the concept of 'the noble' (*to kalon*),[51] though it would seem to be a centrally important concept for him. Early in the *Ethics* he linked together the three notions of what is good, what is noble, and what is pleasant, telling us that *eudaimonia* satisfies all three (I.8, 1099ª24–31).[52] Now we obviously cannot complain that he has left the notion of 'the good' unanalysed, for that was the main topic of almost all of book I, and it was identified with what men ultimately aim for. Subsequently he tells us that there are three kinds of good that men aim for, namely what is noble, what is useful, and what is pleasant (1104ᵇ30–4). But since what is useful must be useful for some further purpose, it seems fair to say that this list of three in fact reduces just to two: what is ultimately aimed for will be either what is noble or what is pleasant (cf. 1110ᵇ9–11), so these are for Aristotle the two main forms of the good.[53] Again, we cannot complain that he leaves the notion of

---

[51] The word *kalon* is untranslatable. In some contexts it clearly means 'beautiful', and in others it clearly expresses moral approval, but it is not perceived as shifting in sense from one context to the other. Thus what is *kalon* is always admirable, and worthy of praise, but also has a kind of aesthetic value. Irwin consistently renders it as 'fine', which seems fair enough to me, but Ross mostly prefers 'noble', and since this has become traditional I have followed him. (But at 1115ᵇ12–13, 1122ᵇ6–7, 1123ª24–5 Ross rather misleadingly has 'for honour's sake' rather than 'for the sake of the noble'.)

[52] Essentially the same passage is used to open the *EE*.

[53] Compare Kenny (1992:3).

pleasure unanalysed, for he has much to say about it later. But we *can* fairly make this complaint about the notion of nobility, for there is really no discussion of this,[54] even though it plays a central role in Aristotle's way of thinking about virtues of character. During the discussion of particular virtues in III.6–IV it is frequently said that the virtuous man acts as he does 'for the sake of the noble',[55] and sometimes this seems to be offered as a general characterization of virtue: all virtue aims for what is noble (1120^a23–5, 1122^b6–7). What, then, is nobility? How do we know what things are noble? And why should we pursue them?

At *Rhetoric* I.9 we find Aristotle's account of common views on nobility, and three points stand out: the noble action is praiseworthy and admired, it is chosen for its own sake, and it benefits others rather than the agent. These points emerge from the *Ethics* too, but the emphasis is mainly on the second: what Aristotle insists upon is that the noble action is chosen for its own sake (and that is apparently taken as explaining why it is praised and admired). It remains true that the actions that Aristotle explicitly picks out as noble are on the whole actions that we would say benefit others rather than the agent, and at one point our text implies this (IX.8, 1169^a6–11), but that is not what is emphasized. Certainly, it is not given as the reason why noble actions should be pursued.[56] What is the reason, then? Presumably Aristotle will say that those actions which a person chooses to do for their own sake simply do, on that account, contribute to that person's *eudaimonia*. But, we ask: does this apply to absolutely *any* action that *any* person—however wicked— chooses for its own sake? Even to those actions that are commonly held to be base and ignoble (*aischron*), and are not admired at all? Here, he must surely say 'no'. It is what the virtuous person chooses to do for its own sake that is noble. Clearly, this brings us round in a circle: the virtuous people are those who do what is noble for its own sake (i.e. because it is noble), and the noble acts are those that a virtuous person does for their own sakes. This gives us no way of discovering which acts are noble, or which dispositions are virtuous.

If there is really no more to the concept of nobility than this,[57] then it cannot help us to determine what the virtues are. Of course, one strongly suspects that there is more to the concept, in particular the thought that noble actions benefit others, which does promise to be of some assistance. But then one is faced with a different question: why should I suppose that my benefiting others will contribute to my own *eudaimonia*? It does not appear that Aristotle has any answer to offer, apart

---

[54] In *EE* VIII.3 there is some discussion of 'being noble-and-good' (*kalokagathia*), in a passage to which nothing in *EN* corresponds. But I do not think that it in fact adds anything useful to what we do find in the *EN*, and so I shall not discuss it.

[55] 1115^b11–13, $^b23$–4, 1116^a27–9, $^b2$–3, 1117^a16–17, $^b9$, 1119^b16, 1120^a23–5, 1122^b6–7, 1123^a24–5. Cf. also 1119^a18, 1120^a12, 1136^b22.

[56] Engberg-Pedersen (1983, ch. 2) disagrees, claiming that (in Aristotle's view) all virtue aims at sharing benefits fairly between myself and others. (He also adds a Kantian account of why 'reason' might require this, for which there is no basis whatever in Aristotle's text.) His attempt (in ch. 3) to apply this idea to each particular virtue only reveals its implausibility. One may hang on to the idea that all virtue aims at nobility, or to the idea that nobility always benefits others, but not to both.

[57] As Annas (1993) would appear to accept on pp. 371–2.

from the (evidently correct) claim that those who are well brought up just will accept that I should take others' interests into account, as well as my own. That is, he must apparently fall back upon education. This tells us what 'the noble' is, and that we should pursue it, but it cannot tell us why. One of course hopes that 'reason' can be brought to bear, but Aristotle does not tell us how.

If this is correct, then it points to a gaping hole, right at the heart of Aristotle's theory. For he began in I.7 with the thought that it is reason (*logos*) that is the specifically human characteristic, and that therefore determines what is to count as a human excellence (virtue). But we are now forced to conclude that it fails to do so. It now appears that, so far as the virtues of character are concerned, it is not reason but mere training and habituation—i.e. absorbing the values already held by those who educate us—that is the source of our 'knowledge'. Plainly, any moral sceptic will reply that this provides nothing that he can recognize as 'knowledge', and he will add, against Aristotle, that it is not even specifically human. For all the 'higher' animals learn from their parents and their herd how to act, and it now appears that humans are in principle no different. Clearly Aristotle could not accept this. He must assume that reason in some form—whether or not it is called 'deliberation' and credited to 'practical wisdom'—*can* provide an answer. But he does not know how to do so himself, and so preserves a discreet silence.

This, of course, leaves us with a question: what *is* the reason why we should help others? Many take this to be the central question of ethics, and certainly more recent theories do not ignore it. But Aristotle, it appears, has nothing to say.

Appendix. Note on 1142ª25–30

I am very unsure of this passage, but here are my speculations.

Text. (i) Retain *en tois mathēmatikois* in ª28. The words are dropped by Bywater in OCT, but for no good textual reason. (Why would the phrase have crept in if it was not originally present?) (ii) Read *hē phronēsis* in ª30. Some manuscripts have *hē*, some have *ē*, and Burnet conjectured *ē hē*. Bywater prefers *ē*. My reason for preferring *hē* is just that it seems to make better sense, but I punctuate so as to place these words in parenthesis. My text for ª28–30 is therefore: . . . *aisthēsis, oukh hē tōn idiōn, all' hoia(i) aisthanometha hoti to en tois mathēma-tikois eskhaton trigōnon; stēsetai gar kai ekei. all' hautē mallon aisthēsis—hē phronēsis—ekeinēs d' allo eidos.*

Translation. 'So *phronēsis* is opposed to *nous*. For *nous* is of the terms (definitions?, *horoi*) for which there is no reason (*logos*), whereas *phronēsis* is of the last thing (*eskhaton*) of which there is no knowledge, but perception. Not the perception of objects special [to each particu-lar sense], but like that [perception] by which we perceive that the last thing in mathematics is a triangle. For [reason, *logos*] stops there too. But this [perception] is more perception—i.e. [the perception that is] *phronēsis*—whereas the other [perception] is of a different kind.'

Glosses. The second sentence says that we have *nous* (rather than *logos*) of the principles from which scientific demonstration starts, whereas we have *phronēsis* of particular things (or situations), and this is perception. The third says that this is not the perception of such things as colours (special to sight), or sounds (special to hearing), but of something else; and it compares this latter perception to a kind of perception that happens in mathematics. (It is called perception at least partly because it is not due to reason (*logos*), and so does not count as something known (*epistēmē*) since it is not grasped by deduction.) The final sentence says that the perception which 'is' *phronēsis* is a more genuine case of perception than is the one that occurs in mathematics. (I understand *hē phronēsis* as a gloss upon *hautē* explaining that 'this perception' means the one mentioned first, i.e. the *phronēsis* and not the one mentioned second. The gloss may be Aristotle's own, or it may be due to a (sensible) scholiast, who saw that this must be what is meant, despite the natural way of reading *hautē*.)

Explanations. The perception that 'is' *phronēsis* is something more complex than just perceiving a colour or a sound; presumably it is perceiving some morally relevant feature of the situation, perhaps 'he is cheerful' or 'he is upset'. The usual way of understanding the comparison to geometry relies on a reference back to something said in III.3, at 1112^{b}20–4, where deliberation is compared to seeking the solution to a geometrical problem. According to the usual account (which itself reads quite a lot between the lines, but very plausibly) Aristotle is there thinking of the geometrical problem of how to construct a particular figure, e.g. a regular pentagon. One proceeds by 'analysing' the figure into its simpler components, until at last one comes upon a component that one already knows how to construct. This last step of the analysis will then be the first step of the construction. To transfer this to our passage, Aristotle will be thinking in particular of a case where (as in the present example)[38] the last step of analysis, and the first of construction, is a triangle of some familiar kind.

There is obviously a great deal of conjecture in this explanation, but it *may* nevertheless be right. However, there are two objections. One is that one does not expect *EN* VI, which is a 'common book', to look back to a point made earlier only in the *EN*, and not in the *EE*. The second is that it does not really suit Aristotle's phraseology. On this account one might have expected him to say 'as in mathematics [i.e. in a certain kind of mathematical problem] one perceives that the last thing is a [constructable] triangle'. But what he actually says is different, namely 'as one perceives that the-last-thing-in-mathematics is a triangle', and this suggests a quite different line of thought.

One reason for claiming that 'the last thing in geometry is a triangle' might be this: rectilinear figures must have *at least three* sides if they are to be figures in the traditional sense, i.e. if they are to enclose a space. In other words '*two* straight lines cannot enclose a space (and nor, of course, can one)'. This formulates a well-known *axiom* for (Euclidean) geometry. Since we do not know what were accepted as the axioms of geometry in Aristotle's day, it is only a conjecture that this was one of them. But it seems to me a perfectly reasonable conjecture.[39]

On the first proposal Aristotle is comparing the 'seeing' that *phronēsis* is (or, better, depends upon) with 'seeing' the solution to a geometrical problem of construction. On the second proposal he is comparing it to 'seeing' the truth of a geometrical axiom. Either proposal would have point in the present context, for (*a*) deliberation does seek for solutions, and it may be said—not too unreasonably—that perception provides them; and (*b*) what is perceived is grasped directly, and without reason (*logos*), as geometrical axioms are. Our grasp of axioms may here be being called a kind of 'perception', as later our more ordinary perceptions are apparently called a kind of *nous* (1143^{a}35–^{b}5); Aristotle is ready to stretch both words until they coincide.

Both of these explanations are highly conjectural, and I would not bet on either. For other

---

[38] See Euclid, book IV, proposition 10, with the commentary of Heath (1908, *ad loc.*).

[39] Irwin's translation (with note *ad loc.*) adopts the interpretation I suggest, but without noting that the thesis in question is (or may be) a geometrical *axiom*.

attempts on this perplexing passage, see e.g. Cooper (1975: 33–41), Kenny (1978: 172), Woods (1986: 162–4), Reeve (1992: 67–70).

Further reading

The most useful account of book VI overall is probably Kenny (1979, chs. 6–9). (But ch. 7 is devoted to the *EE*, and so might be omitted.)

On the particular subject of choice (*prohairesis*), one should certainly see Anscombe (1965), who sparked the controversy. This may be further pursued in Charles (1984), who argues against Anscombe (and others) on pp. 151–5, and in Reeve (1992), who argues against Charles in his section 15.

On the question of how Aristotle conceives of practical wisdom, and how it is related on the one hand to training in virtue and on the other to more intellectual endeavours, it is useful to start with Allan (1953), who argues the case for intellect. On the other side, an essential discussion is Broadie (1991, ch. 4). (But the heart of her position is stated in sects. 3–4 of that chapter, and one might not unreasonably confine attention to these.) See also Sorabji (1973–4), and possibly Irwin (1975), Dahl (1984, pt. I; ch. 6 gives a useful summary), and Reeve (1992, ch. 1). These each defend the cause of intellect, but each in a different way. On the other side, see Woods (1986), who offers a nice clear argument for the importance of perception, and possibly Fortenbaugh (1991). The central issues are well set out by Smith (1996: 56–65), but the 'solution' that he then goes on to offer does not seem to be a 'solution' to me.

On the particular question of the unity of the virtues, which is not addressed in any of the readings so far mentioned, some useful points are raised by Irwin (1988*b*) and by Telfer (1989/90).

I have said nothing in this chapter about what commentators call 'the practical syllogism'. For some reading suggestions on this, see the end of Chapter VI.

Chapter V
Responsibility (III.1, 5)

1. Introduction

ARISTOTLE'S best known-discussion of responsibility is in III.1, and I shall devote most of my attention to this. I then turn to what he has to say in III.5 concerning what *we* call 'the problem of free will', and finally I add a brief appendix on the other treatment of responsibility in our *Ethics* at V.8.

The general title 'responsibility' is mine and not Aristotle's. He introduces the subject by noting that only certain actions are liable to praise or blame, and promising to say which these are. His main thesis is that it is 'voluntary' actions which are thus liable, whereas 'involuntary' actions are not; they should rather be pardoned, and perhaps also pitied (1109^{b}30–5). But before we begin it is necessary to say something about the two Greek words *hekōn* and *akōn*, which in this context are standardly translated as 'voluntarily' and 'involuntarily'. For in non-philosophical contexts they would not normally be so translated.[1]

Usually, to say that I do something *hekōn* is to say that I do it willingly, that I am happy to do it, I want to do it. To say the opposite, that I acted *akōn*, is to say that I acted unwillingly, reluctantly; I did not want to act in that way. But the contrast that Aristotle seems to be aiming for is not this, and is better taken as the contrast between acting intentionally and acting unintentionally. This is certainly a different contrast, for one can do something both intentionally and reluctantly (e.g. when I refuse an offer to participate in the profits of a crime). Indeed, what I do unintentionally I do neither willingly nor unwillingly, for the question whether I wanted to do it or not simply does not arise if I did not intend to do it anyway. The reasons for supposing that it is actually the question of intention that Aristotle means to be discussing are (i) that the analysis which he offers does seem to fit this notion better, and (ii) that it is anyway closer to what is demanded by the connection he draws with responsibility, i.e. with praise and blame. For it is plausible to say that I cannot be blamed or praised for what I do unintentionally, but not plausible at all to say this of what I do not much want to do, but do intend to do. But we shall find as we proceed that even this is not quite right, and in fact Aristotle so twists the word *hekōn* that he uses it to apply to anything that I do, and am responsible for doing,[2] even if it is not something I intended.

---

[1] Ross uses 'voluntarily' and 'involuntarily' throughout, with just one exception ('unwillingly' at 1110^{b}12). Irwin quite often uses 'willingly' and 'unwillingly' (1110^{a}10, ^{a}15, ^{b}12, ^{b}20–3, 1111^{a}17, 1113^{b}14–17, 1114^{a}15–21), as if 'willingly' and 'voluntarily' were interchangeable.

[2] Aristotle is not concerned with things for which I may be held responsible though it was not I who did them. (For example, I am responsible for the damage that my dog does to your garden.)

It will be noted that, in what I have just said, I have been equating 'doing x intentionally' and 'doing x, and intending to do x'. Some might wish to make a distinction. For example, suppose that I intend to assassinate the Queen, and intend to do this by placing a bomb (to be detonated by remote control) in the saddle of the horse that she will ride at a ceremony of trooping the colour. Of course I foresee that if this plan succeeds then the horse will die too. Do I count as *intending* to kill the horse? Some might say 'no', on the ground that my *goal* was to kill the Queen and not the horse, so that and only that was what I intended. But, since I did foresee the horse's death, they might concede that I did kill the horse *intentionally*. On this view, its death was intentional but not intended. But this distinction is irrelevant in the present context, concerning praise or blame, or so anyway it seems to me.[3] To see this clearly, just reverse the example, and suppose that my goal was to kill the horse, and it was merely incidental to this that the Queen would die too. Obviously I would be held responsible for the Queen's death, and could very properly be blamed and punished for it—hung, drawn, and quartered, perhaps—even though I protested that that was not my goal. The distinction between the goal itself, what is a means to it, and what is merely a side-effect, seems to me irrelevant to responsibility and blame; what matters is just whether these effects of my action were foreseen by me.[4] If they were, then I shall count them as both intentional and intended.

2. The analysis of responsibility (III.1)

Aristotle's basic idea is that an action is involuntary (unintended?) if it is done either 'by force' or 'because of ignorance', and that if it is not involuntary in one or other of these ways then it is voluntary (intended?) (1109^b35–1110^a1). But this basic idea is qualified as we proceed.

(a) **Force (1110^a1–$^b17$).** An action is forced when the origin (or cause, *archē* of the action is outside the agent, so that he contributes nothing to it, as for example when the wind carries him off course. (We might say—as Aristotle himself suggests at $^a2$–3—that in such a case what happened was not an *action* at all.) We get an important clarification of this criterion a little later: in a voluntary action 'The origin of the setting in motion of the instrumental parts of the body is in him, and where the origin is in him it is also up to him whether to act or not' (1110^a15–18).[5] The addition is relevant. Thus, suppose I have a cold and I sneeze. Then if you ask where the cause of that act is to be located, the only possible answer seems to be 'in me'. Yet it was not (in the normal case) a voluntary act, because I could not help it; it was not up to me. So what Aristotle evidently means is not just that the cause should be 'in me', but more strongly that it should be 'in my mind'. A very natural suggestion is that its cause should be a choice or decision of mine, and I shall accept this

---

[3] Those who believe in the so-called 'doctrine of double effect' may disagree with me.

[4] Or perhaps foreseeable? I take up this question on p. 110.

[5] Ross translates the phrase 'up to him' as 'in his power'. (I have no objection to this translation, but will mostly use Irwin's more literal rendering.)

suggestion for the time being. Then the basic assumption is that how I choose to act is up to me.

Now Aristotle does not say this himself, and one reason why is that he cannot say it, because *his* concept of choice will not allow him to. He is going on to discuss choice in the next two chapters of this book, and (as I have noted, pp. 39–40) he insists that it always includes some *deliberation* on how to act. But this is not a requirement on voluntary action, as he notes himself:

> Choice seems to be voluntary, but is not the same as the voluntary; what is voluntary extends more widely. For both children and other animals share in voluntary action, but not in choice, and what is done on a sudden we describe as voluntary, but not as chosen. (1111^b6–10)

We, however, need not adopt Aristotle's rather over-inflated view of what choice is, and if we do not then it seems reasonable to suggest that his first condition comes to this: the cause of a voluntary action must be a choice or decision of the agent, and that is how he must 'contribute' to it, for it will then be up to him whether he acts in that way or not.[6]

Unfortunately, the position is muddled by the immediately following discussion of what Aristotle calls 'mixed' actions. These are cases which we might describe as actions done under duress; Aristotle's initial examples are actions done under threat from a tyrant, or in peril from a storm at sea (1110^a4–11). They clearly do not fall under the previous account of 'force'. If I do decide to knuckle under to a tyrant's threats, or to throw my cargo overboard, then this certainly is a decision of mine, so the cause of the act is in the relevant sense 'in me'. This, indeed, is more or less what Aristotle says himself at 1110^a11–18, and yet he persists in calling these actions 'mixed'. Not only are they voluntary, but they are *also* in a way involuntary. In what way? Aristotle's explanation is laconic in the extreme: 'Such actions, then, are voluntary, but perhaps without qualification (*haplōs*) involuntary; for no one would choose them for their own sakes' (1110^a18–19; cf. $^a9$–11). This is wholly unconvincing. The actions are chosen as the best response to the circumstances at the time, and *every* action is a response to the circumstances at the time. Admittedly, the circumstances were not of my own choosing, but that again is almost

---

[6] Irwin (1980) supposes that Aristotle has a 'simple' theory of voluntariness, which is independent of deliberation, but also a 'more complex' theory of 'responsibility' which involves deliberation in this way: 'A is responsible for doing *x* if and only if (*a*) A is capable of deciding effectively about *x*, and (*b*) A does *x* voluntarily' (p. 132). (Decision, which is Irwin's translation of *prohairesis*, is understood by him always to include deliberation.) The obvious objections are (i) that Aristotle shows no sign of having a separate conception of responsibility, distinct from voluntariness (and certainly he has no word for it), (ii) that Irwin motivates his 'complex' theory by the claim that we do not praise or blame children or animals for what they do (p. 125), and yet when one thinks about it this is not obviously correct, nor anywhere endorsed by Aristotle. (Irwin pays no attention *here* to Aristotle's further example of what is done 'on a sudden', but it may influence his final version of the 'complex' theory, according to which 'responsibility' requires only that one be *capable* of deliberation, and not that one has actually deliberated.) I do not deny that Irwin's 'more complex' theory may 'allow us to state more clearly and defensibly Aristotle's arguments about responsibility for action on non-rational desires, for mixed actions, and for character' (p. 142). I do deny that this 'more complex' theory is Aristotle's.

always the case. And surely Aristotle does not want us to conclude that almost all actions are 'mixed'?

I think it is legitimate to suspect that Aristotle is here being drawn by the other, and more usual, meaning of the words *hekōn* and *akōn*. He says that the action is 'involuntary' on the ground that I do not choose it 'for its own sake', which is a way of making the point that there is a sense in which I do not *want* to act in that way. So I act unwillingly, and with reluctance. But at the same time, of course, there is also a sense in which I *do* want to throw the cargo overboard, for (as I see the situation) my life depends on it. So what is 'mixed' about the situation is my wants: in one way I want to do it, and in another way I don't. But there is nothing mixed about my decision, and that is what is relevant to our topic of praise and blame. The decision is simply to throw the cargo overboard, and clearly it is up to me whether to do that or not. So I can be praised for it, if it was the appropriate decision in the circumstances, or blamed if it was not. And this Aristotle himself admits, but with a different example in mind, namely the decision to endure something shameful or painful in order to achieve a great and noble end (1110^{a}19–23), for this too he counts under the general heading of 'mixed actions' (cf. 1110^{a}4–5). As he very reasonably goes on to add, in such cases it can be difficult to decide what the right course of action is, and difficult too to abide by the initial decision (1110^{a}29–^{b}1).

In the course of this discussion he says, 'But in some cases there is no praise [or, we may add, blame], but pardon, when one does something that should not be done, for reasons which overstrain human nature, and which no one would endure' (1110^{a}23–6). From the general run of his discussion so far, it would appear that the 'some cases' considered here are *still* cases of what he calls 'mixed' actions, which are counted as 'voluntary', though also (in an irrelevant sense) 'involuntary'. If so, then he is here saying, contrary to 1109^{b}30–5, that some voluntary actions are not blamed but pardoned. But he gives no example,[7] and it is perhaps better to take him as digressing here to actions which are not to be counted as voluntary at all, since what 'overstrains human nature' presumably cannot be altered by the agent's decision, and it is no longer 'up to him' whether to resist or not. One might think, perhaps, of betraying a secret under torture. If the situation really is that the betrayer *could not* have resisted, then the act is not voluntary, and he should be pardoned; but if he *could* have resisted (and should have done), then the act is voluntary and he should be blamed. That, at least, seems to be the best way of applying Aristotle's principles to this kind of case.[8] But we must of course add that

---

[7] The example that follows (of Alcmaion's reasons for matricide) is an example of what does *not* 'overstrain human nature'.

[8] An alternative view is that, even where resistance would 'overstrain human nature', the action is still voluntary, since the agent does contribute something, namely his decision, although that decision is 'necessitated' (*anankadzetai*, 1110^{a}26, 32). To maintain this view, one must see Aristotle as here distinguishing between 'forced' and 'necessitated', as several commentators do, but it does not seem to me a plausible distinction in the present context. (I think all agree that *EE* II.7–8, which often couples 'force' and 'necessity' in its discussion of this topic, sees no distinction between these terms.)

this question of whether resistance was *possible* (for that particular person, in those particular circumstances) is often not easily resolved.[9]

In any case, I think we may say that Aristotle's general tendency is to paint the picture in black and white: either it was up to the agent whether or not to act as he did, or it was not. In the first case he is open to praise or blame for so acting, and in the second he is not. What one might feel is missing from the discussion is what we call 'mitigating circumstances', where the agent is to be blamed, and perhaps punished, for what he did, but the blame or punishment is to be *lessened* in recognition of the fact that, in the circumstances, it would have been *difficult*, but not *impossible*, for him to have done what is right. Though Aristotle clearly does mention degrees of difficulty in taking the right course, he does not mention corresponding degrees of blame, but only the question whether blame is or is not appropriate. But I do not regard this as at all a serious criticism, for his discussion could surely be expanded to include the topic.

One excuse that Aristotle recognizes, then, is 'I couldn't help it' (or 'it was not up to me', or 'I was forced'). Let us now turn to the other 'I didn't know'.

(b) Ignorance (1110^{b}18–1111^{a}21). As soon as he turns his attention to actions which are involuntary because 'due to ignorance', Aristotle at once begins to introduce qualifications, before he has even stated the basic idea. In fact we at once get three qualifications, at 1110^{b}18–24, ^{b}24–7, ^{b}28–1111^{a}2. But before tackling these it will be useful to pause for a moment on the basic idea.

The knowledge and ignorance that is relevant here is, in the first place, knowledge or ignorance of what I am doing, i.e. it is knowing or not knowing some proposition of the kind 'what I am doing is such-and-such'. There will be many such propositions concerning any particular action, and for any action that I choose to do there will inevitably be some that I do know and some that I do not. For example, I may know that I am pulling the trigger of a gun, but not know that I am pulling the trigger of a gun that is loaded, or again I may know this too, but not know that I am pulling the trigger of a gun that has so far been fired just 102 times. In a sense, then, one *always* acts *both* 'in knowledge' *and* 'in ignorance'. But, in the present context, the knowledge that concerns us is knowledge of those features of the action that make it a good or a bad thing to do, and so potentially a fitting subject for praise or blame. Since actions, and their circumstances, are so various, I doubt whether it is sensible to try to list these, though of course we can make some broad generalizations. (For example, not knowing that the gun is loaded is much more *likely* to be relevant than not knowing how often it has been fired before.) But one can always dream up unusual cases where all kinds of recherché knowledge would be relevant. (For example, perhaps the gun has been designed to self-destruct when it is fired for the 103rd time.) I add further that ignorance of some fact about my action may well be due to ignorance of some other fact that can be described without reference to my action. Thus, to stay with the same example, the

---

[9] In our passage what 'overstrains human nature' is identified with 'what *no one* would resist' (1110^{a}25–6). But the parallel passage in *EE* II.8 says very explicitly that what matters is whether the particular person could or could not have resisted (1225^{a}25–7).

fact that the gun was loaded is not itself a fact about any action, but my not knowing it leads to my not knowing a relevant fact about my action, namely that it is a case of pulling the trigger of a loaded gun. To change the example, my not knowing a general fact about the law, say that it is illegal to drive without third-party insurance, may lead to my not knowing a relevant fact about my action, namely that I am driving illegally. When specifying what the relevant ignorance is, we very often do cite facts of this kind, which in themselves make no reference to action. But what makes the ignorance relevant is that it leads to ignorance of a fact about my present action, and in particular a fact about it which affects its goodness or badness as an action.

Let us sum up in this way. Any action may be described in all kinds of different ways, say as 'a φ-ing' or as 'a ψ-ing'. Then Aristotle's basic thought is that I φ-ed voluntarily if and only if (i) the φ-ing was due to a choice or decision of mine, and so 'up to me', and (ii) I knew that it was a φ-ing. The same action may also be described as a ψ-ing, and it may be that I knew that I was φ-ing but did not know that I was ψ-ing. In that case, I φ-ed voluntarily, but did not ψ voluntarily, even though the φ-ing and the ψ-ing were the same action. And in practice we are going to be most interested in that description, as a φ-ing or as a ψ-ing, that is most relevant to the appraisal of the action as a good or a bad thing to do.[10] This, it seems to me, is a very reasonable description of what it is to φ intentionally, and is the main reason for saying that it is really the notion of an intentional action that Aristotle is seeking to analyse. But so far we have considered only what I have called his 'basic' thought. Let us now turn to the qualifications that he himself begins with, in 1110^{b}18–1111^{a}2. I shall take them in reverse order.

At 1110^{b}28–1111^{a}2 Aristotle says that it is only 'ignorance of the particulars' that makes an action involuntary, namely the particulars 'in which and concerning which the action takes place' (1110^{b}33–1111^{a}2). He proceeds to give a list of these, with illustrative examples (1111^{a}3–19). His contrast is with 'ignorance of the universal', which he says is blamed (1110^{b}31–3), and which he evidently equates with ignorance of what is advantageous (*sumpheron*,[11] 1110^{b}30–1), and with ignorance of 'what should be done and what avoided' (1110^{b}28–9). The contrast, then, appears to be that between general moral rules and facts concerning the particular situation. Why is this relevant? The answer must surely be that he holds that ignorance of the particulars is an excuse, whereas ignorance of the universal is not. (That is why he adds that it is blamed, ^{b}33, and that it is characteristic of the vicious man, ^{b}28–30.)[12]

---

[10] Ackrill (1978) argues that Aristotle does not himself have a firm grasp of how it can be the *same* action that has different descriptions, and so can be wanted or intended under one description but not under another. I agree, though I do not think that it is a serious obstacle to seeing, in any particular case, what Aristotle is getting at. (Charles 1984, ch. 2, sect. B attempts to pin Aristotle down to a definite theory about when actions that are initially given under different descriptions are the same action. I am sure that Aristotle had no such theory as Charles envisages.)

[11] Irwin prefers 'beneficial'.

[12] He is here apparently accepting the Socratic thesis that the vicious man does not know that the goals he is pursuing are wrong; if he did know this he would not be classed as vicious, but as suffering from weakness of will (*akrasia*; see Ch. VI).

We may compare our own legal ruling that 'ignorance of the law is no excuse', and note merely that Aristotle is apparently extending this to 'ignorance of the *moral law*'. I postpone for the moment the reason *why* it is not an excuse.

At 1110^{b}24–7 he introduces an obscure distinction between acting '*in* ignorance' and acting '*because of* ignorance'. His examples are the man who is drunk and the man who is angry. These, he says, act in ignorance, but not because of ignorance; rather, they act as they do because of their drunkenness or anger. Though he never quite says this, it seems clear that his point must be that their actions do not count as involuntary, for it is only acting *because of* ignorance that counts as involuntary action, as he has defined it. Again one asks: why is this supposed to be relevant? And again a very natural suggestion is that the drunk man is blamed for what he does when drunk, even if it is true that at the time he is not fully aware of what he is doing. So this is another example of a kind of ignorance, namely ignorance brought on by drunkenness, which does not count as an excuse.

If this is right, then it is the same thought that lies behind both of these qualifications, namely that there are cases where I do act in ignorance of some relevant feature of my action, and so do not *intend* to do what I do, and yet I can still be blamed for it. Aristotle has given us two cases—ignorance of a general rule, and ignorance due to drunkenness or quick temper—but of course there are many others too. For example, I may plead that I did not know that the gun was loaded, but you may very fairly reply 'You *should* have known; this is the kind of thing that should always be checked'. Later in III.5 we do get a very much better and more general statement of the position, in which Aristotle explicitly recognizes that in some cases ignorance is *itself* culpable. He says:

Legislators punish and exact penalties from those who do what is wicked, unless they have acted by force or because of ignorance *which they are not themselves responsible for*. (1113^{b}23–5)

And again, more fully

Indeed, they punish a person for the ignorance itself, if he seems to be responsible for it. For example, those who are drunk pay a double penalty, for the origin is in the man himself, since it was in his power not to get drunk, and it is the drunkenness that causes the ignorance. They punish too those who are ignorant of something in the laws which they ought to know and which it is not difficult to know. And similarly in other cases where people seem to be ignorant through carelessness, on the ground that it was up to them not to be ignorant, for it was in their power to take care. (1113^{b}30–1114^{a}3)

Clearly this passage both subsumes and generalizes[13] the two qualifications made in III.1 at 1110^{b}24–1111^{a}2. It is also clear that Aristotle wishes to square this point with the general connection that he draws between blame and voluntariness by claiming that one can be ignorant voluntarily, e.g. when one's ignorance is due to carelessness,

---

[13] It also softens the principle that ignorance of the law is no excuse, by allowing that it may be an excuse when the law is 'difficult to know'. We have here a hint of the question 'Could he have been *expected* to know?'

since it is 'up to me' not to be careless. This must lead us to qualify the initial suggestion that it is my choices or decisions that are 'up to me'. For when I am careless, inattentive, negligent, or just plain forgetful, it hardly seems right to say that I intend or choose or decide so to be. It is not (usually) that I consider whether to take care, or to remember, and decide not to, but rather that the thought simply fails to occur to me (though it should have done). Nevertheless, I think we may agree with Aristotle that the situation fits his own formulation: the 'origin' of this carelessness or forgetting can only be located 'in me', and it was 'up to me' or 'within my power' to avoid it (or so, at least, we normally think). To change to a phrase that is nowadays often used, 'I could have done otherwise', and that is why I am to blame for it.

One must admit that it is an odd use of 'voluntary' in English (and of *hekōn* in Greek) to say that when I simply forget to do something which I should have remembered, I do so 'voluntarily'. And I have to add that Aristotle never directly says in these passages that culpable ignorance is 'voluntary', and some deny that he means it.[14] But clearly it is what is required if voluntariness and responsibility are to march hand in hand, and Aristotle is trying to define 'voluntary' in such a way that they do march hand in hand. But has he really succeeded? I have accepted that my forgetting was not 'forced', in that it was 'up to me' to remember, and I could have remembered. But what about the other condition? Can we say that the forgetting was not 'due to ignorance'? More generally, when ignorance is itself culpable, can we say that it is neither forced *nor* due to ignorance? The question is puzzling. The ignorance was due to forgetfulness, inattention, lack of care, negligence, and so on. Are these themselves forms of ignorance? Apparently Aristotle is committed to answering 'no', if he wishes to count such ignorance as 'voluntary'. But it seems better to say that his general analysis here requires an exception. I am responsible for my ignorance if I could, and should, have taken steps to find out. Since I could have taken these steps, the ignorance is not 'forced'. But the question whether this ignorance is itself 'due to ignorance' is one that does not arise, for we do not know how to make sense of it. What one asks instead is 'Could I have been expected to know?' (for which the usual test is 'would the 'reasonable' man have known?'). In any case, let us accept that ignorance can itself be (in Aristotle's somewhat odd sense) 'voluntary', and hence blameable.

A further question which arises is whether, when one subsequently acts in a way which depends upon this ignorance, that subsequent action is *also* 'voluntary', and hence blameable, so that I am blamed twice—once for becoming ignorant, and then again for what that ignorance leads me to do. Since Aristotle apparently approves the 'double penalty' for misdemeanours committed when drunk, one would suspect that his answer is 'yes', though it is difficult to see how there is more than one action here that is 'voluntary'. (For, in the case of the extreme drunkenness that Aristotle seems to have in mind, once I have got drunk, my actions are no longer 'within my power'.) If this is right then Aristotle is prepared to call some-

---

[14] e.g. Curren (1989).

thing 'voluntary' even though it does not itself satisfy his definition of voluntari-ness, but is the result of some previous action—or failure to act—which does satisfy the definition.[15] We shall find a further example of this in the next section, with his claim that character counts as 'voluntary'. But it is not entirely clear that this is how he is thinking here (for the 'double penalty' could certainly be explained differ-ently). And the truth seems to be that, while there are some crimes (notably mur-der, in English law)[16] where 'intent' is crucial, and if I kill you without intending to that is not murder, however negligent I have been, still in most cases negligence is punished more or less severely depending on the seriousness of the result that it leads to. So while it is still the negligence that is punished, it is the subsequent act that dictates the penalty.

Finally, I return to the very first qualification that Aristotle makes, right at the beginning of his discussion (1110^b18-24). What is done because of ignorance, he claims, is never voluntary—recall that the qualifications we have just been discuss-ing are still to come—but also it need not be involuntary either. For it is involuntary only if the agent later regrets what he has done. Thus we have an intermediate category of acts which are neither voluntary nor involuntary, namely those which are done because of ignorance, but not subsequently regretted. Presumably he means that they are not regretted *when* the ignorance is dispelled; if it never is dispelled, then we have to ask instead whether the agent *would* regret them if it were dispelled.

This is the only occasion on which Aristotle suggests that the classification 'voluntary'–'involuntary' is not exhaustive, and I see no rationale for it. If the act was due to ignorance, and ignorance which is not itself blameable, then clearly the agent cannot be blamed for it, whether or not he afterwards regrets it. It should therefore count, in either case, as involuntary. I give two examples. Suppose first that I am walking in the woods, an ant is in my path, and I tread on it and kill it. I did not notice the ant, and I could not be expected to notice it, since it could hardly be seen amongst the other debris in the path. When you point out to me that I have just killed an ant, I may perhaps regret it, or I may feel no regret at all, reckoning that ants are unlovely creatures, since they can bite, and the loss of one is a (minute) gain to the world as a whole. But in either case I killed that ant involuntarily, since I did not know that that was what I was doing. To change the example to something more serious, suppose that I intend to kill you by shooting you, and in preparation I go down to the rifle range for some practice shooting. Unknown to me, some other people who intend your death have kidnapped you, and tied you up behind the target I am aiming at. So I shoot at the target, and thereby kill you. Naturally, I am delighted. My aim has been accomplished, and at no risk to me, since I did not know, and could not be expected to have known, that shooting at the target would result in your death. Again, I kill you involuntarily, and cannot be blamed or

---

[15] Similarly, perhaps, with 'forced' actions. I may plead that the wind blew me and I could not help it. But you may still blame me for what happened, saying: 'You should have paid attention to the weather forecast; you should never have set out.'

[16] But not in Scots law.

punished for killing you, since I did not know that that was what I was doing. The fact that I have no regrets is entirely irrelevant.[17]

Why might Aristotle think that it was relevant? Well, one conjecture is that he is more interested in the agent's character than in the blameworthiness of the particular action, and one who regrets an action with a bad but unintended outcome has a better character.[18] Another conjecture is that he is misled by his own coupling of 'pardon and pity', for the one who has no regrets should indeed be pardoned, as I have argued, but is hardly to be pitied. But what he says himself is just that the act should not count as 'involuntary' simply because the agent is not pained by it (1110^b21-2), and he says quite frequently that involuntary actions are painful (1110^b11-12, $^b18-19$, 1111^a32-3). Now evidently if at the time I do not know what I am doing, I will not be pained by it *then*, but Aristotle no doubt takes this point to be satisfied if later I am pained, when it all comes to light. (And *perhaps* he takes it that the later pain shows that I would have been pained at the time, if only I had known.)[19] But here one must protest that there is in fact no connection between involuntary action and action that pains me, and Aristotle has again been misled by the more ordinary meaning of his words *hekōn* and *akōn*. In the ordinary meaning, to act *akōn* is indeed to act unwillingly and reluctantly, and so it may reasonably be said that it is acting in a way which pains one. But that is the ordinary meaning of the word, and not the rather special meaning which Aristotle here gives it by his own definition, which is (in effect) to do something for which one is not responsible. And really there is no reason why an act for which I am not responsible, whether because of 'force' or because of 'ignorance', should not be one whose outcome delights me. This qualification, then, is that one we should certainly not accept, given that we *do* accept the others, which aim to close the gap between voluntariness and responsibility.

3. Aristotle and the problem of free will (III.5)

In the passages discussed so far Aristotle has simply assumed that certain actions are 'up to us', namely those which we choose or decide to do, and those (failures to act) which it was in our power to avoid. The problem of free will, as we understand it, is set by the fact that there appear to be arguments which show that it is *never* 'up to us' whether to act or not, or in other words that no one ever could have done other than what he did do. For if these arguments are correct then it appears to follow that no one is ever responsible for what he does, and hence that no one ever deserves praise or blame.

---

[17] One may note that a similar distinction may be drawn in the case of forced actions: in some cases I may be upset by what I have been forced to do, and in other cases delighted. It is mildly odd that Aristotle notes the distinction only for cases of ignorance, and not also for cases of force. But in both cases it is irrelevant.

[18] This explanation is offered by Irwin (1988a: 342–3), by Broadie (1991: 126), and by several others. It is pursued at length by Hursthouse (1984), and leads her to a quite unusual account of what Aristotle is trying to do in III.1.

[19] Thus Urmson (1988: 46); but the inference is hazardous.

I briefly mention three such arguments. One is what one might call a 'purely logical' argument, which starts simply from the tautology that whatever will be will be. From this it infers that I cannot prevent what will be, nor bring about what will not be, and from this it is alleged to follow that I cannot prevent or bring about anything. Aristotle is aware of this argument, and he disagrees with it, but his discussion of it is elsewhere (i.e. in *De Interpretatione* 9). Consequently I shall say no more of this argument here. A second is a theological argument, which starts from the premiss that God is omniscient, and so already knows whatever is going to happen. From this it is inferred that he already knows what my choices will be, and it is then alleged to follow that those choices could not have been any other than they in fact will be. The argument depends upon a conception of God which Aristotle certainly does not share, and which is usually associated with Christianity. So neither he nor any other classical Greek philosopher ever made any response to this argument. A third argument, and the one which most bothers us these days, is based upon physical, or causal, determinism. The thought is that the laws of nature include causal laws, according to which the state of the universe at any one time determines its future states at all other times, in such a way that the future is (in principle) predictable, by one who knows all the laws of nature and who knows in full detail some past state of the universe. This argument Aristotle is familiar with, for it can be extracted from his predecessor Democritus,[20] and again he disagrees with it. This is basically because he does not believe that there are causal laws of the kind in question. But once more his discussion is elsewhere (principally in *Physics* II.8–9, but other passages are also relevant, e.g. *De Generatione et Corruptione* II.9). At any rate, none of these arguments is directly discussed in the *Ethics*.

The view that Aristotle *is* aiming to combat in III.5 is mainly the Socratic view that 'no one does *wrong* voluntarily'.[21] But Socrates had not claimed that *no* action is voluntary, and apparently was quite content to allow that right actions are (usually) voluntary. That is why Aristotle so constantly just *assumes* in III.5 that right actions are voluntary, and treats a claim as refuted if it conflicts with this assumption. It is also why his discussion often seems to us to be unsatisfying, for we are more interested in the objection that no actions whatever are voluntary. Nevertheless, some of what Aristotle says can be seen as bearing upon the problem as we understand it, and I shall pay more attention to this part of his discussion. We may divide what he has to say into two main claims, first that actions are voluntary (i.e. if they are neither forced nor due to ignorance), and second that character is therefore voluntary.

(a) Actions are voluntary. At the end of his discussion of forced actions in III.1 Aristotle envisages an objection: 'Suppose someone says that what is pleasant and

---

[20] Whether Democritus himself drew the conclusion about free will is not clear. But, one generation after Aristotle, Epicurus certainly did, and so tried to modify the Democritean doctrine accordingly.

[21] More accurately, 'no one does wrong *hekōn*'; it is not clear that Socrates did mean by *hekōn* what Aristotle means by it, but Socrates certainly did mean that no one does what is wrong *knowing it to be wrong*. His argument will receive more attention in the next chapter. (The Socratic thesis is cited explicitly at 1113^{b}14–15.)

what is noble force us to act, because they are external to us and compel us' (1110 ^{b}9–10). He does not reply, as one might expect (cf. 1114^{a}31-^{b}1), that what leads us to act is *our conception of* what is pleasant or noble, and that is within us, but instead claims simply that *every* action aims either at what is pleasant or what is noble, so on this account every action would be forced, which he treats as absurd (1110^{b}10–15). At the end of his whole discussion in III.1 he envisages a similar objection 'Presumably it is not correct to say that what is done through "spirit" (*thumos*) or desire (*epithumia*) is involuntary' (1111^{a}24–5). Again he gives a similar reply: it would be arbitrary to suppose that only some of these actions are involuntary, but absurd to suppose that they all are, since on some occasions it is *right* to act as one's desires and emotions prompt one to do (1111^{a}27–31) — indeed, we may add that on his own account of virtue it is always right for the virtuous man so to act. So very explicitly here, and as a subtext in the earlier passage too (1110^{b}14–15), he is taking a view to be refuted if it leads to the conclusion that right actions are not voluntary. This line of thought recurs frequently in III.5 (1113^{b}7–11, 1114^{b}12–16, ^{b}19–21), and I shall pay it no further attention. There are, however, two different arguments for the voluntariness of actions which are to be found in III.5.

One claims that the whole practice of blaming and punishing people for their bad actions presupposes that those actions are within our power, and it adds as evidence for this that we take the same attitude even to bodily conditions such as ill health or blindness. For we reproach those who have brought themselves into such conditions, e.g. by refusing to take their doctor's advice, but not those who were born that way, or have suffered from injury or misfortune (1113^{b}21–1114^{a}31). There are two well-known replies to this argument. One (the 'compatibilist' reply) is that blame and punishment are still justified even if no one could ever have acted differently. For they still have a useful role as a deterrent, which may be expected to be one of the factors which is a causal influence on my choices. In support of this it is claimed that we do blame and punish only when this is a potential deterrent, either to me or to others.[22] The other (the 'incompatibilist' reply) accepts Aristotle's claim that our practice does rely on a belief that people often could have acted differently, but it adds that that belief may perfectly well be mistaken. Aristotle might be tempted to reply that the belief is (almost) universal, and that we must assume as a methodological principle that 'what everyone believes is true'.[23] But his opponent may clearly dissent from this principle: in *some* cases, and this may be one of them, a belief that is (practically) universal may yet be false. This is not the place to pursue either of these arguments further.

A second argument that appears in III.5 is this: we cannot actually trace the causes of an action back to factors that lie outside the agent, and if so then we must admit that how to act is 'up to him' (1113^{b}19–21). The first response that one must make to this argument is that it is plainly a *non sequitur*. For suppose that we take it

---

[22] Aristotle himself says that we blame or punish only when the offender will himself be deterred thereby (1113^{b}23–6).

[23] The principle is explicit at 1172^{b}36–1173^{a}2, but often implicit elsewhere. For some discussion see Ch. X, pp. 214–16, 221.

that a person's choices, and hence actions, are caused by, in a broad sense, his desires and his beliefs—in more Aristotelian terms by on the one hand his appetites (*orexeis*), which include both the desires (*epithumiai*) and other emotions of the part of the soul that has feelings, and his wishes (*boulēseis*) stemming from the rational part, and on the other hand by his reasoning or deliberation about how to achieve them. And suppose further, for the sake of argument, that these desires and beliefs (and the reasoning based on them) cannot be traced to any cause outside him. It clearly does not follow that he is in any reasonable sense in *control* of these desires and beliefs (and reasoning), so they are not 'up to him'. But if they cause his choices, then his choices also, and consequently his actions, are equally not 'up to him', even though we cannot trace their causes to something outside him. The general principle that what happens within a person—or even, within his mind—must also be within his control, is clearly false.

As a matter of fact we can, of course, very often point to external causes. Thus my beliefs about the situation I am in (e.g. that there is a glass of water in front of me) are usually caused by the situation itself (the fact that there is such a glass), together with my perception of it.[24] For other beliefs, a more complicated story will be needed, but no doubt it too will trace the belief back to factors external to me. A similar story may be told about at least some of my desires. There are those (e.g. thirst) which arise directly from the state of my body, and others which—on Aristotle's own account—are due to early training and conditioning, which were initiated by others and not by me (e.g. my wish to behave fairly). And even if in some cases (e.g. sexual desires?) all one can say is that they are (largely) due to 'the nature I was born with'—a hypothesis resembling that which Aristotle entertains at 1114^{b}5–12—still that nature in turn has its causes, as we now believe. (It is due to what happened at conception, and the various genes that I then inherited.) In general, then, the modern determinist will be ready to supply external causes for all the beliefs and desires that influence my choices; in broad terms they will be due on the one hand to what I inherit (in my genes), and on the other to the environmental influences that I experience (including what I learn from my parents, my school, my friends, my reading, and so on). It is surprising to find that, in the present passage, Aristotle seems not prepared to recognize *any* of these 'external' factors. But they are perhaps recognized later, at 1114^{b}1–3 and 21–3, where the claim is only that I am 'in a way' a cause, or at least a 'co-cause', of what appears to me to be good.

In any case, the more important criticism of his argument is the first: no matter how these beliefs and desires of mine originally came into being, the fact is that, on any given occasion of acting, I do not *choose* what beliefs and desires then to have. Rather, I just find myself having them, and they are not 'up to me'. So, if there is to be room for something which is 'up to me', it must be in the choice that I then make. One who wishes to uphold the Aristotelian (and, I believe, common-sense) notion of free will must say that these beliefs and desires will of course *influence* my

---

[24] This is Aristotle's own view elsewhere (e.g. *Physics* VIII.253^{a}7–21 and 259^{b}1–20). So perhaps what he means is that my response to such external stimuli is not determined by those stimuli themselves, but by me. I note this view in my next paragraph.

choice, but nevertheless do not *cause* it, in the sense that they do not determine, or necessitate, that choice. For I *could* have had all those same beliefs and desires and yet have chosen differently. That, it seems to me, is the only point where it is at all plausible to locate my own 'control' of what happens. But whether there really is such a control—or whether, in the end, we can even make sense of this idea—is a crucial question which I cannot here discuss further. Suffice it to say that Aristotle's thought that it all happens 'within me' is certainly not enough to give us an answer.

(b) Character is voluntary. The main theme of III.5 is not so much that actions are voluntary as that character is, so that one who is a virtuous or vicious person is 'voluntarily' virtuous or vicious. Since 'voluntary' and 'involuntary' do primarily apply to actions, rather than states of character, this seems to us an awkward use of the word. But we may easily rephrase Aristotle's claim in a more natural idiom as the claim that we are responsible for our virtues and for our vices.

This is the claim with which the chapter opens at 1113^{b}6–7, and it is at once followed by what appears to be a rather careless argument for it. For it is said that being virtuous or vicious just *is* a matter of doing the right or the wrong actions (1113^{b}13–14), so if the actions are 'up to us', the states must be so too. I call this argument careless because—as Aristotle has explained at some length in the previous book—there is more to being virtuous than just doing the right thing. In particular, one must choose the actions 'for their own sake', one must do so with an inner harmony, because reason and emotion are in agreement, and one must have a settled disposition so to act. The problem, then, is that even if it is up to me to choose or not choose this action, it is not obviously up to me to choose to do it *from* a settled disposition (as Aristotle himself says at 1137^{a}4–9). To suppose that it is seems already to presuppose that I choose my own dispositions, and that is the very point at issue.[25]

The argument on which Aristotle mainly relies, however, is not this, but rather that virtuous and vicious dispositions are caused by acquiring the habit of doing the actions they require. (Neither here nor anywhere else does he explain *how* this habituation leads to the extra conditions being fulfilled.) If, then, the actions lead to the state, and the actions are 'up to us', so too is the state that they lead to. One might object that if we are to count as responsible for the states we get into, we must *know* that our actions will cause them in the way that Aristotle claims. Aristotle's reply is that we do know: 'not to know that the dispositions are brought into being by the corresponding particular activities shows complete lack of sense' (1114^{a}3–10). To expand on this a little, one might say that his attitude is that any 'reasonable' man knows this, and that if someone does not then this ignorance itself is culpable: he should know.

But in fact what Aristotle takes us all to know is something we certainly do not know, for we do not believe it. He supposes that habituation leads to a disposition which is then unbreakable. Just as I can choose whether or not to throw a stone, but once it is thrown there is nothing more that I can do about it, and just as I can

---

[25] This problem is well brought out by Ackrill (1978).

choose whether or not to smoke, but once I have got lung cancer I cannot simply choose not to have it, so too with dispositions that are virtuous or vicious. In the beginning I can choose whether or not to behave in an unjust or intemperate way. But if I do choose so to behave, and become habituated to doing so, then it is no longer up to me; my disposition is formed, and there is nothing that I can then do about it ($1114^{a}11$–21).

This leads Aristotle into a peculiar position, which commits him to saying that a present unjust act of mine counts as 'voluntary', and hence punishable, even though I can *truly* say 'I could not help it', since the time when I could have done otherwise is now long past. But of course the truth is that our ordinary and common-sense belief in free will does *not* allow us to say that dispositions, once formed, are thereafter unbreakable. We take it to be possible, though perhaps unlikely, that the man who has behaved honestly all his life should nevertheless be tempted into fraud, and that the hardened criminal should resolve to go straight from now on, and should succeed. No doubt it is difficult to break the habit of a lifetime, but really we do not think that it is impossible, and one can appeal here to Aristotle's own argument about punishment. We punish the habitual criminal for what he does now, not for those long-past actions that first formed his habit. Indeed, we may be led to think that he is hardly to blame for *them*. If all or most of the environmental influences of his early childhood condoned, and indeed glorified, successful crime, then those past acts might perhaps be excused, not on the ground that he could not have refrained from them, but on the ground that he could hardly be expected to know better. But now that he is mature he surely does know better—or anyway he *should* know better—and that excuse is removed. And we evidently do not accept, as an alternative excuse, that it is now no longer 'up to him'.

I conclude that Aristotle here offers to his determinist (and 'incompatibilist') opponent a concession that is very dangerous indeed. It is not in accordance with common sense, and if it is granted then common sense is in deep trouble. But I add that it is clear from elsewhere that Aristotle does not really believe what he here concedes, and so we might take it as a concession made purely for the sake of argument. For example, in IX.3 he will admit that one who was virtuous may cease to be so ($1165^{b}13$–22), and I do not see why he should not say the same thing of vice.[26] Let us, then, take back this concession. It is no doubt fair to say that dispositions, once formed, will *influence* action, even if they do not *determine* it. What, then, should we say of Aristotle's conditional claim that *if* how I act is (usually) up to me, then so too is how I am disposed to act?

The argument seems to me to be persuasive. No doubt there are many factors that contribute to the formation of a disposition, but past behaviour is certainly one of them, and one may be sympathetic to the view that it is the major one. Admittedly there are unreflective people who do not see that the more often they act in a particular way the more likely it is that they will continue to act in the same way, and the more difficult it will be to break their habit. In these cases the

---

[26] Cf. *Categories* $13^{a}17$–31; *Politics* $1332^{b}4$–11.

disposition is perhaps acquired 'in ignorance'. But may we not say that such people *ought* not to be so ignorant? What they should do is to read what Aristotle has said in book II.

4. Two comments

I think we may set aside what Aristotle has to say here on the general problem of 'free will'. He is entirely convinced that how we act is often 'up to us', and he does not really take the opposing arguments very seriously. Moreover, he is clearly right to dismiss the suggestion that, while our good actions are up to us, our bad actions are not; for there is no ground for any such asymmetry. Assuming, then, that how we act is mostly up to us, how good is his account of 'voluntariness', i.e. of those actions that are fit subjects for praise or blame?

Clearly he is right to say that 'forced actions', to which the agent contributes *nothing*, are not in any sense voluntary. But one might complain that there are various ways in which the agent may contribute *something*, even though most of the relevant factors are not in his control. As I have mentioned, it may be one's own fault that one got oneself into a situation where subsequent action was 'forced'; and it may be that while the action was not 'forced', according to Aristotle's strict understanding, still we are inclined to leniency on the ground that resistance would have been difficult, though not impossible. But these are details which one might quite easily accommodate to his general account. Where he does seem to be some-what insensitive, from our point of view, is over the question of a 'compulsion' that is not due to external factors but to an internal disorder in the *psychē*. In so far as he considers this kind of question at all in III.1—e.g. when he notices the thought that my emotions are not 'up to me' (1111^{a}24–5), or that my conception of the good is not 'up to me' (1110^{b}9–10)—he gives it short shrift.[27] But we, who are more familiar with the notion of a psychopathic disorder, may well feel that there is more to be said. (For example, we genuinely do think that the kleptomaniac 'cannot help' stealing, and that this is what differentiates him from the ordinary thief.)[28] In Aristotle's own terms, I suppose that a psychopathic disorder would be a kind of 'madness', but it is interesting to note that even madness does not figure in his discussion of the kind of things that may 'force' one.[29] There would seem to be room here for some extension of Aristotle's own suggestions.

As for his account of ignorance, he is clearly right to say that this can be an excusing factor, and to recognize that it is not always. But here I think he is mis-

---

[27] Perhaps we may count his reference to 'what overstrains human nature' (1110^{a}23–6) as pointing to a 'psychological compulsion'? (Thus Meyer 1993: 98–100.) On the other side, in III.5 he assigns too much 'compulsion' to a different kind of internal factor, namely a disposition.

[28] Some problems that arise here are pursued by Haksar (1964).

[29] It is suggested by Sorabji (1980: 264–6) that Aristotle does not consider insanity because he has said that he is concerned to assist lawgivers (1109^{b}34–5), and in the Athenian courts a plea of insanity would not secure leniency. But I see no reason to suppose that Aristotle's interests are confined to excuses that might pay off in a legal situation, and I add that Plato had considered insanity as an excuse in his *Laws* (864d–e; cf. 881b).

taken in trying to hang on to the idea that whatever is blameable must be 'voluntary'. For there is no reasonable sense of 'voluntary' in which it covers such cases of culpable ignorance as forgetfulness, carelessness, inattention, and negligence. (Certainly, these failures are not 'done willingly' or even 'intended', though it may be right to say that it is 'up to us' to refrain from them.)[30] Moreover, there are many other cases of failure for which I can be blamed—e.g. failing to pass some test or examination, or simply failing to perform well at any task (e.g. a football game)— though again I did not want or intend to fail. For blame to be reasonable one must be able to say that the failure was 'up to me', in the sense that I could have avoided it (e.g. by trying harder, by preparing more thoroughly, by heeding advice on what was needed, and so on); but it is a very odd use of the word, in Greek or English, to say that this makes the failure 'voluntary'. Aristotle would have done better, not to try to twist this word to a wholly new sense, but to recognize that the link between being blameable and acting voluntarily (or intentionally) has exceptions.

Appendix. Comparison with *EN* V.8[31]

In III.1 a voluntary action is defined by two negative conditions: it is not due to force and not due to ignorance. But the discussion of force soon introduced the positive condition that it must be up to the agent whether to do it or not. In the rather different discussion of this topic at *EE* II.7–9 the definition finally reached in chapter 9 is that a voluntary action is one that is up to the agent and not done in ignorance (1225^b8–10). But the earlier discussion of chapter 8 had certainly associated the idea of being up to the agent with that of not being forced. In *EN* V.8 we are offered a curious amalgamation:

As has been said before, I call an action voluntary if (i) it is one of those that are up to the agent, and (ii) he does it knowingly and not in ignorance . . . and (iii) in addition it is neither coincidental nor forced. (1135^a23–7)

Since book V is a 'common book', the reference back to what was said before may well be to *EE* II rather than *EN* III, though neither of them had listed 'up to him' and 'not forced' as *separate* conditions.

   As my discussion of III.1 has shown, the condition 'up to him' is the more important, since it covers more cases, and *EN* V adds a different example which shows this: a natural process such as ageing is due neither to force nor to ignorance, but still it is not voluntary since it is not up to us (1135^a33–$^b2$).[32] One wonders why *EN* V thinks it necessary to make separate mention of 'not forced'. As for 'not coincidental', this is a condition that has not played a part in either of the earlier discussions, though it is briefly mentioned (without explanation) at *EE* II.9, 1225^b6. It appears from Aristotle's next remark in *EN* V that this

---

[30] This point is pursued by Siegler (1968).

[31] This topic is treated in Sorabji (1980, chs. 16–17) and in Kenny (1979, chs. 1–5). I do not pretend to exhaust it here.

[32] The example may come from reflection on *EE* II.8, 1224^b34, which mentions ageing as a natural process, but draws from it no moral concerning the definition of 'voluntary'. (Nothing similar is said in *EN* III.) I add that the example in *EN* V of a clearly forced action that is not up to me, namely when you seize my hand and strike someone with it, is the same as that given in *EE* II, at 1224^b13. *EN* III gives different examples (1110^a3).

condition is supposed to be relevant to ignorance, and his thought is that an act of striking a man may also ('coincidentally') be an act of striking one's father, but one may know that it is the first without knowing that it is the second (1135ᵃ28–31; the context in *EE* II also suggests this explanation).³³ As I have indicated, it would have been helpful if Aristotle had given this point more discussion. In any case, it seems fair to set aside these minor differences in formulation, and to say that the basic idea behind these definitions of 'voluntary' is much the same in all three cases.³⁴ Nevertheless, the way in which *EN* V elaborates this idea is in several ways divergent from *EN* III.

The most striking divergence is that *EN* V presents a distinction, within acts done in ignorance, between those that turn out contrary to reasonable expectation (*paralogōs*)³⁵ and those that do not. The first are called 'misfortunes' and the second 'mistakes' (1135ᵇ16–19). It is natural to suppose that this is another way of trying to draw the distinction between ignorance that is itself culpable and ignorance that is not, a distinction which is clear in III.5 but only hinted at somewhat obscurely in III.1. In the terminology I was using earlier, one might say that this introduces the question 'Ought he to have known?', and suggests as a criterion 'Would the reasonable man have known?' Admittedly, Aristotle does not explicitly say that 'misfortunes' are not blameable, whereas 'mistakes' are, but it seems to be a very natural inference. What is not clear is whether he also supposes that 'misfortunes' are involuntary whereas 'mistakes' are voluntary.

What he does say is that where the result is a 'mistake' the origin of the act was in the agent (though the act does not demonstrate a vicious character), whereas where the result is a 'misfortune' the origin was not in him. (He does not explain where else it was.) Now if we bring in the connection drawn in III.1, that where the origin of the act is in the agent the act is 'up to him', and hence voluntary (so far as this condition is concerned), then we may certainly infer that 'misfortunes' are involuntary, and perhaps also that 'mistakes' are voluntary. But we have reached this conclusion by an odd route, for we would now be supposing that in the case of a misfortune the agent *could not* have done anything about it, and this is surely too strong. There is a difference between what may reasonably be anticipated and what it is possible to find out about. In any case, *EN* V does not seem to accept the strong link that we find in *EN* III between voluntariness and where the action originates. For at one point it says that, when a man acts unjustly because of an emotion such as anger, 'the origin is not in him but in the one who provoked him to anger' (1135ᵇ25–7), and yet at the same time it seems to be implied that he does act voluntarily (1136ᵃ4–5). (No doubt the agent contributed *something*. This, according to the strict doctrine of III.1 would be enough to ensure that the origin was in him, but V.8 does not use the notion so strictly.)

In fact, I think it likely that V.8 wishes to classify both 'misfortunes' and 'mistakes' as

³³ Meyer (1993, ch. 4, sects. 3–8) gives an entirely different explanation, connecting this condition with 'forced' and with 'mixed' actions.

³⁴ The discussion supporting this basic idea is quite different in *EE* II and *EN* III. More than half of the discussion in *EE* II concerns actions that manifest self-control or the lack of it, in order to explain why *both* count as voluntary (i.e. virtually all of ch. 7, and the first half of ch. 8, to 1225ᵃ2). *EN* III pays no attention to this, presumably relying on its generalisation that what is important is that the origin of the action should be within the agent, for this covers all the points raised in *EE* II. (*EE* II does not connect *akrasia* with ignorance, as *EN* VII does.) By contrast *EE* II gives only a very hurried account of the kinds of ignorance that do or do not excuse an action, which is simply tacked on to the end of its discussion, in a quite unsatisfactory way (1225ᵇ11–16).

³⁵ The translation of '*paralogōs*' as 'contrary to *reasonable* expectation' has been disputed by Daube (1969). It is defended by Schofield (1973, n. 5) and Sorabji (1980: 280–1). I do not see what else the word could mean in this context.

involuntary, even though the latter are blameable and the former are not. After all, both are done 'in ignorance', and that is all that is specified in the definition of V.8. As a further indication of this, I note that at the end of V.8 Aristotle does mention (without explanation) the distinction, within acts done in ignorance, between those which are *due* to ignorance and those which are not, and he says that in *each* case the action counts as involuntary (1130ᵃ5–9).[36] He clearly says that those due to ignorance are not to be blamed, but he also says that *some* at least of those due to other causes are to be blamed, namely those due to an emotion which is not 'natural and human' (but presumably 'bestial', as discussed in VII.5). (As to those done in ignorance, but not due to ignorance, because due instead to some 'natural and human' emotion, he does not say whether they are to be blamed or not. But the natural inference from the passage seems to be that they are not.)[37]

The conclusion that I draw is that in V.8 Aristotle is not yet wedded to the thesis that only what is voluntary can be blamed. On the contrary, he thinks that 'mistakes' are involuntary and can be blamed, and that the same applies when a 'bestial' emotion leads to ignorance. This is a noticeable distinction between III.1 and V.8. It can quite plausibly be argued that there are other distinctions too, but that is a topic which I pass over.[38] In any case, the clear-cut views of *EN* III are presumably Aristotle's final thoughts on this topic, even if the earlier (and messier) accounts do seem to contain some insights which are subsequently lost.

Further reading

A nice summary of the account of voluntary actions in III.1, with useful illustrations, is given in Kenny (1979, ch. 3 and the first half of ch. 5, to p. 53). (The remainder of Kenny's discussion in chapters 1–5 is designed to compare *EE* II. 7–9 and *EN* III.1 and *EN* V.8.)[39] An equally clear but also more critical account may be found in Siegler (1968). The problem of 'psychological compulsion' is explored in Haksar (1964). On a relatively minor point, concerning the identity of actions under different descriptions, Ackrill (1980) is good value.

An interesting account of what may fairly be regarded as an 'Aristotelian' view of responsibility is given by Irwin (1980), though (as I have indicated, n. 6 to this chapter) I do not believe that it is Aristotle's own view. Nussbaum (1986: 283 ff.) provides a response to Irwin. There is a book-length treatment of Aristotle's account in Meyer (1993), which is often provocative, and a distinctly unusual interpretation of III.1 in Hursthouse (1984). A well-known and influential modern contribution, which professes (unreasonably?) to be developing Aristotle's own approach, is Austin (1956–7). One might also consult Mackie (1977, ch. 9, sects. 1–2) for a more straightforward modern discussion of what seems to be Aristotle's basic idea, or Glover (1970, ch. 1) for a simple comparison between Aristotle's views and some others.

---

[36] The distinction is also mentioned, but without any explanation of what it is or why it is relevant, at *EE* II.9, 1225ᵇ10. It is only in III.1, 1110ᵇ24–7 that we get some inkling of what Aristotle intends by it.

[37] If this is right, then the account of *akrasia* in VII.3 is in trouble. For in that account Aristotle does appear to think that an akratic act is done in ignorance, but is not due to ignorance, since it is due rather to some desire-or-emotion. But he also thinks that akratic acts should be blamed.

[38] Sorabji (1980) argues that in *EN* V, and in *EE* II, actions which are not 'forced' in the strict sense of *EN* III, but are (as we might say) done under duress, are counted as involuntary. So far as *EN* V is concerned, this argument mainly depends upon reading something into 1135ᵇ4–5 which is not obviously there, but I do not deny that it is a plausible reading.

[39] It may be useful if I note here that Kenny refers to *EN* V–VII, i.e. the 'common' books, as *AE* A–C.

On the argument of III.5, that if actions are voluntary then so are dispositions, there is a complex and sensitive discussion in Broadie (1991, ch. 3, sect. 5–6).

Much has been written on Aristotle and 'the problem of free will'. Two discussions which seem to me to be especially useful, and are not confined to the *Ethics*, are Furley (1977) and Sorabji (1980, ch. 14). As for the problem itself, considered from a modern point of view, the literature is of course immense. But the articles collected in Watson (1982) will form a useful start.

Chapter VI
Self-control (VII.1–10)

1. Preliminaries (VII.1–2)

A⊤ the end of book VI Aristotle has finished his account of the virtues—both virtues of character and virtues of intellect—and their associated vices, and he passes to a new topic. There are, he says, three distinct states of character to be avoided, not only vice but also lack of self-control (*akrasia*)[1] and bestiality (*thēriotēs*)[2] Opposed to these are respectively virtue, self-control (*enkrateia*), and what might be called a superhuman virtue, something heroic and divine. (*We* might perhaps say 'saintliness', but—quite apart from the anachronistic connection with Christian religion—our picture of the saint is surely very different from Aristotle's picture of the hero.) Ordinary virtue and vice have already been discussed. We hear no more of the 'divine' virtue proposed here, which one suspects Aristotle includes only because he wants a contrast with bestiality.[3] Bestiality is treated in what follows (principally in VII.5, but also at 1149^{b}27–1150^{a}8), but is hardly a major theme. The topic that Aristotle wishes to discuss now is the possession and lack of self-control. These are good and bad states respectively, but not as good as virtue or as bad as vice (1145^{a}15–^{b}1).

A word must be said about translation. The pair 'self-control' and 'lack of self-control' are, I am inclined to think, the best translation, and they closely reflect the literal meaning of the Greek words, which is just 'in control' and 'not in control'. (But the Greek does not have built into it any specification of *what* one is or is not in control of; in particular, it does not say 'oneself'.) A popular alternative in modern discussions is 'strength of will' and 'weakness of will'. This has the disadvantage that it suggests to the English reader that Aristotle recognizes the existence of a mental faculty called 'the will', which he clearly does not. But both suggested translations also suffer in a different way. Aristotle recognizes two main varieties of what he calls *akrasia*, namely 'impetuosity' (*propeteia*) and 'weakness' (*astheneia*) (1150^{b}19–22), but to us 'lack of self-control' inevitably suggests the first rather than the second, while 'weakness of will' inevitably suggests the second rather than the first. However, Aristotle means to include both. There seems to be no way out of this dilemma, and that is no doubt why translators have often used 'incontinence', a word which is now almost obsolete, except in the irrelevant sense of 'bed-wetting'. But I prefer to avoid the problem in a different way, by leaving

---

[1] Both Ross and Irwin translate *enkrateia* and *akrasia* as 'continence' and 'incontinence'. I shall discuss the problem of translation shortly.

[2] Ross prefers 'brutishness'.

[3] Hardie, in the revised version of his (1965–6) that is published in Barnes *et al.* (1977), offers some suggestions on how to make sense of this 'divine' or 'heroic' virtue (pp. 42–6).

Aristotle's word untranslated henceforth. An akratic action, then, may be due either to impetuosity or to weakness; what is common to both is that, in some sense, the agent knows that what he is doing is wrong. A major part of Aristotle's discussion, and the part that I shall focus on almost exclusively, is his attempt to determine in just what sense the agent knows this. This occupies the bulk of VII.3.

Before he comes to this, Aristotle sets out very clearly the method he intends to follow (1145^b2–7), which involves a review of what is commonly said on this topic (1145^b8–20), a statement of the difficulties that this gives rise to (1145^b21–1146^b5), and a programme for resolving these difficulties (1146^b6–14). I comment on the method proposed in Chapter X, but I shall also come back to it at the end of this chapter. The particular difficulty that I shall be concerned with is the one set by Socrates, who had claimed that *akrasia* is impossible. For he had argued that *knowledge* cannot be overcome by anything else, and cannot be 'dragged about like a slave'.[4] So, when *akrasia seems* to occur, we must infer that in fact the agent did not have knowledge, but was acting through ignorance (1145^b21–7). It should be noted that Aristotle does not give the Socratic argument that he here refers to, nor does he anywhere subject it to critical discussion, as one might have expected. Instead, he simply takes Socrates' conclusion, and asks what sense can be made of that. Clearly, it conflicts with what appears to us (the *phainomena*),[5] and so requires us to ask how this supposed ignorance could come about, for the agent is not thus ignorant before (or, we may add, after) his *akrasia*[6] (1145^b27–31). He then goes on to observe that some simple distinctions, which one might suggest, do not resolve this problem.

Some say that while it is true that *knowledge* cannot be overcome by anything else, still *belief* can succumb to a desire for pleasure, so it is belief rather than knowledge that is affected (1145^b31–5). To this Aristotle offers two replies. One, given here, is that if the point is supposed to be that belief may be held only weakly, then one might pardon a person for not abiding by it in the face of strong desires, whereas *akrasia* is not pardoned but blamed (1145^b36–1146^a4). The other, given later, when this suggestion is raised once more, is that belief may be held just as strongly as knowledge, and in that case it is equally puzzling to understand how a man may act against his belief (1146^b24–31). The distinction between knowledge and belief is, then, of no help with the present problem, which is simply the problem of how one can act against one's convictions. A further distinction that some might suggest is that it is practical wisdom (rather than theoretical?) which is overcome in a case of *akrasia*, but this too is of no assistance. At any rate, as Aristotle understands practical wisdom, it requires virtue of character, and the virtuous man is not subject to the temptations which generate *akrasia*, so practical wisdom is never overcome (1146^a4–9). These, then, are some initial distinctions that one might make, in the hope of accommodating Socrates' conclusion to our ordinary beliefs, but they do

---

[4] This is clearly a reference to the argument which Plato gives to Socrates in his *Protagoras*, at 351b–358d.

[5] The *phainomena* referred to here are presumably the same as those mentioned at 1145^b3, which are evidently the common opinions on this topic (*endoxa*, $^b5$; *legomena*, $^b20$). See further pp. 214–16.

[6] I take the *pathos* mentioned at 1145^b29–30 to be simply the 'happening' in question, i.e. *akrasia*. But Aristotle may also intend an (ambiguous) reference to the desire that causes it.

not work. We are still left with the problem of how, and in what way, the ignorance that Socrates alleges comes about ($1145^b21–2$, $1146^b8–9$).

One might remark here that both in these initial skirmishes, and in his subsequent treatment of the issue, Aristotle is *accepting* that 'ignorance' is somehow involved; he does not seem to contemplate the possibility that Socrates has got it entirely wrong. And this despite the fact that he has not actually given us any argument at all in favour of Socrates' position. Perhaps he takes it to be too obvious to be worth stating, but you may not feel the same, so I give a *very brief* indication. The leading premiss is simply that we all wish for what is good, or in Aristotle's terms for *eudaimonia*. And from this it seems simply to follow that if we know what is good then we will do it. Consequently, if we do not do what is good, that can only be due to a lack of knowledge on our part. The argument can certainly be embellished in various ways,[7] but this is the heart of it. And apparently Aristotle finds it convincing. At any rate his subsequent discussion of how *akrasia* does in fact occur, to which I now turn, does nothing to upset it.

But one more preliminary remark is in order. Aristotle will argue in VII.4 that the basic case of *akrasia*—what he counts as *akrasia* 'without qualification'—is when one fails to resist a bodily desire (e.g. for food or drink or sex). That is why much of his account focuses on such a desire. But these desires are (for him) a special case of emotion (*pathos*), so there are times when he speaks more generally of failing to resist an emotion (including, for example, anger). As he admits, some people are not akratic in these respects, but can be called akratic in other respects, e.g. money ('he cannot resist a bribe') or prestige ('he cannot resist an opportunity for publicity') or other things. In these cases the desires in question do not stem from the emotional part of the soul, and it is not entirely clear that Aristotle would regard his discussion in VII.3 as applying to them. But perhaps we may agree that these cases are secondary, and the most important task is to deal first with what is primary.

2. Aristotle's account of *akrasia* (VII.3)

Aristotle begins his positive account at 1146^b31 with two preliminary distinctions.[8] First, we must distinguish between on the one hand having knowledge but not using it, and on the other (both having and) using it. It is quite clear that by 'using' one's knowledge Aristotle does not mean acting on it; he means thinking of it, attending to it, having it before one's mind. This is clear both because he quite often invokes the same distinction elsewhere in his writings, where there is no question of

---

[7] The version in Plato's *Protagoras* is embellished thus: we can suggest only one criterion of goodness, namely pleasure, so we cannot say that one's desire for what is good is overcome by one's desire for pleasure. For that is to say that a desire is overcome by *itself*, which is absurd. Presumably Aristotle would not accept this embellishment, for he standardly assumes *two* criteria of goodness, not only 'the pleasant' but also what he calls 'the noble' (*kalon*).

[8] Robinson (1969) describes these distinctions as the first two (of four) 'solutions', but this is to misconstrue the structure of Aristotle's discussion.

acting on the knowledge,[9] and because he here introduces two alternative locutions for 'using' one's knowledge, one of which is 'contemplating' it (*theōrein*,[10] 1146 ^{b}33–5), and the other is 'activating' it (*energein*,[10] 1147^{a}7). While 'activate' might well be taken to mean 'act on', it is quite clear that 'contemplate' cannot be taken to mean this. The distinction is, of course, perfectly fair (and was first introduced, in different words, by Plato at *Theaetetus* 197a–199c): there are many things which we know but for most of the time are not thinking of. Aristotle's comment, which reveals an important feature of the account of *akrasia* that he is leading up to, is that it would be strange for someone to act wrongly while attending to the point that the action is wrong, but not strange if he is not so attending (1146^{b}31–5).[11]

The second preliminary point is that the reasoning that leads to action will employ two types of premiss, a universal premiss (e.g. 'dry food is good for everyone'), and a particular premiss which describes the present situation (e.g. 'these are nuts'). To complete the reasoning there will no doubt be further premisses too (in this case 'nuts are dry food' and 'I am a person', of which *we* would say that the first is universal again and the second particular).[12] In an actual case, the train of reasoning may of course be quite long, and a good number of premisses may be involved, but we may agree that it will always include both a universal premiss (which, in broad terms, states the goal that is pursued, cf. 1144^{a}31–3) and a particular premiss describing the present situation.[13] On the face of it, it would seem that one may 'have but not use' knowledge of *any* of these various premisses, and in each case one would then fail to be aware of the conclusion to which they lead. But it should be noted that, in a way which is again indicative of what is to come, the possibility that Aristotle directs attention to, both in his general comment and in his example, is of failing to be aware of the particular premiss (1146^{b}35–1147^{a}10).

Finally at 1147^{a}10 Aristotle introduces a further distinction which leads (for the first time) to a statement of what *akrasia* is. What he is concerned with is a special subcase of 'having but not using knowledge', which he rather unhelpfully characterizes as 'in a way both having and not having'. But his examples, and his subsequent comment, make it quite clear what he has in mind. The examples are one

---

[9] e.g. *De Anima* 417^{a}21–^{b}1; *De Generatione Animalium* 735^{a}11; *Metaphysics* 1048^{a}34, 1050^{a}12–14.

[10] Unfortunately Ross renders both *theōrein* and *energein* as 'exercising' the knowledge, which does not resolve the ambiguity; Irwin has 'attend to' for *theōrein* which is fair enough, and 'activate' for *energein*.

[11] Broadie (1991, ch. 5), following Mele (1981), argues that when Aristotle speaks of 'using' knowledge in this passage he *does* mean acting on it, and that the same applies to 'contemplating' the knowledge (or 'attending to it', *theōrein* (pp. 292–7, esp. 295–6)). This seems to me to be a clear error, which vitiates her whole account.

[12] Aristotle, I think, is here regarding both as 'particular'; if so, then 'nuts are dry food' is counted as particular because it is *less* universal than the leading premiss. Aristotle quite often uses the word 'particular' in this way, i.e. as meaning '*less* universal', rather than '*not* universal'; e.g. *Categories* 15^{b}1–2; *Topics* 105^{b}13–16; *Posterior Analytics* 79^{a}2–13, 97^{b}26–31; *De Partibus Animalium* 644^{a}28–33, *De Generatione Animalium* 763^{b}15. Similarly here in the *Ethics* at 1141^{b}14–22.

[13] I make some comments on the complexity of actual cases of practical reasoning in the appendix to this chapter.

who is asleep, or mad, or drunk. What these have in common is that the person is in such a *physical* state that, while in that state, they *cannot* attend to the knowledge they may yet be said to have. The sleeper must first be woken, the epileptic cured of his fit, and the drunk sobered up, before it is even possible for them to bring this knowledge to mind. That it is the physical state that Aristotle is concerned with is confirmed by what he at once goes on to say: those who are affected by an emotion—e.g. anger or sexual lust—are altered in body (as well as in mind). If I may add some examples: anger may make one flush with heat, fear lead to cold sweat, love to a racing heart, grief to tears, and so on. So his view must be that emotions, by disturbing one's bodily state, make one incapable of attending to knowledge that one nevertheless 'has'. And *that*, he concludes, is how it is with *akrasia*. The desire or other emotion involved simply blocks one's ability to take in or keep in mind the relevant facts ($1147^{a}10$–18). And if we may extrapolate from the hint we have had already—which will be confirmed, I believe, in what follows—we may add that the facts in question here are particular facts about the present situation.

He at once turns to meet an objection: the akratic may *say* just those things that Aristotle claims he cannot actually comprehend. But that, he replies, is irrelevant. There are many cases of people who can say what they do not understand. The drunkard, the actor, or the young pupil may all be able to recite the verses of Empedocles, but it evidently does not follow that they grasp their meaning. This objection, then, has no force ($1147^{a}18$–24).

This may well strike you as an incredible account of what happens when I take the chocolate éclair that I know I ought not to take. I would agree; it is a wholly incredible account. Nevertheless, it *is* the account that Aristotle gives—or at any rate it is his *first* account, for we have more to come, and the interpretation of what comes next is disputed. I give first an interpretation (which I believe to be correct) according to which what follows is merely a further elaboration of essentially the same account, and then a different interpretation which seeks to find here a new and improved account. But even if the second interpretation is accepted, still it has to be admitted that the account is not *much* improved.

The disputed passage begins, 'we may also look at the cause in this way, as in the study of nature (*phusikōs*)'[14] ($1147^{a}24$–5). This opening certainly suggests an alternative view of the same diagnosis already given, rather than a new diagnosis, but perhaps one should not be too confident of this. (It is not altogether clear how we should understand 'as in the study of nature', or indeed whether this is a fair translation of Aristotle's phrase. I *guess* he means that we shall now pay closer attention to the causality involved, i.e. to just how the body is set in motion, for the study of nature is the study of natural causes. But the point is not important.) This is followed at once by what appears to be a contrast between theoretical

[14] This is Ross's translation (near enough); Irwin has 'referring to [human] nature'. Others offer a rather different account (e.g. Charles 1984: 128 n. 28). I conjecture that this is a (concealed) reference to what Aristotle has (already?) written about how reasoning leads to action in his *De Motu Animalium*, which undoubtedly counts as one of his 'Physical' works.

and practical reasoning.[15] In each case we have both a universal and a particular[16] premiss. When these are put together, and one thing[17] results from them, in the theoretical case one *affirms* the conclusion, but in the practical[18] case one at once *does* it. For example, the reasoning may be '(i) Everything sweet should be tasted, and (ii) this is sweet'; one who completes this reasoning must at once act accordingly 'provided he is able to and not prevented' (1147ª25–31). One notes, incidentally, that Aristotle is here content with a very simplified and unrealistic example, for it is nevertheless enough to illustrate his main point.

He now turns to the application of this general doctrine about practical reasoning to the particular case of *akrasia*, and it seems clear that he envisages two opposing pieces of reasoning like this:[19]

(A) One should not taste anything that is such-and-such
This is such-and-such
So you do not taste it.

(B) Everything sweet is pleasant
This is sweet
So you eat it.

He does not fill in the 'such-and-such' of reasoning (A), nor state its particular premiss. He does comment either on reasoning (B) as a whole, or on its particular premiss—the wording is ambiguous—that 'this is active' (i.e. the agent is aware of

---

[15] Our text does not *say* that one side of the contrast is theoretical reasoning, and the interpretation has been disputed (e.g. Kenny 1966; 1979: 157 n.; Charles 1984: 128 n.). But the parallel discussion in *De Motu Animalium* does introduce just this contrast (701ª7–14), and the only alternative seems to be a contrast between 'practical' and 'productive', which makes no sense in the present context. (See n. 18.)

[16] In the theoretical case, the particular premiss is presumably *less* universal rather than *not* universal (and it is hardly accurate to say, as Aristotle here does, that perception controls it). Cf. n. 12 above.

[17] 'One thing'. Since the word 'one' (*mia*) is feminine, it calls for a feminine noun to supplement it, and the only one that the context supplies is 'one belief' (*mia doxa*). Ross and Irwin translate accordingly. On the other hand Aristotle's point seems to be that in the practical case the conclusion is not a belief but an action, so I prefer an indeterminate translation, and I guess that the noun in Aristotle's mind is 'belief-or-action' (*mia doxa ē mia praxis*).

[18] Aristotle actually says here not 'practical' but 'concerned with production'. This must be a verbal slip on his part, for earlier in VI.4 he has contrasted production with action, and it is quite clear from his example that his concern here is with action generally, and not production in particular. (For similar uses of 'productive' where what is meant must be 'practical', cf. *EE* 1227ᵇ30 and *De Motu Animalium* 701ª23.)

[19] Kenny (1966, retained in his 1979: 159 nn.) is I think alone in supposing that only a single piece of reasoning is envisaged, e.g. '(i) One should not taste anything pleasant, (ii) everything sweet is pleasant; and this is sweet, so (iii) one should not taste this.' Kenny counts (i) as 'the' universal premiss, and all of (ii) as 'the' particular premiss (which in itself is reasonable, cf. 1147ª6–7), but cannot explain the suggested contrast in 'this is active' applied to (ii). So in his (1966) he suggests that the akratic is fully aware of all this reasoning, and explicitly draws its conclusion. In his (1979) he admits that Aristotle must still be supposing that the akratic does somewhere suffer from ignorance, but does not explain how this is possible on his proposed analysis.

it). Presumably this is intended to mark a contrast with something else which is not active. If, as my first interpretation holds, the doctrine here is the same as has been previously hinted at, then the contrast is between the particular premiss of reasoning (B), which is active, and the particular premiss of reasoning (A), which is not (i.e. the agent fails to be aware of it). He adds further that a desire (i.e. for what is pleasant) 'happens to be present', and sums up the situation thus: 'so the one (premiss?) bids one to avoid this, but the desire leads one on'. And he finally remarks (as I believe), 'For each of the parts [of the soul] can initiate motion.'[20] That is, a wish to do what is right, belonging to the rational part of the soul, together with reasoning showing how that wish may be fulfilled, is capable of leading one to act; but so too is a desire belonging to the emotional part of the soul, again together with the appropriate reasoning which shows how it may be gratified (1147^{a}31–5). But why is it that, in the akratic case, it is the desire that does in fact lead to action, and not the rational wish? The only explanation that this passage offers is in the parenthetical comment 'this is active', attached either to reasoning (B) as a whole, or to its particular premiss. But we are given no further explanation here of why reasoning (A) as a whole, or its particular premiss, fails to be active, and one can only assume that the explanation intended is the same as was given earlier: the desire for what is pleasant blocks it from the consciousness.

Aristotle goes on to add some comments. When one acts akratically that is in a way as a result of one's beliefs and one's reasoning, and these are not in themselves, but only 'coincidentally', opposed to 'the correct *logos*'. I suspect that, on this occasion, by 'the correct *logos*' Aristotle means not the whole of reasoning (A), which—after all—he has not stated, but simply the part that he has stated, namely its universal premiss. He wishes to emphasize that one can simultaneously *believe* all of

> One should not taste anything that is such-and-such
> Everything sweet is pleasant
> This is sweet.

For there is no contradiction between these beliefs, and according to my first interpretation the agent does have them all. What he is missing is the belief in the particular premiss of reasoning (A). But according to the second interpretation (which I shall come to shortly) we should add this further belief too, i.e.

> This is such-and-such.

For it is still true that there will be no explicit contradiction between all four of these beliefs, and the agent can hold them all simultaneously. What creates the 'opposition' is (*a*), as Aristotle mentions, the desire for what is pleasant, and (*b*), as

---

[20] *kinein gar hekaston dunatai tōn moriōn.* These words are interpreted as I suggest by Kenny (1979: 159), Wiggins (1978–9: 260), Charles (1984: 130), in all cases without comment. The more usual translation (adopted by both Ross and Irwin) is 'For it [i.e. desire] can set in motion each of the parts [of the body]'. But the 'each' is here decidedly unexpected, and serves no purpose.

he fails to mention, the wish to do what one should. He concludes in any case that *akrasia* does involve reasoning from a universal premiss, and that that is why only human beings can be guilty of it. And he adds finally that the question of how the akratic's ignorance is later dispelled is in principle no different from the question of how the ignorance of one who is drunk or asleep is later dispelled. This completes his account of what it is that actually happens when *akrasia* occurs (1147ª35–ᵇ9). One should note that this last remark—if it does belong with the 'disputed' paragraph, as it certainly seems to—confirms the claim that in this paragraph too Aristotle is still thinking that ignorance is somehow involved, though admittedly he does not here make clear just where, or why.

For this we turn to the concluding remarks, which—at least at first sight—strongly bear out the interpretation suggested so far. It is the 'last premiss' (i.e. the particular premiss) which the akratic either lacks altogether, or 'has' but does not comprehend, and merely 'says' as the drunkard will recite Empedocles. This is a belief based on perception, and is of crucial importance for action, and so his lack of it explains why he fails to act. But just because it is particular, and not universal, his failure to know it does not count as a failure of knowledge, properly speaking. (For, by a prejudice which Aristotle shares with Plato, *genuine* knowledge is always of what is universal, and cannot be otherwise.) So it appears that the result that Socrates was seeking for does actually eventuate: *akrasia* does not occur because what is genuinely knowledge is overcome,[21] and it is not this that is dragged about by emotion, but rather the belief due to perception. That explains how the akratic in a way knows, and in a way does not (1147ᵇ9–19).

The first interpretation, then, says that what Aristotle is claiming throughout is that knowledge of the particular premiss of the reasoning that would lead to correct action is blocked by the operation of an emotion. According to the final paragraph, the akratic may lack this knowledge altogether. But according to the earlier discussion the point is that he may 'have' it—perhaps in the sense that *afterwards* he can fully appreciate it—while at the time it is prevented by the emotion from coming into consciousness.[22] And the fact that he may verbally admit it is taken not to be an objection to this claim, for one may perfectly well say something that one does not in fact understand. It is a temporary ignorance, then, that the emotion induces, but ignorance of the particular and not of the universal, and that is why Socrates was right, at least to this extent: the emotions cannot 'drag about' one's grasp of the universal, but only one's grasp of the particular. That completes the first interpretation.

*

[21] For *parousēs ginetai* (OCT, translated by Ross and by Irwin) I read *periginetai*, as proposed by Stewart (1892), and since adopted by many, e.g. Robinson (1969: 87), Hardie (1968: 290), Wiggins (1978–9, 261). The OCT reading can only be defended in what Hardie justly calls a 'Pickwickian' manner. (But Stewart's proposal is palaeographically improbable. The alternative amendment proposed by Broadie (1991: 311 n. 38) is better on this score, but still has the same overall effect. She suggests reading *pathousēs* for *parousēs* and treating this word as an intrusive gloss.)

[22] One may fairly suggest: the impetuous akratic lacks the knowledge altogether; the weak akratic 'has' it, but fails to bring it to mind.

The second interpretation opens with an obvious criticism of the first. Aristotle does not state in full even the universal premiss of reasoning (A), but describes it merely as a premiss which forbids tasting (1147ᵃ32). One can offer all kinds of suggestions, for example,

> Don't taste anything that will rot your teeth,
> Don't taste anything that isn't yours.

Such suggestions make perfectly good sense of the discussion. But if we ask what is the most likely supplement, given the general context, it is most plausible to suggest

> Don't taste anything sweet.

After all, Aristotle is not concerned that his examples should be realistic, and he has just offered an example which has as its leading premiss 'Do taste everything sweet'. What is more natural than to suppose that he is here using the same example, with a negation inserted? But if we do take this view a contradiction at once appears. For now both pieces of reasoning have *the same* particular premiss, and this same premiss cannot be something which the agent both *is* aware of (to explain why reasoning (B) goes through) and *is not* aware of (to explain why reasoning (A) does not). Moreover, even if Aristotle is thinking of something other than 'sweet' as the filling for argument (A), or even if he is not thinking of anything in particular as its filling—for, after all, he does leave it unspecified—still the same objection arises. For it is quite clear that there can be two pieces of reasoning, opposed as in his example, which do share the same particular premiss. The first interpretation, then, must credit Aristotle with a failure to see an objection easily suggested by his own discussion. This leads us to look for a different interpretation.

   The second interpretation, then, is that what explains the failure of argument (A) is not that the akratic fails to be aware of some of its premisses, but that he fails to put those premisses together so that 'one thing' results from them (1147ᵃ24–5). That is, he fails to see their relevance to one another, and so fails to draw the appropriate conclusion. On this account, there is still ignorance involved, but it is now located in a different place.

   Those who favour this interpretation[23] must say that the disputed paragraph (i.e. 1147ᵃ24–ᵇ9) opens a *new* line of thought, when it begins with its contrast between theoretical and practical reasoning at 1147ᵃ25–31. It is clear that in the earlier remarks Aristotle was interested in the possibility of failing to grasp a premiss, and especially the particular premiss (1147ᵃ3, ᵃ7). But they often do not admit that it is the earlier line of thought that recurs in the closing paragraph (1147ᵇ9–19). For where the first interpretation says that what the akratic fails to grasp is 'the last premiss' (the *teleutaia protasis*, 1147ᵇ9), they would rather say that he fails to grasp 'the last *proposition*', which they take to be the *conclusion* of the reasoning, and not one of its premisses. And it is perfectly true that what Aristotle goes on to say about it—namely, that it is 'a belief concerning what is perceptible', 'of crucial import-ance for action', 'not universal and therefore not [properly speaking] something

---

[23] Charles (1984) is a forceful advocate (ch. 3, sect. ɪɪ), but see also Dahl (1984, ch. ɪɪ).

that can be known'—would certainly fit a belief in the conclusion just as well as a belief in the particular premiss. But I must admit that this does not convince me. For one thing, although the word *protasis* may sometimes be used by Aristotle simply to mean 'proposition' in general, rather than 'premiss' in particular,[24] I do not believe that he could so use it as applying to the *conclusion* of what is being considered as a piece of reasoning. (After all, the word does etymologically mean 'what is held out *in front*'.) But in any case the point of Aristotle's opening contrast between practical and theoretical reasoning is that in the case of practical reasoning the conclusion is *not* a belief, but an action. So even if we do translate 'the *teleutaia protasis*' as 'the last proposition', the fact that Aristotle calls it a belief (*doxa*), and not an action, will still force us to conclude that it is a premiss and not a conclusion.[25]

I conclude that the second interpretation of the disputed paragraph cannot be squared either with what comes before it or with what comes after it. While it remains *possible* that in this disputed paragraph, taken by itself, Aristotle means to offer a different diagnosis of *akrasia*, I do not think it at all probable. The best version of this second interpretation will be, I think, that it applies only to the disputed paragraph, which could be a later insertion (made, presumably, by Aristotle himself), introducing a new idea, but not properly harmonized with its context. But this speculation seems to me baseless. It is introduced in order to save Aristotle from what seems to be a rather obvious objection to his account, but even in this it fails. For the objection clearly is an objection to all the *rest* of his account, even if the disputed paragraph can be interpreted so as to avoid it. And besides, it does nothing to deflect the *main* objection, which is that the akratic need not be suffering from *any* kind of ignorance. Let us turn, then, to this central claim.

3. Objection

As I have said, Aristotle distinguishes two main varieties of *akrasia*, namely 'impetuosity' and 'weakness', which he describes thus: 'the person who is weak has deliberated, but does not stick to the result of his deliberation, because of the feeling; the person who is impetuous is led by his feeling because he has not deliberated' (1150[b]19–22). Now it is fair to say that the impetuous man acts as he does because of a kind of ignorance, a failure of knowledge, for his fault is that he

---

[24] In the *Topics* the word is standardly rendered as 'proposition' throughout, and this creates no awkwardness. But I observe that the 'propositions' in question are mainly those that are proposed to *initiate* a discussion. Elsewhere the situation is less clear; one might cite *Prior Analytics* 24[a]16–17, *Posterior Analytics* 72[a]7, *De Interpretatione* 20[b]24. But the first two may be regarded as ambiguous between 'proposition' and 'premiss', and the third still retains the thought that a *protasis* is something proposed *beforehand*. By contrast, at *Prior Analytics* 42[a]32 the word simply *must* mean 'premiss'.

[25] Charles (1984) will disagree. His ch. 2, sect. C gives a careful analysis of the various passages in Aristotle which bear upon the question whether the conclusion of a practical syllogism is a proposition or an action, and he concludes that the evidence is divided. But I note that even passages which seem to favour the view that 'what is concluded' is not *itself* an action, but leads to one, never say that it is a *protasis*, nor that it is a 'belief' (*doxa*).

does not stop to think, and if he had thought then he might well have acted differently.[26] There is indeed no reason either to say (as the second interpretation has it) that while he is aware of the premisses of the reasoning that would prevent his action, he fails to put them together and so to see what they jointly imply, or to say (as in the first interpretation) that while he is aware of the universal premiss he fails to grasp the particular premiss. Indeed it is really a very strange suggestion that someone might have before his mind a universal premiss that would be relevant to his present situation, but not be aware of the particular premiss that shows how it is relevant. For why should the universal premiss occur to him at all if he does not see how it would be relevant? Perhaps, then, we would do better to modify Aristotle's doctrine in this way: *no* part of the relevant reasoning comes to mind, but that is because the particular premiss which would bring it to mind is blocked by the emotion. (This seems not to be quite what he does say, or imply, but it is only a small modification, and would still allow him the conclusion about Socrates that he is aiming for.) I concede, then, that with this small modification the general account of *akrasia* that Aristotle gives may be seen to fit impetuosity well enough. Admittedly we are given no good reason for saying that when an emotion simply prevents one from stopping to think, it always does so by blocking the knowledge of the particular premiss, rather than the power that such a premiss usually has of bringing to mind the associated universal. But perhaps a reason could be supplied, and in any case I shall not protest further about the account of impetuosity.

But weakness is different, as Aristotle himself acknowledges, and surely his general account does not fit weakness at all. Suppose that I have a strong liking for cream, and someone offers me a chocolate éclair. Suppose also that I *do* stop to think, and what I think is: 'I should not eat what is bad for me; cream is bad for me; éclairs contain cream; this is an éclair; so I should not eat it'. In this way I complete a piece of deliberation. But suppose further, as we all know is entirely possible, that I nevertheless take the éclair and eat it. Then on Aristotle's account this must be because my desire for cream blots out some part of this deliberation (on the first interpretation, the premiss 'this is an éclair'; on the second, the conclusion 'so I should not eat it'). But, as we all know, this is wholly unrealistic, for even as I bite into the éclair I may well be thinking to myself 'I should not be doing this'. And there is no reason at all to say that I fail to understand these words that I think to myself; I do know perfectly well that I should not be doing what I am doing.

So far, I merely appeal to your own experience; you must know that this is how it is. But I now add briefly a couple of arguments (both, in a way, *ad hominem*). For the first, let us consider the opposite case of *enkrateia*, where I control my desire for the éclair and do not eat it. There is, of course, some reasoning which tells me to eat it, perhaps like this: 'cream tastes very nice, and so does chocolate; éclairs contain both cream and chocolate; this is an éclair; so this will taste excellent'. This, coupled

---

[26] Since impetuosity is blamed, it must on Aristotle's principles be 'voluntary' even though it involves ignorance of the particular situation. So he is committed to saying that the ignorance is itself culpable. And even if the impetuous disposition is now so firmly established that (in Aristotle's view) the feeling, say anger, can no longer be resisted, still the impetuous man is to be blamed for having, by his previous actions, allowed this disposition to form.

with the desire to taste what will taste excellent, is a reason for eating it. But in fact I do not eat it. Why not? Well, if we are to conform to Aristotle's leading idea, it must be because some piece of this reasoning—its particular premiss, or its conclusion, or perhaps all of it—is blotted from my consciousness. By what? Well, apparently by the rational wish to preserve my health, for—by symmetry—in the case of *enkrateia* this too must be able to prevent knowledge that I in a way 'have' from coming before the mind, and—if we follow through the details of Aristotle's account—it does this by putting me into such a physical state that I *cannot* properly appreciate this knowledge. So it is not only the emotional part of the soul, but also the rational part, that can block my understanding. But surely this is wholly unconvincing. When my desire for chocolate and cream is defeated, it is not because I have lost my grip on how it could be satisfied. And the same will therefore apply too to the opposite case, when this desire is victorious and it is my rational wish to maintain a healthy body that is defeated. There is no ignorance involved in either case.

The second objection is just that Aristotle himself, in other passages on *akrasia*, admits this fact. For if the reasoning on one side or the other is blocked from consciousness, then of course the agent will not, at the time of acting, be aware of any inner conflict. There will be just the one reason that he is then conscious of, and so naturally he will choose and act accordingly. But Aristotle himself frequently describes *akrasia* and *enkrateia* as involving an inner conflict, for example at I.13 (1102^{b}14–17), at III.2 (1111^{b}13–18), at V.9 (1136^{a}31–^{b}9), at VII.7 (1150^{b}19–25), and at IX.4 (1166^{b}6–9). Indeed, in I.13 he follows Plato in invoking this inner conflict as a reason for saying that there must be two distinguishable parts of the soul involved, one the rational part, and the other the emotional part that can oppose it. It follows that his account in VII.3, which banishes this inner conflict, is not even in harmony with the attitude that he himself usually takes elsewhere.[27]

For this reason, some have attempted to see the 'disputed' paragraph in VII.3 (i.e. 1147^{a}24–^{b}9) as *recognizing* an inner conflict, and providing Aristotle's diagnosis of what happens in a case of weakness, rather than impetuosity. For they suppose that *both* the pieces of reasoning here indicated are supposed to be present to the consciousness. (For this purpose they must set aside the last sentence, i.e. 1147^{b}6–9, which explicitly speaks of ignorance; and they must interpret the comment 'this is active' at 1147^{a}33 as meaning not 'this he is conscious of', but 'this he acts on', despite the earlier indications that 'activity' and 'consciousness' are being equated, 1146^{b}33–5, 1147^{a}2, ^{a}7.) I can only comment (*a*) that this interpretation of the 'disputed' paragraph makes it entirely out of line with everything else that Aristotle says in VII.3, and (*b*) that in any case when the paragraph is thus construed it offers absolutely no *explanation* of why it is the one piece of reasoning that wins rather than the other. But clearly Aristotle *thinks* that he has offered an explanation.

This leads me to a final comment. The explanation that Aristotle offers is that the desire on the one side defeats the reasoning on the other, by preventing it from coming fully to the agent's consciousness. This explanation, however, is wholly

---

[27] Perhaps the thought is that there is an initial period of conflict, which is resolved (by the onset of ignorance) before the action takes place? Well, perhaps. But there is nothing in our text to suggest it.

unconvincing. It would be better to say that the desire on the one side defeats the *desire* on the other, but the reasoning is in each case quite unaffected.[28] It would seem that Aristotle fails to notice the possibility of this description, since he fails to draw attention to the fact that there is what may be called in the broad sense a desire (i.e. an *orexis*) in favour of not eating the éclair. (In his own vocabulary, this is a wish (*boulēsis*) rather than a (bodily) desire (*epithumia*).) But even if we grant this redescription, it does not actually help very much, for there is still no explanation of *why* it is the one desire that wins, and the other that loses. (The determinist will, of course, insist that there *is* an explanation, but he need not suppose that it is an explanation which the agent is conscious of.) I have no explanation to offer myself: sometimes one side wins, and sometimes the other, but we do not know why this is. One final point is perhaps worth making: when a desire is 'defeated' in such a contest, that does *not* mean that it evaporates. Even when I do stick by my initial resolve not to eat creamy things, still I will (usually) continue to desire that éclair. Neither the desire itself, nor the reasoning that supports it, is banished from my consciousness. So I remain in 'inner conflict', until the opportunity is finally removed.

4. Explanations

Why should Aristotle give such a Socratic account of *akrasia*?[29] That is, why should he suppose that, when one fails to act in the way that (in a sense) one recognizes as the right way, the explanation must be a kind of 'ignorance'? This is, of course, a matter for speculation, and I shall consider three such speculations.

First we may note that Aristotle's conclusion is inevitable if we retain his claim that the conclusion of a piece of practical reasoning is an action.[30] For it then follows that where the appropriate action does not result the reasoning cannot have been completed. But this observation merely shifts the question: why should Aristotle have been so attracted to this claim that practical reasoning concludes in action? For example, why should he not have said, as is strongly suggested by some

---

[28] Charles (1984) thinks that this *is* Aristotle's account: I fail properly to 'grasp' the thought that I should not eat the éclair, not in the sense that I do not properly understand it, but in the sense that I do not desire it as I should. (I do not 'accept' it fully, because my acceptance 'is not integrated with the rest of what I accept to form one coherent theory', p. 190. This relies on 1147^{a}22, which says that one who has only recently learned does not yet know 'because it must grow into them, and this needs time'.) A similar account is offered by Dahl (1984, ch. 11). In reply, I simply appeal to the text: when Aristotle speaks of 'ignorance' it is *obvious* that he does not mean 'failure to desire appropriately'.

[29] I cite the question from Wiggins (1978–9: 262). But I shall not consider Wiggins's own suggested answer, which seems to suppose that the ordinary akratic has the same fully worked out conception of *eudaimonia* as does the man of practical wisdom.

[30] It should be noted that he makes this same claim elsewhere too, most emphatically at *De Motu Animalium* 7, 701^{a}8–36. It is a standing feature of his thought, and the occurrence at 1147^{a}26–31 is not a mere aberration. And even if at times he seems to distinguish 'what is concluded' from 'what is done' (as is argued by Charles 1984, ch. 2, sect. D), still he always supposes that the one leads inevitably to the other, which is enough for my purpose.

other things that he does say, that practical reasoning concludes in choice or decision (*prohairesis*), and whether this choice in turn leads to action is a further
question. Usually it does, but in a case of *akrasia* it does not. I make two suggestions
about this.

One is that in our ordinary explanations of why people act as they do we generally cite just their relevant desires and beliefs, thus indicating the reasoning
involved, and we leave it at that. That is, we apparently accept the desires and
beliefs—or, in Aristotle's terms, the desires and the reasoning—as forming by
themselves an adequate explanation of the action. We do not ordinarily add a
further explanation of why the choice or decision to which they lead was acted on.
Now the case of *akrasia* apparently shows that there is scope for further explanation
at this point, and Aristotle hints at this when he says, in our passage, that the person
will then act 'provided he is able to and not prevented' (1147ª30–1). For there may, of
course, be a contrary desire, and contrary reasoning, which one could perhaps
think of as 'preventing' the proposed action. (Or there may of course be other inner
states, e.g. laziness, which 'prevent' it.) But the truth is that we do not think of these
opposing factors as literally 'preventing' the action—i.e. as *necessitating* its non-
occurrence—and generally we cannot in fact explain why, when there is such an
inner conflict, it is one side rather than the other that wins. So, when explaining
actions, we hardly ever do *mention* anything other than the desire and the reasoning
that in fact wins. This no doubt encourages the thought that the desire and the
reasoning are themselves a sufficient explanation of the action, so that once they are
completed the action will inevitably result.[31] One might perhaps explain in this way
Aristotle's standard acceptance of the thesis that the conclusion of a piece of practical reasoning simply *is* an action. But I confess that I do not find this explanation
very convincing, for the obvious counter-example to cite just is the case of *akrasia*,
and so you might certainly have expected Aristotle to show more caution in a
context where he is explicitly considering this very phenomenon. Let us turn to a
different line of thought.

It is an agreeably simple way of distinguishing between theoretical and practical
reasoning to say that the one issues in belief and the other in action. Moreover, one
might well say that Aristotle needs such a distinction, for the contrast between
theory and practice is absolutely fundamental to his way of thinking about the
intellectual virtues. It is true that his own account begins (in VI.1) with a different
way of marking this distinction, namely that theoretical reasoning concerns universals throughout, whereas practical reasoning must also take account of particulars.
But we have already observed that this way of drawing the distinction leaves much
to be desired. And in fact it is evidently the connection between practical reasoning
and action that governs all of his thinking on this topic.

Some philosophers, and I am thinking in particular of R. M. Hare, have had deep
motives for insisting upon this way of drawing the relevant distinction. For Hare,

---

[31] It is natural to take this to be a consequence of Davidson's well-known thesis that desires and
beliefs *cause* actions (Davidson 1963). But, to be fair to Davidson, he does of course allow that there may
also be other causal factors that are relevant.

influenced by Hume's insistence on the distinction between 'is' and 'ought', which he accepts as fundamental for ethics, attempts to explain it in this way: a merely descriptive (theoretical?) belief does not *in itself* have any consequences for action, whereas a prescriptive (practical??) belief—i.e. one that has 'ought' for its copula— is different just because such a belief does 'commit' one to action. It inevitably results that Hare is in difficulties over *akrasia*, as he himself acknowledges.[32] For it seems that in *akrasia* one does hold a relevant 'prescriptive' belief and yet fails to act on it. So, apparently, either the belief was not really prescriptive—but was an 'inverted commas' use of 'ought' (meaning 'most people (*other* than me) think that this is what one ought to do')—or it was not really held: very much as Aristotle suggests, one may pay lip-service to the thesis, while not actually *believing* it, as is shown by the fact that one fails to act on it. Thus Hare, like Aristotle (and like Socrates), denies that what we ordinarily think of as *akrasia* ever occurs. And in his case there is a deep reason: the fundamental distinction between what is descriptive and what is prescriptive can only be explained by saying that the one does not lead to action whereas the other does.

Is Aristotle motivated in a similar way? Of course Aristotle had not had the benefit of reading Hume, and so does not focus on the particular words 'is' and 'ought'. Indeed, he uses many different ways of presenting the leading premiss of a piece of practical reasoning (for example: 'dry food suits everyone' (1147^a5–6), 'one should taste everything sweet' (1147^a29), and from elsewhere 'I need a covering' (*De Motu* 701^a17–18)). But one might suggest that in all cases what makes the reasoning *practical* is that it states a *goal* of the agent (1144^a31–3), and then one can imagine someone of Hare's persuasion asking: 'But what is it that marks a statement as the statement of a *goal*, rather than a plain and straightforward statement of fact?' The suggested answer, then, is that goals lead to action, whereas other beliefs do not, and if this is Aristotle's own answer then he himself must accept that a piece of reasoning which does begin from such a goal, and which is successfully completed, will have to issue in action. *Akrasia* can only be explained, then, either by saying that the alleged goal was not truly a goal of the agent, or by saying that the reasoning to show how it could be achieved was somehow thwarted.

This is an ingenious attempt to assimilate Aristotle to a modern philosopher, and thereby make him 'deeply' committed to the claim that practical reasoning results in action. But again I do not find it at all convincing. For one thing, Aristotle would surely have answered quite differently the question of what makes something a goal, namely by saying that a goal is something the agent desires. (He would not be at all moved by the suggestion that I may recognize something as 'what I ought to do', while in no way desiring to do it. For he would say (*a*) that we all do desire to do what is good, and (*b*) that what is not desired cannot motivate me to action, whether or not I describe it as an 'ought'.) But, more straightforwardly, he surely *need* not have said that the connection between practical reasoning and action is quite so direct as this line of thought requires. As I suggested initially, we might well say that practical reasoning (when completed) leads to choice, without adding that

[32] Hare (1952, sects. 2.2 and 11.2; 1963, ch. 5).

that choice inevitably leads to action. On this view what is characteristic of weakness, as a form of *akrasia*, is just that in this case the choice does not lead to action.[33]

I therefore turn to a wholly different explanation, which focuses on Aristotle's professed *method* in ethics. This is to review the opinions of the many, and the opinions of the wise, and to show (if possible) that they can all be harmonized by drawing the appropriate distinctions. Now it is true that Aristotle does not always preserve either the opinions of the many or those of the wise. (For example, he emphatically rejects the opinion, held by many people, that *eudaimonia* is primarily a matter of (bodily) pleasures, and he also rejects the opinion, held by some of the wise, that pleasure is a matter of (bodily) replenishment.) But his aim, stated clearly at the beginning of VII.1, is to find at least some truth in each of them. The suggestion is, then, that his avowed method predisposes him to find at least some truth in what Socrates said when he denied the possibility of *akrasia*. Of course, Socrates is one of 'the wise' rather than one of 'the many', and his view cannot be accepted just as he put it, for when so put it does evidently conflict with the opinion of 'the many'. But Aristotle will still seek to find some truth in it, and indeed to maximize that truth, for his general method directs him to do so.

There can be little doubt that he does end by finding as much truth in Socrates' view as could be desired. His main point is that the akratic is in a way 'ignorant' when he acts, though in another way he can also be said to act 'with knowledge', for he *has* the relevant knowledge but fails to *use* it. (Indeed, his feelings put him into such a physical state that he *cannot* then use it.) This is very clearly an attempt to harmonize Socrates' claim (which alleges ignorance) with the view of the many (which claims that the akratic knows that he is acting wrongly). And Aristotle goes further: the relevant 'ignorance', he claims, is always ignorance of the particular premiss. But since what is properly called knowledge is of the universal rather than the particular, it is not what is *properly* called knowledge that is 'dragged about' by emotion, and in this way too Socrates was right. But at the same time it must be admitted that something *commonly* called knowledge is 'dragged about', in so far as the particular knowledge is prevented by the emotion from coming into consciousness. Here again is a distinction which seeks to find as much truth as possible in each of the two opposing views. (And in this case that motive is the only explanation for Aristotle's distinction that one can suggest; there is no other reason for supposing that it is particular, rather than universal, knowledge that the emotion defeats.) The general moral is clear: Aristotle's account *does* find as much truth in Socrates' claims as is possible, given that they must also be made to harmonize with the common view that *akrasia* does indeed occur. Since this is what his own professed method directs him to do, it is hardly surprising that it is what he does do.

Yet one may remain sceptical. After all, Aristotle is not always so sympathetic, either to the many or to the wise, as we have noted. Why, then, the particular sympathy to Socrates at this point, when it has already been admitted that his view 'clearly conflicts with the *phainomena*' (1145^{b}27–8)? The answer, I suggest, is that

---

[33] The 'Aristotelian' account sketched by Wiggins (1978–9, sects. I–II) would give him a wider range of options.

Aristotle feels a very general sympathy with Socrates' ethical outlook, even though in truth it is entirely different from his own. Thus in VI.13 he is as nice as he can be to Socrates' claim that virtue just *is* knowledge; while he cannot of course agree with it, he again seeks to maximize the truth in it, on the ground that 'full' virtue does require practical wisdom, even though it is not exhausted by it (1144^b17–32). And then, to one's surprise, he goes on to *endorse* Socrates' claim that one cannot 'fully' have one virtue without having all the others too (1144^b32–1145^a2), though in fact it seems clear that this is a concession he should not have made. There are other places too where he casually endorses a Socratic thesis, as for example when he says that 'every wicked man is ignorant of what should be done and what avoided' (1110^b28–30), though surely he should not have done so. (To simplify somewhat, the wicked man consistently *prefers* what is pleasant—or seems so to him—to what is noble; but it cannot be inferred that he does not *know* what is noble.) Admittedly, he does not always swallow what Socrates says. For example in III.5 he sets himself to argue *against* the Socratic thesis that 'no one voluntarily does what is wrong' (1113^b14–17).[14] But undoubtedly the occasions of agreement, or near agreement, are more numerous than those of explicit dissent. I think it is fair to conclude that Aristotle did have an exaggerated respect for Socrates' views in ethics.

He should not have done, for in fact their basic views are very different. Socrates' view of the soul was wholly intellectual, and he paid very little attention to non-rational desires and emotions. So, on his account, the mere thought that something is good will inevitably lead me to act accordingly. By contrast Aristotle—who has learned from Plato's division of the soul—pays very much more attention to non-rational desires and emotions. Indeed, his account of virtue of character focuses entirely upon these. So, to come back to the present topic, he has an important tool to explain *akrasia* which Socrates lacked, for he is well aware that non-rational desires and emotions are not perfectly under the control of reason. And yet he locates their action in the wrong place; he thinks of them as obscuring the reasoning that would lead to correct action, rather than as opposing the desire (*orexis*) that advocates that action. Indeed, in the whole of his discussion of *akrasia* in VII.3, he never once mentions the fact that there is a desire—in his vocabulary, a wish (*boulēsis*)—that is attached to the 'good' reasoning. But why does he focus so exclusively on this reasoning that is 'defeated', and not on the desire which lies behind it? My suggestion is that he is (without noticing it) accepting a Socratic preconception: the desire for what is good is universal, goes without saying, and need not be considered; so if anything goes wrong it must be with the reasoning and not with the desire behind it. Given this preconception, and given also the methodological principle that one would expect there to be some sense in which Socrates was right, one can perhaps explain why Aristotle should have thought as he did. But of course this explanation does not make his thought any better.

---

[14] Roberts (1989) argues with some plausibility that on this point Aristotle's disagreement with Socrates is more verbal than real.

Appendix. Note on the practical syllogism

The phrase 'practical syllogism' is not used by Aristotle himself, but is a convenient title for his use of the syllogism to illustrate practical reasoning.

The Greek word 'syllogism' (*sullogismos*) simply means 'reasoning' or 'argument' or 'calculation' in a quite general sense, but Aristotle has a theory about deductive reasoning, namely that it can all be cast in the form of what he and we now call 'syllogisms' in a somewhat technical sense. (In this sense a syllogism consists of three propositions, i.e. two premisses and a conclusion. Each proposition is of one of the four forms 'All A are B', 'No A are B', 'Some A are B', and 'Some A are not B'. In the *Prior Analytics* Aristotle gives us a detailed investigation into how three such propositions can be put together to make an argument, and which of these arguments are deductively valid. In the course of this he introduces various bits of technical terminology, for example distinguishing between 'major', 'minor', and 'middle' terms, and between 'major' and 'minor' premisses.) As anyone acquainted with modern logic will know, this theory is in fact mistaken: there are many examples of deductively valid reasoning which cannot in fact be represented either as syllogisms or as chains of syllogisms. But undoubtedly this is what Aristotle's theory was.

It would seem that he also assumed that practical reasoning could be cast into syllogistic form too. For most of the time he does not dwell on this, and his general characterization of deliberation in III.2–4 and in VI.2 makes no mention of syllogisms. But the idea does become prominent in his explanation of *akrasia* in VII.3, and occasionally elsewhere too, where he wishes to go into detail about how reasoning will lead to action, especially in the *De Motu Animalium*, chapters 6–7. Cooper (1975, ch. 1) argues for the unusual view that Aristotle wishes to distinguish two stages in practical reasoning: first there is deliberation proper, which issues in a conclusion stating what *kind* of action should be done, and in this syllogizing plays no part; then at the final stage it is a syllogism that shows how to translate this conclusion into action. But I see nothing in Aristotle's text that would support this view, and there is at least one passage which speaks strongly against it. (1142^{b}21–6 makes use of the technical vocabulary of syllogistic reasoning in what is clearly a discussion of deliberation.)[35] The more natural assumption is that Aristotle is taking it for granted that deliberative reasoning is syllogistic, just as he also supposes that theoretical reasoning is too.

In the *Ethics* his illustrations of practical reasoning in syllogistic form are extremely simplified and artificial. Here is an example from the *De Motu Animalium* which begins to reveal that the actual situation is rather more complex:

(1) I need a covering
    A coat is a covering
    So I need a coat

(2) What I need I ought to make
    I need a coat
    So I ought to make a coat
    (And the conclusion, 'I ought to make a coat', 'is' an action)

(701^{a}17–20)

The first thing that should strike one about this reasoning is that it is in no way logically

---

[35] There is a brief response to Cooper on this issue in Charles (1984: 136–7), and a longer discussion in Miller (1984).

conclusive. The first premiss states that I need a covering, but does not specify any particular type of covering (e.g. a fig leaf? a suit of armour? a roof over my head?). Even if the premiss had been given more specifically, still it seems improbable that only one particular type of covering—a coat—would satisfy my need. The same point may be made about the continuation: *one* way of obtaining a coat is no doubt to make it, but clearly there are other ways too, including buying one, or stealing one. So the first point to make is just this: there will in general be *many* ways of achieving one's goal, and the mere statement of the goal will not dictate any one in particular. The practical reasoner must therefore consider a variety of different means.[16]

A second point is this: when considering the variety of means that are available, I will need to weigh them against *other* goals that I have. For example if, like most people, I wish to stay on the right side of the law, then I am likely to rule out the suggestion that I could acquire a coat by stealing one. And yet, if the situation were sufficiently desperate, and the only other means available seemed to be even worse, I might consider it quite seriously. A piece of practical reasoning, then, will need to consider several goals simultaneously, in order to choose what seems to be the best means to the goal stated initially. And it may indeed conclude that none of the possible means is worth pursuing, for they all conflict with other, more important, goals.

The general point that I wish to make is that practical reasoning is *not* deductive, though Aristotle's very simplified examples make it seem so. On the contrary, it requires inventiveness, in thinking up different ways of achieving a goal, causal reasoning, to follow through the consequences of this or that way of pursuing it, and evaluative judgement on the relative importance of the different goals involved. Consequently a piece of practical reasoning does not *compel* its conclusion, in the way that a piece of theoretical reasoning does. The point may be illustrated even from Aristotle's very simplified examples in the *Ethics*. Suppose that I accept the premiss 'one should taste everything sweet', but suppose that I *also* accept the premiss 'one should not do what will lead to a million deaths', and now imagine a situation in which tasting the sweet thing *will* lead to a million deaths. Clearly, the mere fact that tasting is commanded by a premiss which I accept is not the end of the matter.

There is some disagreement among commentators over whether Aristotle saw the complexity of actual practical reasoning. Some take it for granted that he did,[17] but I am more inclined myself to think that he did not, and one of my reasons is his blithe assumption that it can all be represented in syllogistic form. The complexity of an actual case cannot be represented simply as a chain of syllogisms, beginning with the single goal to be achieved, and ending with the particular action which (in the present circumstances) will achieve it. I also suspect that, if he had seen that the reasoning is not in fact deductively valid, then he would have been less inclined to say that its conclusion 'is' (or inevitably leads to) an action. For while it is fair to say that theoretical reasoning *compels* belief, it is better to say that practical reasoning suggests or supports an action, but there is no reasonable sense in which it *compels* that action. But of course I admit that a piece of practical reasoning will normally lead to action ('if nothing prevents it'), just as other types of non-deductive reasoning will very often lead to beliefs, even though they do not *compel* them.

---

[16] III.3 does admit this (1112[b]16–17), but for the most part Aristotle does not dwell upon it, and none of his illustrations of practical thinking exemplify the point.

[17] e.g. Wiggins (1975–6, sect. 3), Irwin (1988*a*: 336), Broadie (1991, ch. 4, esp. sect. 9); on the other side, see Cooper (1975: 95), and—interestingly—Broadie (1987).

Further reading

The interpretation of VII.3 that I have given may fairly be called the 'orthodox' interpretation. A straightforward exposition is given in Robinson (1969), which I think is well worth reading. The same broad outlines, though with minor variations, are followed by many others (e.g. Walsh 1963; Hardie 1968;, Santas 1969; Cooper 1975; Wiggins 1978–9, Urmson 1988; Woods 1990). Dissenting views are usually motivated, at least in part, by the desire to save Aristotle from what seems so obviously to be an unsatisfactory treatment of the issue. The most interesting of these is probably that given by Charles (1984, chs. 3–4). (But the heart of Charles's position can be extracted just from his ch. 3, sect. B and ch. 4, sect. D.) A more succinct version of essentially the same interpretation is given by Dahl (1984, ch. 11). A quite different dissenting view may be found in Kenny (1966). (This is briefly summarized, and in some points recanted, in Kenny 1979, ch. 14. But for the main arguments one must go back to Kenny 1966.) There is a response to Kenny in Dahl (1984: 168–81). Wholly different again is Broadie (1991, ch. 5); her main dissenting claim is argued in section V of that chapter, and its implications brought out in section VI. There are several whole books devoted to this topic, e.g. Walsh (1963), Milo (1966), Mele (1987), Charlton (1988), but I expect that most readers will feel that they need something briefer.

Turning to the question of what one *should* think about *akrasia*, an essential starting-point is Plato, *Protagoras* 351b–358d, which gives the Socratic argument to which Aristotle is reacting. Influential modern discussions, which also see *akrasia* as problematic, are Hare (1963, ch. 5; with his 1952, sects. 2.2 and 11.2), and Davidson (1970). A useful commentary on all of these at once is Taylor (1980). An outline account which refuses to see any real problem is given by Wiggins (1978–9, sects. I–II). (Wiggins describes his account as 'Aristotelian', and indeed 'more Aristotelian' than what Aristotle says himself, p. 250.)

On the related topic of the form and structure of practical reasoning, especially as it is described by Aristotle, the main passages in the *Nicomachean Ethics* are III.1–4, VI.2, and VII.3. One should also see *De Anima* III.9–11 and *De Motu Animalium* 6–7. (These passages are conveniently collected together and annotated by Ackrill 1973.) Of commentators, one might start with Anscombe (1957, sects. 33–9), with the response by Mothersill (1962). There are many other discussions. My own selection would be Cooper (1975, ch. 1; but pp. 23–58 argue for a peculiar view of his own, which might be omitted); Wiggins (1975–6); Kenny (1979, chs. 10–13). In addition, for the *De Motu Animalium*, see Nussbaum (1985, esp. essay 4); and, for the *De Anima*, Hutchinson (1990).

For some modern treatments, which might be said to begin from Aristotle, but then take off in their own way, one might see Kenny (1965–6a) and Hare (1971, title essay).

Chapter VII

Pleasure (VII.11–14, X.1–5)

1. Book VII and Book X

OUR text of the *Nicomachean Ethics* contains two discussions of pleasure, one at VII.11–14 and one at X.1–5, and neither shows any awareness of the existence of the other. Presumably the explanation is that the account in book VII was written for the *Eudemian Ethics*, not the *Nicomachean* (see Introduction, p. 1), and so, if we are right to suppose that the *Nicomachean* is the later work, Aristotle's intention was that the account in book X should supersede that in book VII. Certainly the book X account is in several ways superior. While it does repeat some arguments from book VII, it presents them more clearly and selects them more judiciously, and it adds extra material of its own. But there is also much of independent interest in book VII, and I shall therefore consider both treatments.

The doctrine of the two treatments is broadly similar, but there are some discrepancies. The central discrepancy is over what pleasure *is*. Both treatments consider and reject a theory found in Plato's *Republic* (at IX. 583b–587a) which claims that pleasure is a process of replenishment, restoring the body to its natural state. (The doctrine is evidently based upon the fact that we enjoy eating when hungry, and drinking when thirsty.) In its place Aristotle wishes to associate pleasure with what he calls activities, rather than processes.[1] But whereas in book VII Aristotle wishes to *identify* the pleasure with the activity (at least, when that activity satisfies certain further conditions), in book X he says rather that the pleasure is not the same as the activity, but is something that 'completes' it. This claim is obscure, and I shall discuss it later. I merely observe here that this divergence affects some other things that Aristotle says, and in particular his treatment of the claim that pleasure is the supreme good. In any case, his own *primary* interest seems to be in the question of whether, and to what extent, pleasure is good. Both treatments open with this question, and give it considerable attention. I shall therefore take this topic first, and only later come back to the question of what pleasure is.

2. The goodness of pleasure

Aristotle begins in VII.11 by setting out three views which are hostile to pleasure, namely:

(1) No pleasure is good.

---

[1] What he means by an 'activity' is not obvious. I shall come back to this.

(2) Most pleasures are not good (though some are).

(3) In any case, pleasure is not the supreme good.

$$(1152^b8-12)$$

He then proceeds to outline some arguments for these views, most of which are arguments which appeal to quite ordinary and common-sense opinions, though the first is clearly a philosopher's argument. It claims that pleasure is a perceived process (*genesis*)[2] towards the natural state, and that what is good about this is not the process itself but the state which is restored at the end of it. This is offered as an argument both for (1) (1152^b12-15) and for (3) ($^b22-3$). The following two chapters then seek to rebut these arguments, claiming that it does not follow from them either that pleasure is not good, or even that pleasure is not the supreme good ($^b25-6$).

A point which is not made clear is whether Aristotle's discussion *merely* aims for a negative result (i.e. that the arguments do not establish their conclusions) or whether he is also aiming for something more positive, namely that the conclusions in question are one and all mistaken. In the third case, it is clear that he does wish to claim something positive, namely that the supreme good is in fact a pleasure (even if other pleasures are not good at all). His argument for this occupies most of chapter 13 (at $1153^b7-1154^a7$), and it relies upon the fact that he has already rejected the doctrine that pleasure is a process (at $1152^b33-1153^a12$) and substituted his own account, which is that pleasure is an activity, namely an activity of our natural state, and one that proceeds in an unhindered fashion (1153^a12-17). So in chapter 13 he can argue that the supreme good is admitted to be *eudaimonia*, that *eudaimonia* is an activity (as he has consistently said),[3] that it is an activity of the natural state, and that it is unhindered.[4] It therefore follows that *eudaimonia* simply *is* a pleasure. So *in a way* pleasure is the supreme good, but we must be clear that this applies to one pleasure in particular, and not to all pleasure. People are dimly aware of this when they say—as almost everyone does—that *eudaimonia* involves pleasure, and when they (wrongly) think that in pursuing the pleasures that they recognize they are pursuing *eudaimonia* ($1153^b25-1154^a7$).

In chapter 13, then, Aristotle is clearly making the positive claim that *in a sense* pleasure is the supreme good (but not in the sense in which this thesis is usually taken). It is also perfectly clear from the latter part of chapter 12 (i.e. 1153^a17-35), where the more straightforward arguments for (1) are rebutted, that he wishes to claim that a fair number of pleasures are good. But what is not very clear is whether he also wishes to claim, more strongly, that *in a sense* all pleasures are good. If so, it can only be by restricting the claim to 'genuine' pleasures, and the way that chapter 12 begins does very much suggest that that is what he is trying to do. Thus he begins

---

[2] Irwin has 'becoming', rather than 'process', which certainly is a better rendering of the Greek, but not so recognizable. See n. 19, below.

[3] Notice that Aristotle here leaves it open whether *eudaimonia* is an activity which manifests just one disposition or all of them. (He must presumably mean not 'all' but 'several'.) This leaves open the question whether it is a 'dominant' or an 'inclusive' goal (1153^b9-14).

[4] I take 1153^b14-25 to be arguing that *eudaimonia* must be unhindered.

with a distinction between what is good without qualification, and what is good for a particular person, or at a particular time (1152^b26–31), and he later applies this same distinction to pleasures too (1153^a2–7). So it would seem that the thesis that he has in mind is this: what is a pleasure without qualification is also a good without qualification, and what is a pleasure to a particular person at a particular time is also a good for that person at that time. He goes on to introduce a further thought (borrowed from Plato),[5] which tends in the same direction but is even more implausible; some things that seem to be pleasures are not actually pleasures at all (even to the person, at the time?), and he gives as an example the pleasures involved in being cured[6] (1152^b31–3, 1153^a2–7). These would certainly seem to be moves designed in defence of the thesis that all (genuine) pleasure is good, and to be what allows him to sum up at 1153^b1–7 that since pain is admittedly something bad, and to be avoided, pleasure must be good—all pleasure, apparently.

But still, as I said initially, it is not clear that Aristotle does mean to endorse this thesis in propria persona, for he may instead be taken merely as showing how, for the sake of argument, one could defend the position against those who attack it. And in fact this seems to be the better interpretation, for in chapter 14 we find a different (and more sensible) view which surely is his own opinion. This is that while some pleasures are wholly good, the bodily pleasures of food and drink and sex are good only 'up to a point'. For this kind of pleasure can be indulged to excess, and this is the mark of a bad man (1154^a8–21). He then goes on, in the rest of the chapter, to explain why, when most people think of pleasure, they think only of such bodily pleasures, and how this leads to mistaken views about the nature and value of pleasure. There is no repetition here of the suggestion that since the pleasures of the glutton are pleasant (to him, at the time) they must also be good (for him, at the time), nor of the suggestion that these pleasures merely seem pleasant but are not really so, even to him.[7] The considered view of book VII is, then, that some pleasures are just bad, though most are good, and on this point book VII and book X are largely in agreement. But they do not agree on the claim that there is a pleasure that is the supreme good. For book VII surely does endorse this claim, whereas book X cannot, since book X no longer identifies a pleasure with an activity. Let us move on, then, to book X.

As in book VII, the account in book X opens by observing that very different things are said about the goodness of pleasure: some (notably Eudoxus) say that pleasure is the supreme good, while others say that it is entirely bad (1172^a27–8), and these views are then discussed in chapters 2 and 3. It is clear that Aristotle endorses neither of them.[8] He grants that when Eudoxus argues that all things pursue

---

[5] Republic IX, 583b–586a; Philebus 31b–55c.

[6] I shall say more of these 'curative' pleasures in Sect. 5.

[7] The idea that 'curative' pleasures only seem to be pleasant is in a way repeated (1154^a34–$^b2$, $^b17$–19), but I take it that the glutton's pleasures are not 'curative'. (On the question whether Aristotle seriously thinks that something may seem pleasant, to a person at a time, without really being so, even to him then, some useful points are made in Gottlieb 1993.)

[8] Gosling and Taylor (1982, ch. 14) argue that Aristotle in the end defends Eudoxus. They take a view of the significance of X.5 that is rather different from mine.

pleasure (1172^{b}9–15), and shun its opposite, i.e. pain (^{b}18–23), he has a good point. For one must agree that what everything pursues is good (1172^{b}35–1173^{a}5), and one cannot accept the alternative (proposed by Speusippus)[9] that pain and pleasure should *equally* be avoided, on the ground that it is the neutral state that is good. For this alternative cannot explain why we do in fact pursue the one and shun the other (1173^{a}5–13). But on the other hand one of Eudoxus' own arguments can be turned against him. For he had said that when pleasure is added to something already good, e.g. to just or temperate action, that makes it even more worthy of choice (1172^{b}23–5). But we may respond, as Plato did (at *Philebus* 20d–22c), that when wisdom is added to a life that is already pleasant that makes *it* even better (^{b}26–34). This, clearly, is the point that Aristotle mainly relies upon against Eudoxus, for he repeats it at the end of chapter 3. While pleasure is one of the things that we value, it is only one; for example, we also value sight, memory, knowledge, and virtue. But also, we do not value *all* pleasures, e.g. not those enjoyed by small children, or those that are shameful (1174^{a}1–8). So, he concludes, while pleasure is (often) a good thing, we cannot count it as *the* supreme good, for that is *eudaimonia*, and we cannot say that *eudaimonia* simply is pleasure. (The claim of book VII, that *eudaimonia* is *a* pleasure, is simply not mentioned in book X.)

Turning in chapter 3 to arguments intended to show that *no* pleasure can be good, Aristotle here omits the more or less popular arguments that had occupied much of book VII (i.e. 1152^{b}15–22, 1153^{a}17–35), and gives only some more philosophical considerations. Some (apparently)[10] had argued that only qualities can be good, and that pleasure is not a quality. Here Aristotle simply denies the leading premiss (1173^{a}13–15). An argument found in Plato's *Philebus* (27e, 31a) claims that what admits of degrees cannot be good, and pleasure admits of degrees. Here again Aristotle simply denies the leading premiss (1173^{a}15–28). But the argument to which he pays most attention is the one that opens his treatment in book VII, namely that pleasure is a process, and in particular a process restoring one's body to its natural state. From this it is held to follow that it is not the process itself—i.e. the pleasure—that is good, but the state that results from it. This argument is here given an extended refutation (which I discuss later), at much greater length than in book VII (1173^{a}29–^{b}20). The conclusion, then, is that we have no reason to deny that some pleasures are good.

Yet it also seems that some are not, namely the 'disgraceful' ones. Aristotle canvasses three replies. The first reintroduces the initial ideas of VII.12, namely that we might say that these so-called pleasures are not in fact pleasant (sc. without qualification), even though they are pleasant *to* the person in question (*at* the time) (1173^{b}20–5). So it appears that Aristotle has not entirely given up this idea, though it also appears that he puts little trust in it. At any rate, it does not recur in the summary of the discussion at 1174^{a}10–11, though the other two suggestions do.[11]

---

[9] The attribution to Speusippus is in book VII, at 1153^{b}4–7. (For some discussion of the evidence on what Speusippus' views actually were, see Gosling and Taylor 1982: 225–6.)

[10] This argument is not familiar from elsewhere.

[11] However, it *does* recur in ch. 5, as we shall see.

These are that we might admit that such pleasures are bad, either (*a*) because they arise from [activities] that are bad, or (*b*) because pleasures differ in kind from one another, and the pleasures arising from different [activities] are simply different pleasures. In that case we need not hesitate to say that some are bad while others are not (1173^{b}25–31). A brief and wholly unconvincing defence of this last suggestion is added here (1173^{b}31–1174^{a}1), but it receives much more extended consideration in chapter 5, to which I now turn.

The thesis that pleasures differ from one another in kind is introduced with this simple example: 'We cannot have the pleasure of the just man without being just, or the pleasure of one who is musical without being musical' (1173^{b}29–31).[12] This seems so obviously right that no more argument is really needed. But Aristotle does go on to argue at some length that whereas the pleasure arising from a particular activity will enhance that activity, and improve one's performance in it, it will not enhance other activities, and will in fact detract from one's enjoyment of them, if it obtrudes. (For example, one who is fond of the flute will be distracted from a conversation if he hears a flute being played.) Again, these points seem to be undeniable, and they do lend appropriate support to the thesis. The pleasures arising from different activities are indeed different from one another and have quite different effects (1175^{a}21–^{b}24). If I may repeat Urmson's rather nice illustration:[13] 'one could not chance to get the pleasure of (say) reading poetry from stamp-collecting'. Now it does not actually follow from this that the goodness or badness of the pleasure simply follows from the goodness or badness of the activity it is derived from (for, after all, in book X Aristotle does not *identify* the pleasure with the activity; indeed he says at 1175^{b}30–6 that this identification would be 'absurd'), but it does clear the way for it. In any case, Aristotle firmly asserts the connection, and this is his final word on the goodness and badness of different pleasures: pleasures arising from good activities are good, and those arising from bad activities are bad (1175^{b}24–1176^{a}3). The result is, then, that pleasure has no ethical value of its own, for its value is simply dependent on that of the associated activity.

This would have been a good place to stop, for it is a major and important claim, and one that is highly relevant to all subsequent discussions of pleasure ever since. Unfortunately, Aristotle goes on. He notes that different species of animals have different pleasures (1176^{a}3–9), and this leads him to suppose that there should be certain pleasures which are proper to the human species in particular. So which are they? He proposes that they be taken as the things that appear pleasant to the *good* man, for he should be accounted the judge of what really is pleasant. And then, reverting to an idea which we had hoped was now abandoned, he infers that what seems pleasant to the bad man should not be counted as a pleasure at all (except, *to him*). The 'real' pleasures, then—i.e. the 'really human' pleasures—are those

---

[12] 'Musical' is an (Anglicized) transliteration of Aristotle's word *mousikos*. The meaning is probably more general, i.e. 'one who has been educated in the works of the Muses'. This includes poetry (or, more generally, literature) as well as music.

[13] Urmson (1967: 324).

associated with the one or more activities[14] that constitute the *eudaimonia* of the supremely good man (1176ᵃ10–29).

This last paragraph is best ignored. *All* pleasures are pleasures *to* someone, and the best reason for dropping the qualification 'to so-and-so' is when what we are speaking of is a pleasure to the vast majority of people. But presumably the good men do not form the vast majority. Besides, Aristotle appears to be presupposing that all good men share the same pleasures, and there is surely no reason to agree with this. For example, some may prefer listening to Bach, some to Bartok, some to Bernstein, and some to the Beatles. (One may object that these activities would not qualify as activities that constitute *eudaimonia*. But then the consequence is that they are not 'really human' pleasures, which is surely absurd.) In any case, as I say, I shall pay no further attention to this thought. But what is important is the thought that pleasures differ from one another as the corresponding activities differ, and the (alleged) consequence that their goodness is simply dependent on the goodness of the corresponding activities. To bring out the significance of this thought, I offer a brief contrast with classical utilitarianism.

The utilitarian recommends the maximization of happiness, and takes the measure of happiness to be the total sum of pleasures less pains. It has long been recognized that we do not in practice know how to calculate this sum, for it requires us to rank all pleasures on an additive scale, and no such scale seems to be available. Aristotle's claim that pleasures are different in kind from one another is at least part of the explanation of why we do not know how to construct such a scale. It is not just, as Mill suggested, that some are 'higher pleasures' and some are 'lower pleasures', for the truth is that *all* of them are different.[15] But it is Aristotle's second claim that creates the most direct conflict, for on his account some pleasures are simply *not* good, and so should not be given any positive value in the utilitarian calculus.

Bentham famously said 'quantity of pleasure being equal, pushpin is as good as poetry'. (Pushpin is a children's game; to update the example we might substitute snakes and ladders.) Would he have been equally happy to say 'quantity of pleasure being equal, gang rape is as good as poetry'? On his own principles, he should have been, for his claim is that *every* pleasure is a positive gain, no matter how it arises. Moreover, it seems quite possible, on utilitarian principles, that in this example quantity of pleasure should be equal. No doubt the victim is in much distress, but each of the gang gets a great kick out of it, and they are many while she is one. It follows that the total sum of pleasures less pains could be positive, and might therefore be equal to that got by some amount of poetry-reading. But Aristotle will have none of this. On his view, gang rape is itself bad, so any pleasures arising therefrom are bad too. They have *no* positive value, and so could not be invoked to show that the action was (overall) good, and therefore justifiable. On an example such as this common sense would, I am sure, side with Aristotle and against the

---

[14] Note again 'one or more', and compare n. 3, above.

[15] The significance of this point should not be exaggerated. After all, I can and do *prefer* one activity to another (say going to the cinema, rather than reading my book) simply on grounds of the pleasure expected from each. But the utilitarian sums need more than a mere (ordinal) scale of preferences, specific to a particular individual.

utilitarian. That is perhaps sufficient illustration of why I say that his views on this topic are important, and deserve serious consideration. No doubt there is more to be said,[16] but I shall not pursue this topic any further. Instead, I turn to the vexing question of what, according to Aristotle, pleasure is.

3. Process and activity

In both treatments Aristotle associates pleasure with activities, and contrasts activities with processes,[17] so we must begin with some elucidation of this distinction. I take first the notion of a process, which is described in some detail at the start of X.4 (i.e. 1174^{a}13–^{b}14).

It is necessary to start with a word on translation. The Greek word *kinēsis* is often translated 'movement', and it is so translated by Ross throughout X.4;[18] a better translation would be 'change'; but Irwin consistently uses 'process', and I shall follow him. The relation between these concepts is that movement is one kind of change, namely change of place, whereas there are also other changes too, e.g. change of colour, of size, of temperature, and so on. Aristotle very often uses *kinēsis* to cover all kinds of change. But he also has a special theory about changes, which is that they are all what we may reasonably call processes. That is: (i) they take time; any change lasts over some stretch of time, and no change is instantaneous. (ii) A change is always 'from something' and 'to something'; that is, it has both a starting-point and a finishing-point, and these indeed form part of the definition of that change, specifying what change is in question; they are part of what Aristotle calls its 'form' (1174^{b}5). (The definition will also specify the manner of the change, e.g. 'by flying' or 'by walking', or 'by jumping', so this too is part of its 'form', 1174^{a}31.) Hence (iii) a change is not 'complete in form' at any time at which it is going on, for at that time it still has not reached its finishing-point. A change is therefore complete only when it is finished, which is to say when it no longer exists. These three conditions may be taken as defining what it is to be a process, and Aristotle's own view is that all changes are processes in this sense. For present purposes, then, it seems simplest just to translate his word *kinēsis* as 'process', for that best represents what he is actually thinking of.[19]

I add here a further point, not because it could reasonably be taken to be part of

[16] For example, listening to Bach is presumably not, in itself, a 'good' activity. Would it follow that if the activity were banned then those who enjoyed it would not be being deprived of any good? (For some other problems, see e.g. Annas 1980.)

[17] In fact it is *not* obvious, from the *Ethics* itself, that Aristotle means to claim that no process is also an activity. But the point is generally accepted by interpreters, and I have argued it at length in my (1988: 256–61), so I here assume it. (For those—discussed below—who interpret Aristotle as saying that the *same* occurrence may be described either as a process or as an activity, my assumption here is that no process-description is also an activity-description.)

[18] Also at X.3, 1173^{a}29–^{b}20. But in VII.12 he too uses 'process', which the reader will surely find confusing.

[19] Quite often in these discussions Aristotle uses *genesis* (literally 'generation' or 'coming into being') synonymously with *kinēsis*. When that occurs, I shall translate it in the same way. (Cf. n. 2, above.)

the *definition* of a 'process', but because it is something Aristotle believes, and is relevant to his discussion here. A process, as we have said, always occupies a stretch of time. We would normally suppose that one could say, of any instant during that stretch, that the process *is occurring* at that instant. But Aristotle holds that one cannot say this, since contradictions then result; a process cannot, even in this way, be ascribed to an instant, but only to a stretch of time. (I do not wish to suggest that this doctrine is defensible—nor the more general doctrine that all changes are processes—but this is not the place to discuss these issues.)[20]

Let us now turn to the other side of the contrast, namely 'activities'. Here again it is necessary to begin with a word of caution on the translation. The Greek word in question, *energeia*, appears to be a word of Aristotle's own coinage, so we get no help on its meaning from other contemporary writers. Moreover, his use of this word is in fact extremely puzzling. Sometimes (as in the *Ethics*) it is contrasted with 'process' (i.e. *kinēsis*) and here the translation 'activity' is traditional. But in other places (e.g. in the *Metaphysics*, but often elsewhere too) it is contrasted instead with 'potentiality' (or 'capacity' or 'ability' or simply 'what could be', i.e. *dunamis*), and here it is usually translated 'actuality'. I have not seen any good explanation of why Aristotle should have introduced a new word of his own, and then used it in these two apparently very different ways, which force translators into different renderings in each case. Certainly, it can be confusing.[21] But at least for present purposes we can say this: in the *Ethics* the word is contrasted only with 'process' (*kinēsis*) on the one hand, and with 'state' or 'disposition' (*hexis*) on the other,[22] and the one word 'activity' seems to work well enough in both these contrasts.

But one should I think be aware of this nasty possibility: perhaps it is again true that Aristotle means one thing by the word when it is contrasted with 'state' or 'disposition', and another thing when it is contrasted with 'process'. There is, indeed, some reason to suspect this. For when 'state' and 'activity' are contrasted, the 'activities' referred to would seem to be the *actions* which express or manifest that state. But actions would seem (very often) to be *processes*, according to Aristotle's account of what a process is.[23] For an action has a goal, and is not completed until that goal is attained, and the action is finished. Moreover, this seems to apply to many cases of actions which manifest the virtues. (Think, for example, of an action manifesting generosity, which is the giving of a gift to someone; or an action manifesting justice, which is a distribution of goods; or an action manifesting courage, which aims at the defeat of an adversary.) But when 'process' and 'activity'

---

[20] Aristotle elaborates his claim that changes are processes in *Physics* VI. I have discussed this in my (1991). (For the particular point about change 'at an instant' see 234ª24–ᵇ9.)

[21] For example, in *Physics* III.1 'process' is *defined* as a special kind of *energeia*, but presumably this is *energeia* in the sense of 'actuality', not in the sense of 'activity'. (I note here that Gosling and Taylor 1982 translate *energeia* as 'actualization' throughout, but distinguish processes (*kinēsis*) as 'incomplete actualizations', which is something that *Physics* III does say.)

[22] Exceptions: there are contrasts with *dunamis* at 1098ª5–6, 1103ª26–8, 1153ª25, and 1170ª16–18; but the translation 'activity' still seems perfectly apt.

[23] *Metaphysics* Θ.6, 1048ᵇ18–23 (discussed below) explicitly says that some actions are processes and some are activities. Cf. Engberg-Pedersen (1983: 32–6).

are contrasted, it appears to be crucial that the activity has no goal beyond itself, and the prime examples that Aristotle gives us are not what *we* would call 'activities' at all, but are seeing (1174^b14–16), or more generally perceiving (1174^b14–20, and *passim*), and contemplating (1175^a1), or more generally thinking (1174^b21–3). Indeed, in book X it is *only* for these cases of 'activity' that he explains (at 1174^b14–23) what a pleasant activity is. At the moment I simply draw this moral: while, in the *Ethics*, 'activity' would seem to be an acceptable translation of *energeia* (for the word means, etymologically, something like 'being in working'), still we must be very wary of supposing that what Aristotle actually has in mind is at all the same as what we would naturally call an 'activity'. I shall pursue this point further what follows.

The contrast between processes and activities is mainly explained here by the claim that an activity is 'complete in form' at any time at which it is occurring (1174^a14–17), from which we may infer that it does not go 'from something to something'. Aristotle also seems to say that it need not take time, but can occur 'in an instant' (1174^b7–9). If this is taken in the natural way, it seems a very strange idea, for surely there cannot be just one instant at which I am seeing, whereas all earlier instants and all later instants are instants at which I am not seeing. (I observe here that Aristotle very clearly does construe the notion of an instant with complete strictness; it is a *point* of time, and not any kind of stretch, however small.) Indeed, *we* are in a position to say that this must be wrong, for we now know that there is a (very short) interval of time which is the minimum threshold for seeing, and what happens in less time than this simply cannot be seen. Even if Aristotle did not have our knowledge, still—as I say—the idea of a strictly instantaneous seeing seems very strange.[24] But I suspect that this was not what he meant at all, and the explanation is quite different. As I have noted, he does think that one cannot say, of any instant, that a given process *is occurring* at it, whereas he here means to say that one *can* say this of an activity: an activity will no doubt last over some stretch of time, but of any instant during that stretch we can say that the activity is occurring then. If this interpretation is right,[25] then we need pay the point no more attention, for the *truth* is that there is nothing to prevent us from speaking in this way in both cases. Finally he says that since an activity is 'complete' at any time, and therefore 'a whole', it has no parts. Consequently it cannot be said to come to be, by which he means (as he always does) that there is no *process* of coming into existence for it. For such a process would have to take time, and would bring into existence first one part of the thing and then another. The same applies to anything else that has no parts, for example to points (1174^b9–14).[26] This (standard) doctrine of his about

---

[24] One might defend the idea in this way. Some things happen instantaneously, for example an object in motion reaching a particular position in the course of its journey. We can apparently see this happening, and the perception of what is instantaneous must itself be instantaneous. But one cannot attribute this thought to Aristotle, for he really does believe that *nothing* happens instantaneously (and he discusses this very example at *Physics* VIII.8, 262^a28–$^b22$).

[25] It is accepted by Gosling and Taylor (1982: 311–12).

[26] For the doctrine in general, see *Physics* VI.6, 237^b9–22. For the application to points, see e.g. *Metaphysics* B.5, 1002^a32–$^b11$.

coming to be is not particularly plausible in itself, and is very unconvincing as applied both to seeing and to pleasure. For surely there can be a stretch of time during which one comes to see something (for example, as the eyes focus on it more clearly), or comes to enjoy it (as one's appreciation of its beauty improves). Indeed, in the case of pleasure what he says here contradicts what he has said only a little before in X.3, where he has admitted that there is a process—which may be quick or slow—of coming to be pleased (1173^a32-$^b4$). In any case, I remark that this point, even if admitted, would not *distinguish* activities from processes, for Aristotle's own doctrine is that a coming-to-be *is* a process, but there is no coming-to-be *of* any process whatever.[27]

I conclude that the only worthwhile guide to activities that we are given here is that an activity is supposed to be different from a process because it is 'complete in form' at any time at which it is occurring. This is not a very helpful characterization, and one naturally looks for help from elsewhere. Apparently, help is forthcoming. At any rate there is a passage in *Metaphysics* book Θ, chapter 6, which looks as if it is trying to draw in a different way the same distinction between processes and activities as we find in the *Ethics*.

The *Metaphysics* passage makes two contrasts. First, it says that some actions are processes and some are activities, explaining that where an action is an activity its goal (or completion, *telos*) is simply itself, whereas where it is a process it has a goal (or completion, *telos*) beyond itself, namely the state which comes into being when it is finished (1048^b18-23). Here Aristotle *may* be thinking of the agent's motive: is he aiming for some result other than the activity itself? But there is certainly no *need* to take him in this way, for he very often speaks of a process as having a goal (or completion) when that process is not something done by a conscious agent at all. (For example, when an acorn grows into an oak tree, the 'goal' or 'completion' of that process is the mature oak tree that results; but this is not to say that any conscious agent intended that result.) The second contrast is what one may call his 'tense-test': if φ-ing is a process, then 'x is φ-ing (at t)' is incompatible with 'x has φ-ed (at t)'; but if φ-ing is an activity, then 'x is φ-ing (at t)' entails 'x has φ-ed (at t)' (1048^b23-35).[28]

Now both of these ways of distinguishing between processes and activities make the distinction relative to the way that the action in question is described. This is particularly clear with the tense-test, for on this test 'he is walking' describes him as engaged in an activity (since it entails 'he has walked', at least some of the way), whereas 'he is walking to Thebes (for the first time)' describes him as engaged in a

---

[27] This is argued in *Physics* V.2.

[28] Greek does not distinguish the so-called 'continuous' present tense 'x is φ-ing' from the plain present tense 'x φs', so we should ignore this distinction when applying the test. (But it does distinguish between two past tenses, 'x was φ-ing' (imperfect) and 'x φ-ed' (aorist), both different from 'x has φ-ed' (perfect).) My supplements 'at t' are naturally taken as indicating an *instant* of time (e.g. 'at 4 p.m.'), and I believe that when they are so taken they fairly represent Aristotle's thought. Admittedly his official doctrine is that one cannot legitimately speak of a process as being in process *at* any given instant, as I have noted, but he does not in practice conform to this (quite unnatural) restriction.

process (since it entails 'he has not yet walked to Thebes'). But obviously the action described as walking to Thebes (for the first time) may *also* be described simply as walking. To generalize (somewhat rashly?) from this one example, it looks as if every action which has a process-description, according to this test, will also have an activity-description. Could this be what Aristotle intended? One cannot believe it. After all, he himself gives 'walking' as an example of a process in both passages, without the slightest hint that it may *also* be counted as an activity, and (according to the tense-test) should be so counted under the very description that he is using himself. This cannot be how he meant his test to work out, for his basic thought is just that walking *does* go 'from somewhere to somewhere', whether or not we put the 'to somewhere' into its description. I conclude that the tense-test was simply a mistake on his part. It does not actually pick out the distinction that he really has in mind.[29]

We find a somewhat similar situation with the first test proposed in the *Metaphysics* passage, concerning whether the action has a goal or completion beyond itself, at least if we take this to be a question about the agent's aims and objectives. Someone may be building a sandcastle because he wants to see that sandcastle completed (or he may be walking to B because he wants to get to B); in that case he would be engaged upon a process, according to this way of understanding the distinction. But someone else may be building a sandcastle just because he enjoys building with sand (or may be walking just because he likes walking, and is not aiming to get to anywhere in particular); then on this account he is engaged upon an activity. For it depends simply upon what the agent is aiming for. (Thus, the second sand-builder would not be at all perturbed if you destroyed every castle that he was building before it was finished, so long as you allowed him to continue building on another.) This, admittedly, is not so implausible as the results that we get from applying the tense-test in a literal manner, but even so I do not find it at all convincing. The first and most obvious objection is that, as I have mentioned, we *need* not suppose that when Aristotle speaks of the 'goal' or 'completion' of an action he means to be talking of the aims of the agent, and in the present instance there is good reason to suppose that he is not.[30] For (*a*) it is quite clear that the *Ethics* passage does not rely on any such consideration, and (*b*) nor does the tense-test (which is no doubt *intended* to elucidate the same distinction, even if—as I have just argued—this intention fails). But, more importantly, (*c*) there is the plain statement in both passages that walking and building *are processes*; there is absolutely no suggestion that they *can* also be activities, if one engages in them with the appropriate aims in view. But Aristotle could not have meant this, and yet allowed it to go completely unsaid.

I conclude that the *Metaphysics* passage in fact sheds no useful light on what we already have in *Ethics* X.4. It is probable that it means to be discussing the same, or at least a very similar, distinction, for the examples given are very similar: in both, walking and building are cited as examples of processes, and seeing and thinking as

---

[29] Heinaman (1995) discusses the point at length. I am broadly in agreement with his conclusions.
[30] Cf. Ackrill (1965: 137–8).

examples of activities.[31] And the main theme in both is similar too: processes are 'incomplete (in form)' at any time at which they are going on, whereas activities are not. But where the *Metaphysics* appears to add further information, this must be discounted. Its tense-test is simply a mistake, for this test actually delivers results which are not at all what Aristotle intended, and the passage should not be interpreted as saying that the agent's aims are what matter, for there is no warrant for this interpretation. (Nor would it improve Aristotle's doctrine.)[32] Some interpreters have fastened on the *Metaphysics* passage because it seems to rescue Aristotle from the utterly paradoxical claim that one cannot enjoy a process—for example, one cannot enjoy walking to Thebes, or building a temple, or quenching one's thirst, and so on for hosts of further examples. They hope to be able to say that his view is that one *can* enjoy these things, so long as what is enjoyed is appropriately described (under an 'activity-description'), or appropriately desired (for its own sake, and not for its further goal).[33] But this simply is not Aristotle's own view, and we must face the paradox: he really does think that one *cannot* enjoy a process. But what could have led him to such an extraordinary opinion?

I will tackle this question head-on in Section 5. But first, I must beat a little more about the neighbouring bushes.

4. The relation between pleasure and activity

How did Aristotle take pleasure and activity to be related? The doctrine of book VII is straightforward. Suppose we take an activity which is enjoyed, say viewing a picture or listening to a piece of music. Book VII claims that the activity will be enjoyed if and only if (*a*) it is an activity of the natural state (for example, one's eyes or ears are not in some abnormal condition), and (*b*) there is nothing that hinders it (for example, bad light, or extraneous background noise), so let us suppose that these conditions too are fulfilled. Then, according to book VII, the pleasure in question just *is* that activity. (It cannot be *another* activity, that is somehow present 'in addition', for if it were then Aristotle could not argue as he does in VII.13 that *eudaimonia is* a pleasure.) But clearly book X disagrees on this point. It says explicitly that pleasure cannot *be* thought or perception, though it is so closely associated with them that the two cannot be separated (1175^b30–5). So, on its account, pleasure is something else. Many commentators suppose that the point of the initial passage in X.4 (i.e. 1174^a13–$^b14$) is to claim that pleasure nevertheless *is* an activity, though a different one.[34] But I protest that Aristotle never says this. He

---

[31] Other examples given in the *Metaphysics* passage are: slimming, learning, and becoming healthy as processes; living well and being *eudaimōn* as activities. (The last two are also given as activities in the *Ethics*, and I shall comment on them later; becoming healthy is no doubt connected with what the *Ethics* calls 'curative processes', and these too I discuss later.)

[32] There is no plausibility whatever in the view that I can enjoy walking only if I am not trying to get to anywhere in particular, that I can enjoy drinking only if I am not trying to quench my thirst, and so on.

[33] e.g. Gosling and Taylor (1982, ch. 16); Broadie (1991, ch. 6, sect. V).

[34] e.g. Urmson (1967: 323); Broadie (1991: 344).

certainly says that pleasure is *like* activities, in that it is complete at any time at which it occurs, and from this one can infer that it is not a process (1174^a16–19). But it seems rash to infer further that it *is* an activity. (Obviously, we cannot infer from the fact that pleasure is *like* a point, in that neither has parts, that pleasure *is* a point.) And when Aristotle does come to a positive characterization of what pleasure is, he does not say that it is an activity but that it 'completes' (or 'perfects') an activity (1174^b14–1175^a3). There is absolutely no indication that his view is that what, in this way, 'completes' an activity must itself be *another* activity. Indeed, my own feeling is that the idea of one activity 'completing' *another*, simultaneous with it, would strike him as in some way absurd. But what, then, is this 'completion' which he now identifies with pleasure?

First, it must be observed that the notion of completeness is here being used in a different way from that employed earlier. Earlier it was claimed that *every* activity is complete, in the sense that is 'complete in form' at any time at which it is occurring, for this is what distinguishes activities from processes. But not every activity is enjoyed. On the contrary, Aristotle here tells us that perception and thought are enjoyed when (i) the faculty of perception or thought is in good order, and (ii) the object perceived or thought of is also 'the finest' (most *kalon*) of those that fall under that faculty. (That is, in the case of sight 'the most beautiful', and in the case of thought 'the most noble', according to my standard translation.) These activities he calls 'most complete', and also 'most pleasant', and it is the pleasure that (in this sense) 'completes' them (1174^b14–23). Since this clearly is a new sense of the same Greek word (*teleios*), one might naturally mark it by shifting to a new, but equally legitimate, translation, namely not 'complete' but 'perfect'.[15] Some activities, then, are perfect activities, and pleasure is what perfects them. I take this to be an equally good account of what Aristotle says. So, in my comments, I shall allow myself to use either 'perfect' or 'complete' interchangeably, but when translating I shall stick with the word 'complete', since that is certainly the usual translation in this context.[16]

As to how it is that pleasure completes (or perfects) an activity, Aristotle apparently offers us two comparisons, but they are both obscure. The first is

The way in which pleasure completes [an activity] is not the same as the way in which the perceiving faculty and the perceived object do, by both being good, just as health and the doctor are not in the same way the cause of being healthy. (1174^b23–6)

Now it is not at all clear that we are intended to take this as a detailed parallel, but if we are then presumably the details must work out in this way: the goodness of the perceiving faculty and perceived object make the activity 'complete' in the same way as the doctor makes one healthy, i.e. by being in each case what Aristotle calls the 'efficient cause' of the result; by contrast, the pleasure makes the activity 'complete' in the same way as health makes one healthy, which in Aristotle's terminology

---

[15] Gonzales (1991) takes me up on this point, maintaining that we do not need to see any change of sense here: all activities may be complete, though some are more complete than others (pp. 151–2). This leads to the odd view that the 'more complete' activities are completed by pleasure, whereas the 'less complete' ones are completed by nothing (or perhaps by pain?).

[16] It is used by both Ross and Irwin, except that Ross slips into 'perfect' at 1174^b15–16.

is by being the 'formal cause' of being healthy. That is (roughly speaking)[37] the explanation of what it *is* to be healthy is given by reference to health, and similarly, then, the explanation of what it *is* to be (in this sense) complete is given by reference to pleasure. I cannot pretend that this is at all clear, but let us press on at once to the other comparison, which presumably must be intended as explanatory:

Pleasure completes the activity, not as the state that is in [it?] does, but as a sort of end that supervenes, as their *hōra* [supervenes on] those at their *akmē*. (1174^{b}31–3)

Here 'the state that is in [the activity?]' must presumably be intended as a reference to the goodness of the perceiving faculty and perceived object; if it is not, then I have no idea what is referred to.[38] Assuming this, then Aristotle is again saying that in *one* way it is this good state that makes the activity a complete (or perfect) one, but in *another* way it is pleasure, and pleasure makes it complete 'as a sort of end (goal, completion, *telos*) that supervenes', i.e. happens in addition (*epiginetai*).[39] But the crucial question is what to make of the comparison that is intended to illuminate this, which I have left untranslated.

Ross offers as a translation 'As the bloom of youth [supervenes] on those in the flower of their age'. (And Irwin more or less follows him.) First I offer a criticism which is not in itself at all important. In standard Greek thought the *akmē* of a man is not usually his youth but his manhood (and the doxographers took this to be the age of 40; compare our use of the Latin 'floruit'). So we should rather translate 'those in the prime of life'. But second, and much more important, the two words *hōra* and *akmē* and particularly their derivatives *hōraios* and *akmaios*, can easily be seen as *synonyms*. So one may perfectly well take it that Aristotle has said 'as primeness supervenes on those at their prime', except that he has used two different words, both of which mean 'prime'. On the other hand one could also suppose that he has genuinely different concepts in mind, translating in some such way as 'as flourishing supervenes on those at their prime', and interpreting 'at their prime' as simply a way of referring to a particular age (say 40), but 'flourishing' as a way of referring to the good state of a person that (normally) occurs at that age. This leads to two quite different interpretations of what Aristotle's theory is.

The first[40] plays down the implication that pleasure is something *extra* (which

---

[37] This is not the occasion for an exposition of Aristotle's doctrine that there are four types of cause (i.e. explanation), namely the efficient, the final, the material, and the formal. (His own explanation is in *Physics* II.3 and 7; but this explanation is often desperately obscure, and particularly on the 'formal' cause.) Note that health is also given as the 'formal cause' of being healthy at 1144^{a}3–5.

[38] This translation follows Irwin (near enough). Ross has 'not as the corresponding permanent state does, by its immanence', which I take it is intended to have the same sense. (It might be more natural to suppose that this state is in 'the man', rather than in 'the activity', and the Greek would certainly allow this. But in that case, if my interpretation is right, we have to suppose that not only the perceiving faculty but also the perceived object is 'in the man'. This is not, of course, impossible. See p. 164 below.)

[39] 'An end that supervenes' is Ross's translation. Irwin has 'a consequent end', which rather leaves out the idea of something 'in addition'.

[40] This interpretation is proposed by Gosling (1973–4). A revised version is in Gosling and Taylor (1982, chs. 11 and 13, esp. sects. 11.3 and 13.4).

epiginetai), and points to the synonymity of hōra and akmē and to the comparison (which it takes to be intended) 'as health causes one to be healthy'. On this account, to say that an activity is pleasant just *is* to say that it is complete (i.e. perfect), and nothing 'extra' is really intended. Of course the two words 'pleasant' and 'complete' are different words, just as the two words hōraios and akmaios are different words, but the thing indicated is the same in each case. It follows that the apparent change of view between book VII and book X is not a real change of view at all, but only an improved way of stating what is essentially the same view. Books VII and X admittedly differ on when an activity is perfect. The condition of book VII that the activity be one 'of the natural state' can be thought of as more or less equivalent to the condition of book X that the faculty of perception or thought involved should be 'in good order'. But the condition that the activity be unhindered can hardly be equated with the condition that the object perceived or thought of should be one of 'the finest' available to that faculty. So here there is indeed a change of view (and a still better view would incorporate both of these conditions, and perhaps some others too). But, setting this aside as of merely minor importance, the position in both accounts is that we have a pleasant activity when and only when we have a perfect one. In book VII it had been said that, when we do have a perfect activity, the pleasure in question just is that activity. In book X this is rejected as oversimple, but what is substituted for it is simply this: the pleasure in an activity is not the activity itself but the perfection of it. It is what 'makes' an activity perfect, in the same way as primeness 'makes' a person at his prime, and in the same way as health 'makes' a person healthy. That is to say, the pleasure just *is* the perfection. What, in our sense, *causes* the perfection is something different, perhaps that the activity is unhindered, or perhaps that its object is one of 'the finest', or perhaps something else. But *what* it causes is perfection, *and* what it causes is pleasure, so these must be the same as one another.[41] The pleasure in a perfect activity, then, is not a *further occurrence*, over and above that activity itself, but is the *further fact* that the activity is perfect. And that is simply what pleasure *is*, according to this first interpretation.

The second interpretation[42] stresses the fact that Aristotle does describe pleasure as an end—something we seek for—that happens *in addition*, which it takes to mean 'in addition to the activity itself'. Consequently, it stresses the fact that hōra and akmē can be taken to have different meanings (related roughly as 'bloom' to 'youth'), and it sees no relevance in the point that health is, in one sense, what makes one healthy. This is offered simply in illustration of the claim that there are different senses of 'cause', which in turn is offered simply as an analogy to the claim that there are different senses of 'to complete', but no detailed parallel is intended. So on this account Aristotle does indeed, in book X, think of pleasure as something else, that happens in addition to the relevant activity, but one has to admit that he gives *no* explanation of what this 'extra something' is. Educated as we are, separated from Aristotle by more than twenty centuries, the thought that inevitably occurs is

---

[41] Pleasure is 'added' only in the sense that what is caused is 'added' to what causes it, i.e. to 'the state that is in [the activity]'.

[42] This interpretation is explicit in Owen (1971–2), but I think implicit in several other discussions.

that pleasure (like pain) should be thought of as a certain kind of *feeling*, that accompanies those activities that we find pleasant. But one has to admit (*a*) that there is nothing that Aristotle says that suggests this,[43] (*b*) that it is evidently an entirely wrong account of most of the pleasures that concern him—for example, there is no special 'pleasure-feeling' that accompanies one's enjoyment of a view, or a symphony, or a geometrical proof—and in any case (*c*) this view surely threatens his claim that pleasures differ in kind from one another. For example, one might well say that the *feelings* one has in any exciting contest—a chess match, or a yacht race, or a game of squash—are essentially the same (due, in each case, to an excess of adrenalin in the bloodstream). But Aristotle will insist that the pleasures are nevertheless different. The best version of this interpretation, then, will be that he does regard pleasure as something 'extra', that is 'added' to the activity, but that he tells us nothing about what this 'extra' is.[44] The interpretation is attractive, for it seems to represent exactly *our own* position: we too think of pleasure as something 'extra', but cannot say what this 'extra' is.

Nevertheless, I am inclined to think that the first interpretation is correct, and I offer two reasons. (i) One of *our* grounds for thinking of pleasure as something 'extra' is that we suppose that essentially the same activity will sometimes give pleasure and sometimes not. But this is certainly not what Aristotle thinks. For he claims very explicitly that when an activity satisfies his conditions for being 'perfect' then it simply *must* be pleasant (1174^b29–31, 1174^b33–1175^a3). This is also a reason for supposing that the 'bloom of youth' interpretation of 1174^b31–3 must be wrong. For presumably it is possible for one to be a youth and yet have no such 'bloom' (or, more exactly, for one to be at the right age of manhood and yet not be 'flourishing'), but apparently this is *not* Aristotle's view about pleasure. (ii) On the second interpretation there would appear to be a real and important change of view between books VII and X, whereas on the first interpretation such change as there is is more verbal than real. So which is the more likely? Of course, any thinker may change his opinions quite radically, from one occasion to another, but there is, over all Aristotle's (surviving) works, an amazing consistency. For that reason too, an interpretation which minimizes the change of view seems to me preferable to one which maximizes it. But this brings me to Owen's interpretation in his (1971–2) of the difference between books VII and X. For on his account the second interpretation of book X is correct, and yet there is *no* change of view from one book to the

---

[43] In so far as he has a general word for feelings it is *pathos*, and his standard examples are ('bodily') desires and emotions. He does say that these are *followed* by pleasure and pain (1105^b23), but (so far as I am aware) he never says that pleasure, or pain, *is* a *pathos*. (But perhaps 1105^a3 is an exception?)

[44] Urmson (1967) gives good reasons for saying that it is not a *feeling* that is added (except, he suggests, in the case of Aristotle's pleasures of touch). But as to what *is* added his proposal is just: 'We might say that [enjoyment] adds *zest* to the activity' (p. 324). Owen (1971–2) cites some more definite claims from chapter 5: 'the pleasure . . . augments the activity in that people who engage in the activity with pleasure are more exact and discriminating (1175^a30–$^b1$); the stronger it is the more it prevents them from attending to other activities (1175^b1–13), and the longer and better the activity goes on (1175^b14–16)' (p. 99). But none of this seems to tell us *what* it is that is added, but only the effects of adding it.

other. For they do not give different answers to the same question; rather, it is the *question* that changes.

The word 'pleasure' (in Greek as in English) has two distinct uses. If I enjoy playing chess then I may say 'chess is one of my pleasures' or 'chess is a pleasure to me', or I may say instead 'chess gives me pleasure'. In the first use chess *is* (a) pleasure, and in the second chess *gives* pleasure, and one must presumably suppose that what it gives is not simply itself. Owen's suggestion is that in book VII Aristotle aims to analyse the first use, and so correctly says that (a) pleasure just is the activity that is enjoyed, whereas in book X he aims to analyse the second, and that is why he no longer identifies the pleasure and the activity, but (correctly?) says that the pleasure is something else which happens in addition. In other words, book VII is concerned with what is enjoyed, whereas book X is concerned with what it is to enjoy something, i.e. with the enjoying of it. Both questions can be given the same title, 'What is pleasure?', but they are different questions, which evidently explains why we get different answers in the two places.

Now one cannot deny that this is a very attractive suggestion, and when the two discussions are viewed in this light they do each make very much better sense. But, if I understand him correctly, Owen means to suggest too that Aristotle is *aware* that he has changed the question from one discussion to the other, and that seems to me very improbable. First, if we are right in supposing that book X is the later account, then on this hypothesis you would expect Aristotle somehow to warn the audience there that his question could be taken in two ways, but it is clear that there is no such warning. But second, and more importantly, there seems to be a strong indication that he himself is not aware of it, in what he says about the theory he rejects, i.e. the theory that pleasure is a (perceived)[45] process of replenishing a bodily lack. This is clearly a theory about *what* is enjoyed, and no doubt Aristotle understands it in this way in book VII. However, when he argues against it in book X he does there appear to understand it as a theory about what enjoying—or being pleased—itself is. For he argues that one cannot *be pleased* quickly or slowly, whereas every process can be qualified as quick or slow, from which it follows that being pleased is not a process ($1173^a31–^b4$). (But it would not seem to follow that *what* pleases one is not a process.) Again he argues that since it is the body that is replenished it should, on this theory, be the body that is pleased, but that is not what we believe ($1174^b7–11$). So once more it follows that being pleased is not a bodily replenishment, but it does not seem to follow that *what* pleases us cannot be a bodily replenishment.

In one way, then, these arguments support Owen's suggestion, for it is clear that Aristotle is in book X concentrating on what being pleased is, i.e. on what enjoying itself is, and is claiming that *it* is not a process, and in particular not a bodily replenishing. But at the same time they *also* suggest very strongly that Aristotle is not aware of the distinction that Owen draws, for if he were he would have to have seen that these arguments are beside the point. The theory that he is opposing is

---

[45] The qualification 'perceived' was seen to be necessary by Plato (*Republic* IX, 584c). The book VII discussion includes this qualification, but the book X discussion omits it.

most naturally seen as a theory about *what* one enjoys, namely that it is (always) a process, and (often)[46] a process by which the body is 'refilled'. As he himself says in the next sentence, the theory arises because it certainly seems that we do enjoy eating when hungry and drinking when thirsty, and it generalizes from these (1173ᵇ13–15). But he takes it that his arguments have refuted *this* theory, and that is why he feels entitled, in what follows, to take it for granted that *what* we enjoy is never a process but always an activity. In his own vocabulary, it is only an activity, which is already in one sense 'complete' (i.e. 'complete in form', at any time at which it is occurring), that can be further 'completed' (perfected) by pleasure. But we have had no good arguments for this claim except the two just mentioned,[47] and they could only seem to support it if in fact Aristotle has failed to notice Owen's (perfectly good) distinction. In fact I think that this failure is absolutely central to his whole way of thinking about pleasure.

5. Pleasure and activities[48]

Can Aristotle *really* have believed that what we enjoy is always an activity and never a process? What I shall claim is that he really did believe that what we enjoy is always something mental, in particular it is either a perception or a thought, and he somehow takes it for granted that these mental occurrences will count as 'activities'. But as to *why* he should have believed that only a mental occurrence can be enjoyed, the only explanation that I can see is that he failed to distinguish clearly between the thing enjoyed and the enjoyment of it. The *latter* is clearly 'in the mind', so he assumes that the former must be too.

The claim that what we enjoy is always either a thought or a perception is a very surprising claim, for we so frequently speak of enjoying all kinds of other things too, so one must hesitate before ascribing it to Aristotle. Indeed, if we look at some of his own examples in X.1–5, it must seem that he cannot have believed it. It is true that when he explains in X.4 what it is for an activity to be 'completed' by pleasure, he does mention *only* the activities of thinking and perceiving (1174ᵇ14–1175ᵃ3), and *most* of the examples that he has given earlier in X.3 (1173ᵇ16–20), and will give later in X.5 (*passim*) are either pleasures of thought or pleasures of perception. But there do certainly appear to be exceptions.

First he says at the end of X.4 that we enjoy living (or life, or being alive), and he describes this as an activity (1175ᵃ10–21), but it would not seem that being alive is *just* a matter of thinking and perceiving, for surely it involves acting too? But here

---

[46] In the version which Plato certainly *seems* to endorse, in *Republic* IX, there are both bodily and mental pleasures, and mental pleasures are 'fillings' (of the soul, with knowledge) but not 'refillings' (585a–b). (But I observe incidentally that his 'theory of recollection', which is not mentioned in the *Republic*, would allow him to assimilate the two cases even more closely.)

[47] In both books Aristotle does point out that what we enjoy is *not always* a process of replenishment, preceded by a painful emptiness (VII.12, 1152ᵇ36–1153ᵃ2; X.3, 1173ᵇ15–20). But this still allows us to claim that *sometimes* it is.

[48] Most of this section summarizes the argument I develop more fully in my (1988).

we can reply, surprisingly, that Aristotle *is* prepared to identify living (for a human being) with perceiving and thinking, for he explicitly does just this in an unexpected argument at IX.9, 1170^{a}16–^{b}8. This apparent counter-example, then, may be set on one side, and at the same time I think we may set aside the apparent counter-example of *eudaimonia*, for this too is a matter of *living* well (and it has consistently been described as an activity *of the soul*). It is more of a problem that on several occasions in X.3 what Aristotle says reminds us of his claim that it is characteristic of the good man that he enjoys 'the activities in accordance with virtue' (1173^{a}15, ^{b}29–31, 1174^{a}6–8). It may seem comprehensible that when he is thinking in very general terms about what living is (for a human being) Aristotle should overlook the fact that people display life not only in thinking and perceiving but also in acting, yet it seems very improbable that *this* phrase is intended to point to a purely mental activity.

In any case, the problem is not confined to this. As I have remarked, many of these 'activities in accordance with the virtues' would seem in fact to be processes rather than activities, given Aristotle's own way of making this distinction. Moreover he does sometimes explicitly talk of enjoying what is, on his account, a process. Thus in X.5 we hear of those who enjoy building (1175^{a}34–5), but in X.4 building is also cited as a star example of a process (1174^{a}19–29). We hear too of those who do not enjoy writing or calculating (1175^{b}16–20), in a way which surely implies that it is possible to enjoy writing and calculating, yet these also would appear to be processes and not activities. And in any case it seems absurd to count building or writing as cases of either thinking or perceiving or both. But the star counter-example is of course eating and drinking, which Aristotle himself counts as bodily processes (i.e. of replenishment). But surely these *are* things that people enjoy? Let us concentrate on this case, since it was evidently important in Aristotle's own thinking. What does he believe actually happens when one *seems* to enjoy eating or drinking?

The quite lengthy discussion in X.3 is unrevealing. It admits that 'one might be pleased while replenishment is going on' (1173^{b}12), but does not explain what this pleasure is. There is more information in book VII, which consistently characterizes these bodily pleasures as 'curative', since they repair a lack, which is painful (1152^{b}31–3, 1154^{a}26–31, ^{a}34–^{b}2, ^{b}17–19). But it also says that these 'cures' are *only incidentally* pleasant (1152^{b}33–5, 1154^{a}34–^{b}2, ^{b}15–19), and in explanation of this it says that 'the activity (in such cases)[49] is an activity of the remaining state and nature' (1152^{b}35–6), or again that one is cured 'when the remaining thing that is healthy is acting, and that is why they seem to be pleasant' (1154^{b}18–19).[50] The doctrine, then,

---

[49] I construe *en tais epithumiais*, literally 'in the desires', as meaning 'in *cases* which involve desire [i.e. a preceding and painful lack]'. For a defence of this suggestion see my (1988: 265–6).

[50] Ross has, not 'when the remaining healthy thing is acting' but rather '*because as a result* people are cured, *through* some action of the part that remains healthy' (my italics). On this account, what is actually pleasant is the result reached at the end of the process, and not some activity simultaneous with it. But, while I do not deny that the Greek *could* be taken in this way, (*a*) it gives an implausible account (for it does not explain why the drinking should seem pleasant while it is going on, not after it has finished), and (*b*) it is not consistent with what (apparently) Ross himself takes to be the correct interpretation of 1152^{b}35–6.

must be this: when one is 'cured' (e.g. of thirst) it *seems* to be the process of replenishment that is pleasant, but this is not actually what is happening, for what is actually pleasant is an activity that is going on at the same time as the process, and this activity is an activity of something that is *not* affected by a lack. So we naturally ask: what *is* this 'other thing', still in its 'natural (healthy) state', whose activity it is that is *really* what is pleasant? The plain truth is that Aristotle nowhere says, so we are reduced to conjecture.

The obvious conjecture, it seems to me, is this: when the body is affected by a lack, it is *the soul* which remains unaffected, and it is therefore some activity of the soul which is what is really enjoyed. As for what activity this is, the answer must be obvious: it is the *perceptions* that take place as the replenishment is going on. For example, I can taste the drink, I can feel its cooling touch in my throat as it goes down, and I also have a certain bodily sensation which can only be described as the feeling that it is doing me good, as my thirst disappears. These are perceptions, and they take place in the soul and not in the body, and they are to be counted as activities, whereas what happens in the body is a process. To put the point in a summary way: we all speak, of course, of enjoying drinking; but this is a loose way of speaking, for what is actually enjoyed is not the drinking itself but the perception of it; and this perception takes place in the mind, not in the body. A sign of this is that if I am drinking but not perceiving it—e.g. if I am anaesthetized, but nevertheless taking liquid, through the mouth, into the stomach—then I do not enjoy it. The point generalizes very naturally to *all* those cases where, in our common way of talking, we speak of what is enjoyed as if it were something non-mental—not only eating and drinking, but also building or writing (or, say, playing squash), and indeed to the 'activities in accordance with the virtues'. For in all cases we would not enjoy them if we were not aware of them, from which Aristotle infers that it is not these processes or activities themselves that are enjoyed, but our awareness of them. And awareness is a mental activity, which can be characterized broadly as either perception or thought.

It has to be admitted that in the *Ethics* Aristotle never explicitly states this thesis that I am attributing to him. I take it that the explanation is that it has been a feature of his way of thinking about pleasure for so long a time that he no longer recognizes that it needs an explicit statement.[51] But it is possible to point to one other discussion in the *Ethics* that does offer quite strong confirmation that this is how he naturally thinks. In III.10 he gives us an account of the 'self-indulgent' man, who is to be roughly characterized as one who pursues pleasure to excess. But we should ask what pleasures in particular are relevant to 'self-indulgence', and here he begins—as often elsewhere—by distinguishing pleasures into two main kinds, namely those of the mind and those of the body (1117^b28–13). Setting aside those of the mind as irrelevant in this case, he proceeds to run through those of the body, and we hear successively of the pleasures of sight, of hearing, of smell, and finally of taste and touch. (It is these last that are relevant to 'self-indulgence'.) The point to

---

[51] It may be claimed that he *does* state it explicitly in what is presumably a very early discussion of pleasure, namely *Physics* VII.3, 247^a6–19.

notice is that, without stopping to explain or defend this view, Aristotle automatically classifies bodily pleasures by the kind of *perception* that is involved, and this is
just what one would expect him to do if he thinks that all bodily pleasures *are*
perceptions. Moreover, he recognizes only two basic kinds of pleasure, those of the
mind, which are pleasures of thought, and those of the body, which are here taken
to be pleasures of perception. There is, for example, no separate category of the
pleasures of action.

I conclude that Aristotle really meant it when he apparently implied in X.4 that
what we enjoy is always either a thought or a perception, and the fact that he gives
no other examples there of activities that are enjoyed is not an accident. For he
really does think that there are *no* others.

Is this view defensible? Certainly, it is strange and unexpected. For of course we
ordinarily think that we can enjoy such things as walking, gardening, fishing, playing squash, and a whole host of other things, which *we* might naturally call 'activities', though Aristotle does not. And he does not call them activities partly because
they satisfy his definition of 'processes' as opposed to 'activities', but more basically
because in any case they do not take place 'in the mind'. But when he is thinking
on *this* topic[52] he will not accept that there are any activities other than mental
activities. What *reason*, however, do we have for accepting this unexpected view?

The one reason given in his text—and even that is never made fully explicit—is
that *enjoying* takes place in the mind. But clearly it does not follow that *what* is
enjoyed must also take place there, for the two are different. The difference is
obvious if we are allowed our usual view that the one (the enjoying) is in the mind
while the other (what is enjoyed) usually is not. But even where what is enjoyed is in
the mind still the two must be distinguished. Consider, for example, enjoying a
thought—say, one suddenly realizes that a certain philosophical doctrine, which
one has always mistrusted, can in fact be refuted by an entirely compelling argument. First, let us note that the fact that such a realization is pleasant cannot be
explained by its fulfilling any or all of Aristotle's conditions for an activity being
'perfect' (i.e. that at the time the mind is in good working order, that it is not
impeded by competing considerations, that the philosophical doctrine—or its
negation—is a 'good' object of thought). In fact these conditions are often quite
irrelevant to the pleasantness of an activity, and I do not believe that any alternative
set of conditions would be any better, as this example shows. For here it is not the
nature of the thought itself that makes it pleasant, since that very same thought may
be pleasing to one thinker and displeasing to another. (It will be pleasing to one
who has always suspected the doctrine, displeasing to one who has believed it.) So
pleasure is indeed something 'extra', and not to be explained by the nature of the
activity that pleases. Anyway, to come back to the point that I am arguing, the
enjoying and the thing enjoyed must certainly be distinguished from one another,

---

[52] As I have noted, I think it quite possible that in *other* contexts he thinks rather differently about
what constitutes an 'activity', e.g. when he contrasts the 'activities' that manifest a virtue with that
virtue itself.

even where both do occur in the mind, just because the same mental occurrence may be enjoyed by one person but not by another. And once this is granted it is clear that there is no general argument from the premiss that enjoying takes place in the mind to the conclusion that what is enjoyed must be there too.

But there is also another line of thought, which I guess was important to Aristotle's thinking. This starts simply from the premiss that I do not enjoy what I am not aware of, and concludes that it must be the awareness itself which provides the enjoyment. We might develop it in this way. Contrast two cases in which (a) I am (as we would say) enjoying a game of squash, and (b) the alleged 'game' is a merely 'virtual reality', and the fact of the matter is that I am not playing squash at all, but being fed the appropriate sensations by a 'virtual reality' machine (or, more simply, I am just dreaming). Then the argument is that my enjoyment is the same in each case, and this is because my awareness is the same in each case, whether or not it is a real game of squash that is being played. To be sure, there will be differences afterwards. When I realize that I have not actually beaten Jahangir Khan, but only been given the sensations of doing so (or have dreamed it), my enjoyment will certainly be diminished. But can we not say that at the time the enjoyment was the same, just because the thoughts and (apparent) perceptions were the same, and does not this show that the 'real' object of enjoyment was merely the thoughts and (apparent) perceptions then occurring?

Clearly, this line of reasoning is familiar, for it is just the same reasoning as is employed in order to convince us that we never do perceive an 'external reality'; rather, the object of my perception is always 'in me'. It seems to me that the reasoning is no better, and no worse, when applied to enjoyment than it is when applied to perception, but this is not the place for a proper discussion of it. Instead, I content myself with an ad hominem observation: Aristotle himself should be perturbed by the analogy between perception and enjoyment that I suggest. At any rate, he is usually interpreted as being in a broad sense a 'realist' about perception, and certainly it is true that he pays absolutely no attention to such phenomena as hallucination, which are so much the stock in trade of those who wish to claim that we never do perceive an 'external reality'.[53] (Very roughly, his account in the De Anima[54] is that in perception the sense organ takes on the same form as the object perceived already has. So in one way what I perceive is 'in me' (i.e. in my sense organ), though in another way it is also 'outside me', for it is the same form that is in these two different places.) Yet his account of enjoyment is that we never do enjoy an 'external reality'. There seems to me to be no good reason for taking one view in the one case and a different view in the other, for the essential arguments are just the same in each case. That is why I say that he should find my analogy perturbing, but there I must let the issue rest.

Finally, I must add a brief word on the claim that perceptions and thoughts are activities rather than processes. What Aristotle is thinking of, when he makes this

---

[53] This is not because scepticism about perception had not yet found its way into philosophy. Most of the standard sceptical moves are given in the first part of Plato's Theaetetus.

[54] De Anima, principally II.12, III.1–2, and IV.8.

claim, is that such a state as seeing a tree is not a process, for it does not go 'from something to something', but is 'complete in form' at any time during the period in which it occurs. The same might be said of contemplating a thought—i.e. holding it steadily in 'mental view'—and perhaps too of remembering something pleasant, or anticipating it (1173^{b}18–19). But Aristotle needs more than this. For he wishes to include too a thought which is a calculation (1175^{b}18), or the working-out of a proof (1175^{b}6), and he should add such things as working at a crossword puzzle. But these thoughts *do* go 'from something to something', and they surely satisfy his definition of a process. More importantly, the relevant perceptions must also include perceptions of processes—not only perceptions of eating and drinking, but also such perceptions as listening to a particular piece of music being played. But these perceptions too are surely *themselves* processes, for again they go 'from something to something', and they are not 'complete' until they are finished. What has happened, I believe, is that he has noticed that the best examples of occurrences that are, in his sense, 'activities' are certain thoughts and perceptions, and he has rashly generalized from there to the conclusion that *all* thoughts and perceptions are 'activities', but in fact the generalization cannot be accepted. No doubt he really does think that what is *most* enjoyable is the activity of contemplation, a topic that I pursue further in Chapter IX. But, if his general account of enjoyment is to retain any plausibility, it must include other thoughts and perceptions too, and in these other cases his distinction between activities and processes is not in fact of any relevance.

If I am right, then, what one usually thinks of as Aristotle's main claim on what pleasure is, namely that it pertains to activities and not to processes, is actually a red herring. And what really is central to his view, namely that only thoughts and perceptions can be pleasant, is a thesis which he himself never states at all. Nor is there any good way of defending it, except by arguments which have awkward consequences for his views elsewhere. I conclude that there is little to be said in favour of his account of what pleasure is (though I add that the same verdict would seem to apply to all other accounts too). He has more to offer us in his views on the value of pleasure, which are of course very much more germane to what *we* call 'ethics'.

Further reading

Most of those who write about Aristotle's discussions of pleasure are more interested in his views on what pleasure is than in his views about its goodness. But a pleasing exception is Annas (1980), whom I recommend as taking this topic rather further than I do.

As for what pleasure is, I would suggest starting with Urmson (1967), who does not really tackle what I see as the main problem, but who represents an influential line of thought. Then one must see Owen (1971–2), on the relationship between the two accounts. Owen's view is attacked by Gosling (1973–4), but perhaps it is better to see the later exposition of this interpretation in Gosling and Taylor (1982, ch. 11, esp. sect. 3, and ch. 13, esp. sect. 4). These readings do not pay very much attention to the distinction between activities and processes, and on this topic one must start with Ackrill (1965), who gives a careful account of what

Aristotle says (not only in the *Ethics* but also elsewhere), and notes some obvious problems. The distinction is differently interpreted by different commentators. A view very different from mine will be found in Gosling and Taylor (1982, ch. 16). My own view is elaborated and defended at somewhat greater length in Bostock (1988), and I could mention Heinaman (1995) as broadly in agreement with me, for at least some of the way.

Our best evidence for the theory that Aristotle is attacking is to be found in Plato. One might start with his *Republic* IX, 583b–586a, but the task of chasing up what he says elsewhere is much lightened by consulting A. E. Taylor's commentary on Plato's *Timaeus*, the long note on 64d.

Several modern treatments of pleasure are influenced by (what is taken to be) Aristotle's account. One might here mention Ryle (1954, ch. 4), Kenny (1963, ch. 6), and possibly Gosling (1969).

Chapter VIII

Friendship (Books VIII–IX)

1. Introduction

WHEN listing the various virtues of character in II.7, Aristotle mentioned three which fall under the general heading 'social virtues', and one of these is standardly called 'friendliness' by translators and commentators (1108ᵃ26–30). In fact Aristotle himself regards it as nameless,[1] and in the later discussion at IV.6 he says that it is like friendship in a way, but only in a way, and that it is displayed both to those who are friends and to those who are not (1126ᵇ20–8). He describes it as a mean between being on the one hand obsequious or flattering, and on the other hand quarrelsome and surly. The friendship that is discussed in books VIII–IX is not this virtue, and indeed Aristotle never says that it is a virtue (but only that it involves virtue), and certainly he does not describe it as any kind of mean between two extremes. But it is nevertheless an important topic for him, because friendship is in his view an essential ingredient in the good life. Indeed he gives it more space—i.e. two whole books—than he gives to any other topic in the *Nicomachean Ethics*.

Much of his discussion is concerned in one way or another with the different varieties of friendship, and I shall skate rather quickly over this (in Section 2), partly because I am somewhat sceptical of the value of Aristotle's taxonomy, but more because it is only one particular variety of friendship that he insists must play a part in the good life, and so it is natural to concentrate on this. I shall mainly be concerned (in Section 4) with the arguments that he offers in IX.9 for saying that this particular kind of friendship is essential to the good life, and (in section 3) with what may reasonably be called his main discussion of the apparent conflict between egoism and altruism, which is in IX.4 and IX.8. This problem very naturally arises for him in the discussion of friendship, for on the one hand his overall position (as sketched in I.7) is that each man will seek *his own eudaimonia*, but on the other hand he thinks that in the best kind of friendship one will often act for one's friend's sake, and so apparently *not* for one's own sake. His treatment of this dilemma is worth attention.

By way of preliminary it must be said that 'friendship' is not a good translation of Aristotle's term *philia*, though no other English word would be any better, for under this title Aristotle groups together a much wider variety of social relationships

---

[1] In II.7 the virtue is apparently called *philia*, i.e. friendship, at 1108ᵃ28, but both Ross and Irwin (very reasonably) translate the word there as 'friendliness'. In any case, it is more significant that Aristotle has prefaced his account in II.7 by saying that these social virtues are often nameless (1108ᵃ16–19), and in IV.6 he insists emphatically that this applies in the present case.

than we would. Some of them we would more naturally call 'love', such as the love of a mother for her child, or the erotic passion of a lover for his beloved;[2] some are friendships in our usual sense; but some seem more to be mere business relationships, as when I trust and rely on my regular butcher or greengrocer to supply wares of good quality, and they trust and rely on me to provide payment. The translator's problem is further exacerbated by the fact that the associated verb *philein* is standardly rendered as 'to love', and again it is difficult to see any alternative that is more suitable. Yet it seems to us quite absurd to say that if I and my butcher trust one another in this way then we thereby count as *loving* one another. There are, then, features of Aristotle's terminology which are somewhat strained even in Greek, and still more so in English. We must simply recognize that in his discussion the noun 'a friend' and the verb 'to love' have a much wider application than is usual.

2. The varieties of friendship

In VIII.1 Aristotle introduces the topic, indicates its importance, and mentions some problems that are raised about it. In VIII.2–4 he argues that there are different forms of friendship, and in fact he distinguishes three. What he calls 'complete' or 'perfect' friendship is described at 1156^b7–32. Its main characteristic is that each person wishes good to the other for his own sake ($^b9$–10), though we should also add that each knows that the other so wishes (1156^a2–5), and that they act accordingly (1166^a3). Let us call a friendship that satisfies this condition a 'true' friendship. But Aristotle also makes it part of his definition of 'perfect' friendship that it can hold only between two people who are each good people, and equal in virtue (1156^b7–9). I shall keep these notions separate, since it is by no means obvious that only good people can be 'true' friends, as defined. This, then, is 'perfect' friendship, which is also called friendship 'based on character' (1164^a12). He adds that friends of this sort will spend time with one another (1157^b18–24),[3] so he is clearly thinking of what we might call 'close' friends, and that such friendships are not easily dissolved (1156^b17–19).

In contrast to 'perfect' friendship Aristotle recognizes in VIII.2–4 two 'secondary' forms of friendship, namely friendships based either on usefulness or on pleasure. Thus two people may be friends because each is useful to the other, or because each gives pleasure to the other, or because one of them gives pleasure to the other, and in return the other is useful to the first.[4] There is some dispute over whether Aristotle means to imply that, in each case, it is still required that each wishes (and does) good to the other for his own sake, if this is to count as a 'friendship'. It must

---

[2] Aristotle (like Plato) has in mind the homosexual passion of a man for a youth; he does not describe a heterosexual relationship as 'erotic'.

[3] Aristotle's phrase is that they will 'live together' (*suzēn*) but he is not thinking of living under the same roof. (The relationship of husband to wife is excluded from 'perfect' friendship, as we shall see.)

[4] Erotic passion is one of his examples. The lover will find pleasure in his beloved, but often the beloved will find his lover merely useful (1157^a3–14, 1164^a2–12).

be admitted that, at first sight, the text seems to be inconsistent on this point. VIII.2 appears to lay it down that this is a necessary condition for *all* types of friendship ($1155^{b}31$–$1156^{a}5$). But immediately in VIII.3 the condition is weakened to: 'those who love each other wish goods to each other *in that respect in which* they love each other' ($1156^{a}9$–10); this is explained as meaning that, in a friendship based on usefulness or pleasure, I do not love the other for himself, but only for the useful-ness or pleasure that he provides to me ($1156^{a}10$–19). It is very natural to take this as meaning that I wish my useful friend to prosper only in so far as I wish that his utility to me should prosper, and that is how I shall understand it.[5] Later in IX.5 Aristotle goes so far as to say that well-wishing simply does not arise in friendships for utility or pleasure ($1167^{a}13$–14; similarly *EE* $1241^{a}4$–10), but that is perhaps a mild exaggeration.

To illustrate the situation as I see it, let us return to the relationship between me and my regular butcher. We each wish the other well, but only in the sense that I wish his business to continue to prosper, and he wishes the same for my financial affairs, just so that we may continue to be useful to one another. It may be that I hope that he enjoys his job, and take some small steps to contribute to this, e.g. by smiling when we meet, and chatting pleasantly to him as we complete a transaction. But if I do then my rationale is simply that his business is more likely to continue if he does enjoy it. For I do not value *him*, as a person, but only his usefulness to *me*. Similarly he may think that if he is nice to me in turn then I shall remain his customer, and so continue to be useful to him. As I understand it, that is all that Aristotle requires of this form of 'friendship', and that is why he adds that it will naturally cease if its utility ceases, e.g. if the butcher retires ($1156^{a}10$–30). Of course, this is hardly what we would call a 'friendship' of any kind, but it is a recognizable form of social relationship, and an essential part of any community. We are more likely to recognize as 'friendships' what Aristotle describes as friendships 'based on pleasure', where each enjoys the company of the other, and he himself admits at least this difference between the two: friends 'for utility' do not spend time with one another, whereas friends 'for pleasure' do ($1156^{a}27$–30, $^{b}4$–6). Nevertheless, his over-all position assimilates these two to one another, as each in its own way a deviation from 'perfect' friendship, because in each case I love the other not for his own sake but for the sake of something that he gives to me.

We are offered two ways of explaining this deviation. One is that there are three kinds of things that are lovable, namely what is good, what is pleasant, and what is useful, and the three forms of friendship arise from these ($1155^{b}17$–19). But this threefold distinction is at once upset by two further remarks, (*a*) that the good and the pleasant are alike in each being loved for their own sakes, whereas the useful is loved only for the sake of what it is useful for, which will be something either good or pleasant ($^{b}19$–21); (*b*) that one should distinguish between what is, and what appears, either good or pleasant *to a given person* (at a given time), and what is either good or pleasant unconditionally ($^{b}21$–7). However, the many further

<hr/>

[5] For the contrary view, see Cooper (1980, esp. 309–16). I am in sympathy with the response to Cooper by Price (1989, ch. 5). (Pakaluk 1998 attempts an adjudication on his pp. 61–3.)

subdivisions of friendship that these remarks appear to suggest are not further explored here, and I shall not pursue the topic.[6] The second explanation is that a 'perfect' friendship is at the same time one that involves both pleasure and utility, since good men both give pleasure to one another and are useful to one another. So the inferior forms of friendship each preserve *one* feature of 'perfect' friendship, though not the others (1157^a25–36, 1158^b1–11). Whether this rather tenuous similarity is enough to entitle each of the inferior forms to be called 'friendships' may be debated, but in any case my focus from now on will be on what I am calling 'true' friendships, in which each loves the other for his own sake.

In VIII.7–12 Aristotle's main topic is friendship between people who are unequal. He has said earlier that 'perfect' friendship holds only between equals, meaning by this that each is equally virtuous (1156^b7–8), but having in mind too that each is of equal status, for example in sharing the same political rights (1158^b1). His main examples of friendships between people of unequal status are family relationships, e.g. that of father to son, husband to wife, mother to child, and brother to brother. As I understand him, these are supposed to be examples of 'true' friendships, rather than friendships based on pleasure or utility, and yet they too deviate from 'perfect' friendships, just because the participants are not of an equal status. That is why they deserve a special treatment.

The main feature of this special treatment is that, because the status is unequal, the degree of love to be expected of each is in inverse proportion to the status. For example, the father is superior to his sons, because he has given them something more valuable than anything that they could give to him, namely their existence (1161^a15–21); consequently they should love him more than he loves them. In a similar way, since a husband is superior to his wife—in this case, just because men in general are superior to women in general[7]—she should love him more than he loves her. By contrast, the status of two brothers is more or less equal, though with a small priority to the elder, so in this case the love borne should be more or less equal too. No one nowadays is likely to feel much sympathy with these propositions, and I shall not dwell upon them. Nor shall I comment on the comparisons in VIII.9–11 between family relationships and the political relationships that obtain in this or that kind of constitution (e.g. kingship, aristocracy, timocracy, and their 'deviations'). From our point of view, the main interest of this part of the discussion is just that it shows how narrow Aristotle's conception of 'perfect' friendship is. First, it must be a 'true' friendship, in which each loves the other for his own sake; but second, it must also be a friendship between equals; and third, it can obtain only between good people. For to love someone for himself is, as Aristotle sees things, the same as to love him for his goodness, that is to say his virtue.

---

[6] EE VII.2 does pursue it, and by identifying some apparent alternatives (e.g. by identifying what is good unconditionally both with what is pleasant unconditionally and with what both is and appears good to the good man) it manages to end with the same threefold classification as the EN.

[7] Aristotle here gives no reason for this alleged 'superiority' of men to women, but he does have reasons, which, however, are too complex to be given here. (They include his metaphysical claim that form is superior to matter, and his biological claim that in reproduction the father provides the form and the mother provides the matter.)

There is a very general problem lurking here, which has occasioned much philosophical literature: how can one treat as equivalent the two ideas that someone is loved *for himself* (or, of course, herself),[8] and that he is loved *for* some property, or combination of properties, that he may well lose while remaining himself, and may well share with others who are not himself? I shall not enter this general dispute,[9] but I remark only that a person's virtue is not obviously the right property to pick out when trying to explain what it is to love someone *for himself.* Aristotle offers the justification that the properties of being useful to another, or pleasant to him, are less stable than the property of being virtuous (1156^a21–4, $^a31$–$^b4$, $^b10$–12), and this is no doubt relevant.[10] But, philosophically speaking, it seems not to be enough. For he also admits that people can fall off in virtue, in which case he thinks that they cease to be lovable (1165^b13–22), though presumably they do not cease to be themselves.[11]

Perhaps for our purposes it is best simply to forget about the awkward notion of loving someone 'for himself', and to say just that Aristotle recognizes three reasons for loving someone, namely for their goodness (i.e. their virtues), for their utility to oneself, and for the pleasure they give one. This prompts one to ask whether there might be other reasons too, of a like kind or perhaps of a very different kind, and I shall (very briefly) return to this topic at the end. But let us turn now to Aristotle's own argument in IX.4 for saying that only good people can be 'true' friends, each wishing good to the other for his own sake.

---

[8] I trust that it is obvious that this qualification applies to my whole discussion.

[9] Whiting (1991) offers a treatment which is related to Aristotle's text though not confined to it.

[10]
> If thou must love me, let it be for naught
>   Except for love's sake only. Do not say
>   'I love her for her smile . . . her look . . . her way
> Of speaking gently, . . . for a trick of thought
> That falls in well with mine, and certes brought
>   A sense of pleasant ease on such a day'—
>   For these things in themselves, Belovéd, may
> Be changed, or change for thee,—and love, so wrought,
> May be unwrought so. Neither love me for
>   Thine own dear pity's wiping my cheeks dry,—
> A creature might forget to weep, who bore
>   Thy comfort long, and lose thy love thereby!
> But love me for love's sake, that evermore
>   Though mayst love on, through love's eternity.
>
> (E. B. Browning, *Sonnets from the Portuguese*, 14)

The negative message is clear: do not love me *for* this or that, that I may lose. But the positive proposal is utterly opaque: what could it be to love a particular person 'for love's sake only'?

[11] Aristotle has said that being useful or pleasant to someone is a mere 'accident' of a person (1156^a16–19), so in terms of his usual metaphysics one might expect the contrast that being virtuous is part of the 'essence' of a person. But (*pace* Reeve, 1992: 175) Aristotle never says this, no doubt for the very good reason that it evidently is not true that all men are virtuous, or that anyone who is virtuous must always remain so.

3. Friendship and altruism

In a surprising discussion, which begins in IX.4 and continues in IX.8, Aristotle claims that one ought to love oneself, and that love of a friend 'comes from' this love of oneself, since a friend is 'another self'. His argument for this claim is interwoven with the theme that only good men can enjoy the primary kind of friendship.

(a) The argument of IX.4. The claim is announced at the beginning of IX.4: 'Friendly relationships towards others who are close, i.e. those relationships by which friendships are defined, seem to have come from one's relationships to oneself' (1166^{a}1–2). In support he at once proceeds to list some features which people propose as defining friendship, namely (i) that the friend wishes good to his friend for his own sake, and does good to him for his own sake; (ii) that he wishes his friend to live, again for his own sake; (iii) that he spends time with his friend; (iv) makes the same choices as his friend, and (v) shares his joy and his sorrow (1166^{a}2–10). (I note in parenthesis that these proposed criteria would seem to be offered as defining what I am calling 'true' friendship, for the extra criteria that Aristotle requires for 'perfect' friendship—namely that the friendship is between good men of equal status—are not included.) But all these relationships, he observes, also hold between a man and himself—at least, that is what one expects him to say, but in fact he qualifies it: they hold between a *good* man and himself ('and for others insofar as they think themselves good', 1166^{a}10–11), but not—as he goes on to argue—between a bad man and himself. This is then offered as a ground for saying that only good men can be true friends of one another, namely because only they can be true friends of themselves (1166^{b}28–9).[12]

His reason for distinguishing in this way is that the good man is in *harmony* with himself, whereas the bad man is not. More precisely, in the good man the fully rational part of the soul is in harmony with the part that has desires and emotions, so his wishes (*boulēseis*, stemming from the fully rational part) and his desires (*epithumiae*, stemming from the semi-rational part) coincide with one another. But in the bad man his wishes and his desires pull in different directions, so he is perpetually in a state of inner conflict. Moreover, each person *is* (1166^{a}17), or *is most* (^{a}22–3), the rational part of his soul, so when a man acts for its sake he can be said to be acting for his own sake, but when he acts against his rational wishes he is not acting for his own sake.

The rest of the chapter works out these ideas at some length (1166^{a}13–^{b}29), and I pick out only a few salient points. The good man does wish good things for himself, and act accordingly, because he acts for the sake of his rational part. He also wishes

---

[12] Aristotle of course admits that wicked people can be friends of one another for utility or pleasure. Just occasionally he appears to concede that they can also be true friends *for a little while*, but the texts here are brief and inexplicit (1157^{b}7–10, 1172^{a}8–10).

his life to be preserved, for his life is a good for him. He enjoys his own company, since he enjoys his memories and anticipations, neither regretting what he has done nor fearing what he will do. And he shares his own joys and his own sorrows, since he remains of one mind about what is a cause of joy and a cause of sorrow (^{a}13–29). By contrast, the bad man recognizes that his life has been (and will continue to be) hateful, since he frequently fails to do what he wishes, so his life is not a good for him, and he does not wish to continue it. He does not enjoy his own company, because he is ashamed of what he has done, and fearful of what he will do, and for that reason he seeks to forget himself in the company of others. He cannot even be said to share his own joys and sorrows, for the two parts of his soul pull him apart, one part rejoicing in his actions and the other regretting it (either simultaneously, or at any rate in close succession). The bad man is full of regrets (^{b}6–25). So Aristotle concludes that, while the good man does bear to himself those relation-ships which are characteristic of friendship (^{a}29–^{b}2), the bad man does not (^{b}25–9), for he is not satisfied with himself, even though he may think that he is (^{b}3–7). And he infers without further argument that a good man can be a (true) friend, whereas a bad man cannot (^{a}29–33, ^{b}28–9).

Clearly this argument is quite unconvincing. An immediate point to make is that the claim that the wicked suffer from a perpetual inner conflict is just what Plato had urged in his *Republic*, but is not consistent with Aristotle's usual view elsewhere in the *Ethics*.[13] For usually he holds that inner conflict is characteristic of the *inter-mediate* states of self-control (*enkrateia*) or the lack of it (*akrasia*), whereas both the virtuous man *and* the vicious man have an 'inner harmony'. In the case of the vicious man, his goals are wrong, and that is what makes him vicious. But he is usually portrayed as being single-minded in the pursuit of them, not ashamed of his conduct, and unaffected by regret; that is why he, by contrast with the akratic man, cannot be cured (1150^{a}21–2, ^{b}29–31). It has been suggested[14] that the intemper-ate or self-indulgent man, who thinks it right always to gratify his bodily desires, will *afterwards* regret that such-and-such *was* at the time what he most desired. In accordance with his general policy, he pursued it at the time in a single-minded way (by contrast with the akratic, who was at the time internally divided), but later he comes to regret it, perhaps as its harmful consequences become apparent. (Com-pare: 'after a little he is grieved that he was pleased, and he would have wished that such-and-such had not been pleasing to him', 1166^{b}23–4.) But (*a*) Aristotle does not elsewhere suggest that this is a feature of self-indulgence; (*b*) it is only a feature of thoughtless or uncalculating self-indulgence, and there is surely no reason to suppose that all self-indulgent people are uncalculating; (*c*) it is not at all clear how the same thought would apply to other vices (e.g. miserliness, or various forms of injustice). It seems to me very implausible to suppose that one who has the wrong

---

[13] Principally *Republic* VIII–IX, *passim*, but see also I, 351a–352a, and IV, 442d–444e. (This objection is raised by Annas 1977: 553–4. She takes it to be a reason for saying that Aristotle's argument here is borrowed from Plato's *Lysis*, which makes this point at 214c–d, and infers that Aristotle's discussion of friendship is a separate and early composition, not designed to fit his *Ethics*.)

[14] Price (1989: 127–9); Pakaluk (1998: 177); cf. Irwin (1988a, sect. 203).

goals must always afterwards regret that he had them, and usually Aristotle concurs. Here he seems to be supposing that such a person is not *really* satisfied with himself, though he professes to be (1166ª11–13, ᵇ2–7), but I see no good reason for agreeing.

A second objection to the argument is that it makes an illegitimate use of the idea that a person *is*, or *is most*, the rational part of his soul. There is a metaphorical sense in which this identification may perhaps be accepted (and I come to this shortly), but in any case it will not serve the purpose here. First, suppose we take quite literally the thought that a person just *is* his rational part. In that case we cannot accept the idea that when he acts against the wishes of his rational part he is not acting 'for his own sake', because *he* cannot act against the wishes of his rational part. For this would be to say that his rational part (which is what, by hypothesis, he is) acts against itself, and this is clearly not what Aristotle is thinking of. To put this in another way, if the person simply *is* his rational part, then there is no disharmony within *him*, for the disharmony that Aristotle is drawing attention to is not within the rational part, but between that part and another. So we must distinguish the person from the rational part, and the obvious thing to say is that the person contains both a rational and a non-rational part. But this still will not give Aristotle the conclusion that he wants. To illustrate, let us just consider the last suggested condition for (true) friendship: the friend shares his friend's joys and his friend's sorrows (1166ª7–8). Of course it must be true—indeed a mere tautology—to say that any person, no matter how internally divided, has exactly the same joys and sorrows as he himself has. What Aristotle is thinking of is that the one part does not have the same joys and sorrows as the other has. But to turn this into 'he does not share his own joys and sorrows' we must take the one occurrence of 'he' to stand for the one part, and the other for the other (in either order; and it does not in the slightest matter whether 'he' is *more* one part than the other, for he has to be *equally* identified first with one and then with the other). But this is quite clearly illegitimate. The same applies to the other conditions of friendship that we began with; they all hold between an internally divided person and himself, so long as 'he' and 'himself' are held to the same interpretation throughout (either as one of the two parts—whichever you please—or as something that combines them both). So even if we grant, for the sake of argument, that the wicked are internally divided, it still will not follow that they do not bear to themselves the relations which are said to characterize true friendship.[15] But, even if it did, how would this help Aristotle's overall aim in this chapter?

He has opened with the surprising claim that a true friend's love for his friend 'has come from' his love for himself. This appears to mean that the one has developed from the other, but we could equally take it as saying that the one is explained by the other: it is because the good man loves himself that he also loves

---

[15] An exception: perhaps we cannot say of *everyone* that they wish themselves to live, but there is no reason to think that all (or only) the wicked suffer from suicidal tendencies.

others.[16] But admittedly the phrase is not entirely clear, and one would hope for elucidation from the argument that is given for it. Yet it seems that we do not actually get any argument at all, for what happens is that Aristotle first lists a number of features which characterize the relationship of true friendship, and he observes that the good man bears these relations to himself. The bulk of the chapter then argues that it is only the good man who loves himself in this way, and the bad man does not. Suppose that, for the sake of argument, we grant this too. What follows? Aristotle himself apparently infers that only the good man can be a true friend, but clearly this inference *depends upon* the thesis he states at the outset; it does nothing to *prove* it. So a better way of looking at the argument would seem to be to take it as *already* established that only a good man can be a true friend (on the ground, stated earlier, that true friends love one another for their virtue, from which it follows that each must be virtuous). Then we can see this chapter as an inference to the best explanation: it is argued here that only the good man loves himself in the requisite way, so the thought is that if we assume that love of a friend 'comes from' love of oneself, we then have an explanation of why only good men can be true friends. But clearly this is not the only way of connecting these two (alleged) facts in an explanatory relation. (For example, someone else might suggest that, in so far as the good man does 'love himself', this love 'comes from' his antecedent love for others, thus reversing Aristotle's own thought.) So if we look at the chapter in this way then it does provide some argument, though not a particularly strong argument, for the surprising thesis with which it opens.[17]

I suspect that Aristotle has in his mind a further argument which is not stated here, and which depends upon taking very seriously the metaphorical thesis that a friend is 'another self'. (In our chapter, the point is stated only once, almost in parenthesis, at 1166ª31–2. But, as we shall see, it does actually play a large part in Aristotle's thought.) Given this, it may seem very natural to suppose that my attitude to my friend, my 'other self', will be a kind of extension of my attitude to myself (my 'first self', as it were). So naturally I will wish for him the things that I also wish for myself, and so on. Yet the obvious difficulty is, as I stated earlier

[16] The perfect tense (*elēluthenai*) at 1166ª2 suggests development, but the suggestion is not specially strong. However, the tense surely does rule out the suggestion that Aristotle is concerned only with a point about *definition*, i.e. that what is to *count* as friendship is settled by the relations that a person bears to himself. (Such a claim would be very foolish, for there are all kinds of relations which I bear to myself and which have nothing to do with friendship—e.g. being born at the same hour as, or having the same colour eyes as. In any case, Aristotle does not make it. He says that all the relations which define friendship are relations which a (good) person bears to himself, but he does not claim the converse.)

[17] Pakaluk (1998: 163–4) sees that the chapter contains no obvious argument for its opening thesis, and so suggests that the succeeding chapters 5–7 are designed to supply one, by giving an account of how self-love develops into love of 'another self'. But in fact there is nothing in these chapters which would support or explain the opening thesis. (There is just one sentence in ch. 5 which might seem to indicate a concern with development, namely the claim that goodwill, which is not by itself sufficient for friendship, yet 'seems to be the origin of friendship', 1167ª3–4.) Rather, ch. 5 and 6 aim to explain why goodwill and concord, often regarded as marks of friendship, were not included in the conditions of friendship listed in ch. 4; ch. 7 begins the discussion, which continues to the end of the book, of various problems about friendship which arise out of ch. 4.

(p. 167), that there appears to be room here for a conflict of interests. It may be that what I wish for him is actually not compatible with what I wish for myself, or in other words that I cannot both act for his sake and at the same time act for my own, since the situation prevents it. (This situation can arise even if we are both equally virtuous.) So I now turn to the argument of IX.8, where Aristotle offers a different line of thought that seeks to evade this problem.

(b) The argument of IX.8. IX.8 opens with the question 'Should one love oneself more than others, or vice versa?', and it proceeds to give preliminary arguments on each side. On the one hand the term 'self-lover' is commonly used as a term of reproach (corresponding to our word 'selfish'), and the good man is distinguished from the self-lover, on the ground that he disregards his own interests and acts rather for the sake of what is noble, and for the sake of his friends. But on the other hand there is the reasoning of chapter 4, which claims that the good man's love for others comes from his love for himself. Moreover, it is agreed that one ought to love most those who are closest to one, and the person most close to one is oneself ($1168^a28–^b12$).

To resolve the dispute we must, says Aristotle, distinguish the different meanings of 'self-love'. As commonly used, the word is applied to those who assign to themselves more than their share of money, or honours, or bodily pleasures, since these are the things that most people desire, for they aim to gratify the non-rational part of the soul ($1168^b12–25$). So in common usage the word is not applied to those who seek always to do what is just, or temperate, or in general virtuous, i.e. those who are concerned always to secure for themselves what is noble ($^b25–8$). Yet the truth is that such people are more properly termed 'self-lovers', for they assign to themselves what is most noble and most good, and thereby aim to gratify the most controlling part of themselves (*kuriōtaton*), i.e. the rational part, and each person *is* this part, or is so *most* ($1168^b28–1169^a6$).

To support this last claim Aristotle briefly mentions the phenomena of *enkrateia* and *akrasia* ($1168^b34–5$), though he does not spell out the argument that he has in mind, and at first sight there is no argument to be found here. For *akrasia* occurs when the rational part—the part that ought to be in control—in fact fails to control some desire or emotion stemming from the non-rational part. In English we call this 'lack of self-control', implying apparently that the person fails to control *himself*, thereby identifying the person both with his rational part (which fails to control) and with his irrational part (which fails to be controlled). But (as noted earlier, p. 123) the Greek words simply signify being or not being in control, and do not explicitly specify what one is or is not in control of, so we can if we wish supply 'one's desires and emotions'. In any case, I think it is fair to say that, in a case of *akrasia*, I do tend to identify myself *more* with my rational part. For example, I may say '*I* didn't really want to do that; it was my craving (my anger, my jealousy) that *made* me do it'. Here one is, as it were, distancing oneself from one's own desires and emotions. In a reflective mood, I may recognize that I do have desires and emotions, or indeed just habitual responses, which I wish I did not have. For example, I may wish that I did not desire to smoke, or I may wish that I did not feel so obliged to keep my promise to finish writing this book by such-and-such a date—for, after all, reason tells me that no harm will actually be done if I am late. (In the first case it is a 'bodily' desire that I wish I did not have, in the second it is a feeling instilled by habituation and training.) In all such cases it is quite natural to identify *oneself* with the 'part' that wishes the desire or feeling were not there, rather than the 'part' that has this desire or feeling. So we might not unreasonably generalize and say: we identify ourselves with the part that ought to be in control, most noticeably when in fact it is not in control, but presumably also when it is. To this extent, then, one can sympathize with Aristotle's claim that I am *more* to be identified with my rational part than my irrational part, though in either case the identification is of course figurative and not literal, for the fact is that both parts are equally parts *of me*.

To come back to the main argument, what we have here is a noticeable variation on the theme of chapter 4. For that chapter had argued that the good man does love himself, because he is in harmony with himself (i.e. the different parts of his soul are in harmony with one another). But this chapter argues rather that the good man should love himself, for that is to love his rational part, and so, since his rational part aspires to what is noble, it is in effect to aim for, and secure for oneself, what is noble. And this is to be praised, not to be reproached (1169ª6–8). (I note, incidentally, that chapter 8 thus escapes the objections raised to chapter 4. For we may now admit that bad men too may be in harmony with themselves, but still say that they should not love themselves, for in them the rational part does not aspire to what is noble.) As for why it is praiseworthy to aim to do what is noble, Aristotle here gives *two* reasons, (i) that other people benefit thereby, and (ii) that the agent also benefits, since he thereby secures for himself the greatest of all goods, namely virtue (1169ª8–15). Later, when arguing for a different point in X.7, he will stress the first of these reasons rather than the second; but here he is bound to stress the second, for that is why he thinks that in acting nobly one is acting for one's own sake, and thus manifesting self-love.

This then brings him face to face with the point that noble actions may apparently require self-sacrifice, which hardly seems to manifest self-love. I quote his response at length.

It is true that the good man will do many things for the sake of his friends, and for his country, and will die for them if necessary. He will sacrifice money, and honours, and in general the contested goods, in order to secure for himself what is noble. For he would rather choose intense pleasure for a brief while than mild pleasure for a long time, and choose to have lived nobly for one year rather than for many in whatever way it may turn out, and choose one great and noble action over many small ones. This is presumably how it is with those who die for others; they choose for themselves something greatly noble. In addition he will sacrifice wealth in order that his friends may benefit, for in this way the friend has the money but he has what is noble, so he assigns to himself the greater good. The same applies to honours and offices, for he will sacrifice all these for his friend, since this brings praise and nobility to him. So it is right and proper that he should be accounted good, as he chooses for himself what is noble, above all else. It is also possible that he should give up to his friend the opportunity for actions, since it may be more noble to be the cause of his friend's acting than to act himself. So in all the things that men are praised for, the good man clearly assigns to himself the greater share of what is noble.

It is in this way, then, that one should love oneself, as has been said, but not in the way that most people do. (1169ª18–ᵇ2)

Here Aristotle very clearly says that altruism is a kind of self-seeking, for it is seeking to secure *for oneself* what is of most value, i.e. that which is 'noble'. There is of course a different way of looking at the situation which claims that what one *seeks* is the other's good, for this is the primary aim, but it so happens that as a (foreseeable) result one does secure nobility for oneself, as a kind of side-effect. This extra was no doubt foreseen, but still not what one was aiming for, not what one intended to achieve, and the action would still have been worth doing even without this extra. I shall exploit this line of thought in a moment, but it certainly does not seem to be the line that Aristotle is taking here. He speaks of acting *in order to* secure for oneself what is noble (1169ª21), of *choosing* for oneself what is greatly noble (ª26, ª32), of *assigning* to oneself the greater good (ª28–9, ª35–ᵇ1), and so on. His account is that securing nobility for oneself is the primary aim, and—at least at first sight—it appears that we must take him in this way if we are to see how the account explains that the action 'comes from' self-love.

I remark that it is not unreasonable to take this passage as a discussion of altruism in general, i.e. of acting in ways that benefit *others*, whether they are friends or not. As one expects from the overall context, Aristotle is for the most part concentrating on acting for the benefit of a friend, but he does himself widen the discussion at one point, namely where he notes that one may lay down one's life 'for the sake of one's friends, *and for one's country*' (1169ª19–20). Clearly, I cannot count all my countrymen as (true) friends of mine, as Aristotle himself would agree,[18] yet if I die nobly in battle there need be no one countryman in particular whom I am trying to benefit. So I think we may reasonably say that the account covers all noble actions which (as *we* are inclined to say) benefit others and not the agent, whether or not those others are friends. But I add that the account does not cover the whole range of altruistic actions, for many of our very ordinary everyday actions are in fact altruistic, but in a very trivial way, and surely do not count as 'noble'. For example, if in a crush I deliberately avoid stepping on someone else's toes, then (usually) I do so for his sake and not for mine, but it would be absurd to call this 'noble'. Or, to give an example which does concern friendship, if my friend happens to be short of a bus fare, and I have enough to pay for both of us, then of course I will pay his fare, and neither of us will expect this to be repaid. But again there is nothing 'noble' about this. We can only apply the account, then, to cases where some quite large sacrifice is involved, and the concept of 'nobility' is not entirely out of place. For the sake of argument, let us grant this restriction. What then should be said of Aristotle's suggestion?

First, Aristotle nowhere explains why the 'nobility' he speaks of is of such value

---

[18] Aristotle allows a somewhat extended sense in which the whole community is (or should be) a community of friends (VIII.9), but he also insists that one can enjoy only a few 'perfect' friendships (IX.10).

to the good man, or how it contributes to his own *eudaimonia*. (That it contributes to the *eudaimonia* of others is plain enough, and is something he will emphasize later in X.7, but that is not what is supposed to be relevant here.) We must, then, take it as a kind of axiom that one who counts as 'good' just does place more value on the nobility of his own actions than on anything else. But now we may contrast with this 'good' man someone else, who places more value on the *eudaimonia* of his friends (or perhaps, of mankind in general) than he does on the alleged 'nobility' of his own actions. Each may act in exactly the same way, but apparently their motives are different, since one is seeking nobility for himself, and the other the *eudaimonia* of others. Then the axiom states that only the first is 'good' (and hence 'virtuous'), but could Aristotle really accept this contrast? And, if he did, could he reasonably expect us to agree that the first deserves to be called a 'better' man than the second? Surely, the answer to this must be 'no', and he must try to eliminate the contrast. In his own terms, he must try to show that each is equally motivated by 'self-love'— but, of course, 'self-love' of the best sort. We are familiar with the kind of argument that he might use, roughly on these lines.

Our rival candidate is described as one who does not care much, if at all, about the 'nobility' of his own actions, but does care for the welfare of others. Nevertheless it is *he* who cares. In Aristotle's language, it is his rational part that sets these values, and in acting on them he is acting 'for the sake of' his rational part. But he *is* (or *is most*) his rational part, so this is to say that he is in fact acting for his own sake. In a word, it is self-love, i.e. love for one's rational part, that motivates love for others. For this line of argument to work, it does not really matter whether or not we bring in the concept of 'nobility'; what is important is just the thought that if *I* want something then it is my love for *myself* that motivates me to achieve it. *All* rational actions therefore result from 'self-love', however 'altruistic' they may seem to be.

This is a well-known argument for psychological egoism, and I must leave its assessment to you. But I note that, even if the argument is granted, there is still a further step that Aristotle's overall position requires, namely the step from egoism to eudaimonism. For even though (let us suppose) I always act from love of myself, it certainly does not at once follow that I always act for my own *eudaimonia*. Aristotle himself attempts to bridge that gap by means of the concept of nobility, but as we have seen this is a mere red herring. The point that it draws attention to is just that we all set a high value on actions which significantly benefit others to the detriment of the agent's own interests, for these are the actions which Aristotle calls 'noble'.

4. Why one who is *eudaimōn* needs friends (IX.9)

Aristotle opens IX.9 by telling us that it is disputed whether someone who is *eudaimōn* will need friends, and he at once sketches an argument for a negative answer. Someone who is *eudaimōn* is supposed to be 'self-sufficient', which means that he needs nothing further, and yet the role of a friend, who is other than oneself,[19] is to provide for one what one cannot provide oneself (1169^{b}3–8).[20] He nowhere gives a direct response to this argument, but we may easily supply one for him, namely that the argument misuses the notion of self-sufficiency. For if friends genuinely are needed, then one who lacks them simply is not 'self-sufficient'.[21] (As Aristotle has said in I.7, 'we ascribe self-sufficiency not to one who is alone, living a solitary life, but to one who has parents, children, a wife, and in general friends and fellow citizens, since man is by nature a political animal', 1097^{b}8–11.)

Instead, he at once proceeds to some straightforward arguments on the other side. (i) Friends are thought to be the greatest of external goods, so it would be strange not to include them in the *eudaimōn* life; (ii) it is characteristic of the good man to do good to others, and better to do good to friends than to strangers, so the good man must have friends; (iii) it would be absurd to suppose that the *eudaimōn* life is solitary, since no one would choose such a life. For man is naturally a political animal, and the same still applies to one who is *eudaimōn*. He therefore has all the goods that are natural to man, and accordingly he will spend his time with friends who are good, rather than with strangers or chance acquaintances (1169^{b}8–22). Aristotle adds a diagnosis of why the opposing argument has appealed to some, namely that they mistakenly take *all* friendships to be friendships based on utility or pleasure. He concedes that one who is *eudaimōn* will have no need of the former, and only a minimal need of the latter, for he already has what is useful for him, and his own life is already a pleasure to him. So there is after all some truth in their view, but it does not cover all cases, for it fails to consider the central case of 'perfect' friendships (1169^{b}22–8).

There, one might have thought, the matter could rest. But Aristotle is not prepared to leave it at that, perhaps because—as he has said initially—the question was in his day a matter of genuine dispute. So he proceeds to three further arguments aiming to show that friendship is a necessary part of the good life. These become progressively more abstract, and more difficult to interpret.

First, he invokes the premiss that *eudaimonia* is an activity (I.7, 1098^{a}5–18), and that the good man's activity is both virtuous and pleasant in itself (I.8, 1099^{a}7–21). Next, he repeats his point that the fact that something is one's own makes it pleasant (VIII.12, 1161^{b}18–29), and adds the new observation that 'we can contemplate our neighbours and their actions better than ourselves and our own actions'. From these premisses he infers without more ado that contemplating the actions of a friend who is good is a pleasure to the good man (1169^{b}28–1170^{a}4). The thought appears to be (*a*) that the good man finds pleasure in contemplating his own activities, since they are both 'pleasant in themselves' and 'his own'; so (*b*) he will also find pleasure in contemplating his friend's activities since they too are both

---

[19] I read *autou* rather than *auton* in ^{b}7. (Reeve 1992: 178, noticing the (mild) awkwardness that *auton* introduces, is led to give a wholly impossible translation of this sentence.)

[20] A similar passage at *EE* 1244^{b}1–21 strengthens the argument by adding that God is truly self-sufficient, and really has no need of friends. But *EN* does not mention God in this context.

[21] To respond in a similar way to the strengthened version of *EE* (previous note), one might naturally say that man is not God, and should not try to resemble him in this way, for—unlike man—God is *not* a political animal. But this response would make difficulty for Aristotle's later argument in X.7–8.

'pleasant in themselves' and—in an extended sense—'his own'; moreover (c) he will find *more* pleasure in the latter, since one can contemplate another's actions *better* than one's own. Here it is difficult to keep claims (b) and (c) in harmony with one another. For (b) apparently relies on the thought that my friend *is* me (he is my other *self*), since that explains why I think of his actions as 'my own'; while (c) apparently relies on the thought that he *is not* me (he is my *other* self), since that allegedly explains why I can contemplate his actions 'better' than my own.

Now (b) by itself can in a way be defended. For it is true that if my close friend (or my wife, or my son, etc.) does something which I very much admire, then that will give me a special pleasure, and more pleasure than I would feel if it had been someone else entirely, quite unconnected with me, who had done that very same thing. I am in a way proud of their action, though it was their action and not mine, and though I would not wish to suggest that I myself influenced that action in any way.[22] This need not be because what they have done is what I would have done myself. (Indeed, it is probably fair to say that we mostly admire actions which we think we would not have done ourselves.) Nor does it follow that if I had done it myself I would have been proud of it, for perhaps I would have just shrugged and said 'it was the obvious thing to do in the circumstances, and nothing to be proud of'. Perhaps *here*, then, we can bring in Aristotle's third thought. He does not explain why he says that I can judge the actions of others better than my own, and it is natural to think that his point is that one can easily be blind to one's own faults, and so judge too leniently in one's own case.[23] But my suggestion is that what he actually has in mind is the opposite: one can easily be blind to one's own merits too, and think of one's action as nothing very special, when in fact it deserves to be admired. That, at any rate, seems to be the best way to make sense of this extremely cryptic argument. Nevertheless, the tension I noted initially still remains. To change the metaphor: it is those who are especially *close* to me whose actions I can see as, in a way, 'mine'; but it is those more *distant* whose merits (and demerits) I can see more clearly; and my friend must be both close and distant if the argument is to work.

On this reconstruction of the argument it aims only to show that the good man gets more pleasure from his friend's good actions than he gets from his own; it is not suggested that he also and thereby obtains a better view of his own actions.[24] But it is quite tempting to suppose that Aristotle wishes to add this step, since one rather obscure sentence of the corresponding discussion in the *Eudemian Ethics* does perhaps suggest it (1245ᵃ35–7), and the thought is very clearly developed in the *Magna Moralia* (1213ᵃ10–26), which *may* reflect Aristotle's own thinking. The author of that work explicitly speaks of one's friend as a 'mirror', in which one can see oneself and find out about oneself, and it claims that such a 'mirror' is essential to

---

[22] Compare *Rhetoric* 1385ᵃ1–3, which notes (on the other side) that we feel shame for what our ancestors did, or others closely related.

[23] The *Magna Moralia* version of this (?) argument, which I discuss in a moment, explicitly gives this explanation (1213ᵃ16–20).

[24] I note that Kraut (1989: 142–4) agrees with me that this is the intended conclusion of the argument, though he gives a rather different elucidation of the premises.

self-knowledge. So the sequence of thought suggested is: 'I see that what my friend has done is admirable; but I now note that I myself have done similar things, and in his situation I would have done the same thing; so I may conclude that my actions are admirable too'. The obvious problem with this argument is to see why it matters that the admired action was done by a friend, and Cooper (1980: 320–4) suggests a solution. As he sees it, the awkward point is to explain why it is any easier to be confident that one would have done the same thing oneself than it is to see the merits of one's own actions directly. And here the suggestion is that it is important that I know that it is my friend who did it, and know that my friend is 'another self', and therefore someone who closely resembles me. As to how it is that I know that this other person is 'another me', it may be suggested that the *next* argument supplies the answer: a friend is someone one knows well, because one has been engaged with him on many co-operative activities. But I protest that this seems to be an extraordinarily roundabout way of coming to know that I would have done the same thing myself, and surely it is not essential. So, if Aristotle did intend this further step to his argument (which uses in a quite different way the thought that the admired action was done by a friend), then one can only comment that it is not particularly convincing.

Aristotle's next argument is too cryptic for any serious elucidation to be possible. It is that the good man's activity will be 'more continuous' (and hence better?) if it is pursued 'with others and towards others' (1170^a4–13). If the activities in question are what we would call 'good actions', then many of them, of course, are only possible 'towards others', but it would not seem that these others have to be friends.[25] As for 'with others', a natural suggestion might be that Aristotle is thinking of co-operative endeavours, where you and I act jointly to secure a result that neither of us could have achieved individually. That may be right, but there is no special reason to think so, for the points that he makes himself are just that activities pursued with others are easier ($^a5$–6) and more pleasant ($^a4$, $^a7$), and that living together with good people provides a kind of training in virtue ($^a11$–13). The last appears to be strictly irrelevant, since we are asking why one who has already been trained in virtue still needs friends, and it is simply not explained why the first two should make virtuous activity 'more continuous'. *Perhaps* his thought is that if I am constantly in the company of others, who will notice any backsliding on my part, then my backsliding is less likely, so good deeds will occupy me for more of my time. Or *perhaps* he is thinking that if several of us are in the habit of discussing our good deeds over dinner, this will act as a stimulus to greater effort. But there are all kinds of things that he *might* be thinking, and speculation here seems to me profitless.[26]

Finally he gives a long and involved argument (1170^a13–$^b19$), which is introduced as one that is 'more in keeping with the study of nature' (*phusikōteron*, $^a13$). He means, it would seem, that he wishes to go back to very basic principles about the

---

[25] Aristotle thinks that it is *better* to benefit a friend than a stranger (1169^b12), and this may be what lies behind his thought here. But it is not clear how it would support his claim for greater 'continuity'.

[26] At X.7, 1177^a21–2 it is not what we call 'good actions' but rather sitting and thinking in one's study that is said to be the 'most continuous' of human activities.

nature of life itself, for at any rate that is how the argument starts. If I may summarize it somewhat baldly, it goes like this.[27]

For living things, to exist is to be living. Now for animals in general to live is to be able to perceive, but for human beings in particular it is to be able both to perceive and to think.[28] But when we perceive or think, in either case we are aware that we do so. So we are aware that we live (^{a}16–19, ^{a}29–^{b}1). Next, living is (for us) something that is good, and pleasant 'in itself' or 'by nature', which is shown by the fact that we all desire it. So being aware that we live is being aware that we possess a good, and this is by nature pleasant, and hence a part of *eudaimonia* (^{a}14–16, ^{a}19–29, ^{b}1–5). But the friend is another self, so being aware that he lives is also a pleasure in the same way, or very nearly (^{b}5–10). And this awareness is achieved by living together with him, i.e. by sharing in discussion and thought, since that is what 'living together' means in the case of a human being (^{b}10–14). From all this Aristotle concludes that the one who is *eudaimōn* must have friends. In outline his thought is that having a friend affords pleasure, namely the pleasure of being conscious of his existence. This is a pleasure because being conscious of one's own existence is a pleasure, and a friend is 'another self'. Moreover, it is something that is 'by nature' a pleasure, and hence a pleasure to the good man, and hence a pleasure that should be included in the fully *eudaimōn* life (^{b}14–19). (Aristotle's own exposition frequently insists that it is the good man he is speaking of, for in his case his life is specially a pleasure to him, and the life of his friend is similar (cf. IX.4). But this would not appear to be important to the argument, which needs to invoke the notion of goodness only at its second step, in order to explain why something that is a pleasure 'by nature' must be included in *eudaimonia*.)

This argument is surprising in several ways. First, it begins by claiming that I have a special and immediate way of knowing that I myself am living, because I am automatically conscious of my own thoughts and perceptions. But we must set this aside as irrelevant to the argument that follows, since I do not know in the same special and immediate way of anyone else's thoughts and perceptions. The point then can only be that I do know (by whatever means) of my own thoughts and perceptions, i.e. of my existence, and this is a pleasure to me. So apparently if I also know of anyone else's existence, that too should be a pleasure to me, at least so long as that other person counts as 'another self'. But how is this to be judged? A suggestion that one might take to be implicit in the argument is that someone else is 'another self' if I know (by whatever means) what their thoughts and perceptions are, and then the idea is that in practice I will know this only by 'living with' them, i.e., as Aristotle says, by sharing with them discussion and thoughts. But this suggestion clearly will not do. For some of those with whom I live, and whose thoughts and perceptions I therefore know about, may well be people whom I regard as my

---

[27] Ross, in his translation, quotes from Burnet (1900) an analysis involving twelve distinct syllogisms. Cooper (1980: 318) reduces it to four propositions. The account that I give here is intermediate.

[28] It is the doctrine of the *De Anima* that what distinguishes animals from vegetables is the ability to perceive, and that what distinguishes humans from other animals is the ability to think. But we may add further activities as distinctive of human life—e.g. choosing and acting—without upsetting the argument, which is simply that humans are conscious of those activities which constitute their living.

enemies rather than my friends. So their existence is not in any way pleasant to me, even though I am well aware of what they think. If the argument is to be successful, then, it must build more than this into the notion of 'another self'. But what else can we add, other than the very point at issue, that 'another self' must also be one whose existence gives me pleasure? This, of course, would at once introduce a vicious circularity, and the argument would collapse. One can only conclude, I think, that Aristotle is once again relying too easily on the metaphor that a friend is 'another self'. Without any real justification, he is apt to build into this metaphor whatever happens to be needed to yield the conclusions about friendship that he aims for.

5. Conclusion

Aristotle's long discussion of friendship in the *Ethics* receives rather little attention in modern moral philosophy, and I would say that this neglect is deserved. I do not mean to imply that the topic of friendship is itself unimportant for moral philosophy. On the contrary, if one is at all inclined to think, with Hume, that morality depends strongly on a 'sympathy' that people very naturally have for other people, then there is evidently something to be said for starting with the case where this 'sympathy' seems to be at its strongest, i.e. (as we say) with 'our nearest and dearest', and then going on to ask how it may be extended more widely. But clearly this is not Aristotle's focus of interest.

He is no doubt right to say that there are different forms of friendship, but his own choice of the central case seems to me to be perverse. It is perhaps fair to focus first on what I have called 'true' friendship, whose salient characteristic is that each wishes (and does) good to the other for the other's sake. But it was surely a mistake to suppose that this form of friendship can exist only between good people of equal status. We all know from our own experience of close friends who are not especially good people. It is true that this might be accommodated to Aristotle's position in the way that Cooper suggests (1980: 306–8), for on his account people of only few or moderate virtues may still be 'virtue-friends' of one another, each loving the other for such virtues as he does possess, meagre though they may be. But is this really the only ground for 'true' friendship? There are, after all, partners in crime, and they may surely love one another because each admires the criminal tendencies in the other. Or, to make a point *ad hominem*, Aristotle himself suggests one reason for true friendship which has nothing to do with virtue. For he observes that mothers love their children, even if the love is not reciprocated, and says in explanation that this is a special case of the general proposition that all people love what they themselves have produced (1159^a28–33, 1161^b16–31). This explanation does not rely either on the virtues of the mother or on the virtues of the child, which is surely as it should be. And it seems to me that true friendships may arise for many other reasons too, which are not specially connected with goodness or virtue.

I hesitate to offer on my own behalf a rival taxonomy of friendships,[29] but here is

---

[29] For an interesting attempt in this direction, see Telfer (1970–1).

at least a possible starting-point. In *many* cases friends enjoy one another's company, and this seems to me a central idea, but it may come about for all sorts of reasons. One may love someone for their beauty, or for their wit, or for their wisdom, but for all kinds of other reasons too. (Cf. 'for thine own dear pity's wiping my cheeks dry', note 10.) A mutual love may come about because two people share the same goals, and co-operate in achieving them, and for this point it does not matter whether those goals are good or bad (e.g. two who share the aim of using bloodshed to create anarchy). But it may equally come about because their goals are very different (e.g. a theist and an atheist who enjoy disputing with one another). There is no limit that I can see to the possible reasons for which one might enjoy another's company, nor any useful generalization about which of these will lead one to 'wish good things to the other, for his or her own sake'. But some of the associations which Aristotle reckons as 'friendships' need not depend in any way upon enjoying the other's company. He says himself that friendships for utility need not (1156^a27–8), and we might add that family ties equally need not. (For example, I may hate my brother, but still acknowledge the tie between us.) Whether such associations are best regarded as species of the same genus 'friendship' is open to debate, but I shall not pursue the issue further. I merely remark that, *whatever* are taken to be the central cases of friendship, one may no doubt use the metaphor that a friend is 'another self', but it is only a metaphor, and no conclusions can be drawn from it that could not equally have been drawn without it.

I have suggested that Aristotle is influenced by this metaphor when he claims in IX.4 that love for another 'comes from' love for oneself. That may be wrong, for admittedly the metaphor plays no role in the argument that he actually gives. But there are so many objections to that argument that one does naturally look for something else to bolster it. In any case, with or without the metaphor, the main claim carries no conviction at all, for each of us can say, looking into our own experience, that self-love had nothing to do with any reasons that we may have had for coming to love another. In IX.8, where the problem is to explain how one can act for the loved one's sake, and not one's own, it would be quite natural to expect a further appeal to the idea that a loved one is 'another self', but in fact that is not what we find. Instead Aristotle argues that I am 'really' acting for my own sake after all, since I thereby secure for myself the greater good, namely nobility. But (*a*) this could only cover a small fraction of cases, for many things that I do for the one I love could hardly qualify as 'noble', and anyway (*b*) it is a quite unconvincing explanation even in those cases that it might seem to cover. For, again, we can say from our own experience that 'nobility' is not part of our motivation; what moves me is love for the other, not for anything that might accrue to me. So the only way to defend Aristotle is to fall back on a very general argument for saying that, *whatever* I do, I must be doing it 'for my own sake', and I do not think that anyone should be convinced by this argument.

I add here that *we* see a particular interest in this chapter, because it seems to be the one place where Aristotle faces up to what we see as a crucial problem for him, namely how to reconcile altruism with his opening claim that each man acts always for his *own* good (his own *eudaimonia*). But I think we can be sure that Aristotle himself did not see it in this light, (*a*) because he notices the problem only in this context of friendship, whereas it is actually a much more general problem, and (*b*) because he notices it only as one of a series of puzzles or problems that arise out of what he has said in chapter 4, and he gives it no special pre-eminence. (I might add (*c*) that earlier, in the *Eudemian Ethics*, he had seen no need to discuss the problem at all, for *EN* IX.8 is one of the very few discussions of *EN* VIII–IX that has no counterpart in *EE* VII.[30]) Elsewhere he seems quite content to say, without further explanation, that to be *eudaimōn* one must be virtuous, and that virtue aims at what is noble, and then to cite as noble actions those that benefit others. Another line of thought that leads to a similar conclusion is that all the virtues of character are summed up in 'justice', taken in its general sense, and that this is a matter of doing what the law commands, and that the law aims to promote the common good (V.1–2).[31] In a similar vein Aristotle first tells us that the man of practical wisdom is the man who deliberates well about how to secure his own *eudaimonia*, but then adds that this is really no different from good deliberation about the *eudaimonia* of one's household, or of the community at large (VI.8). In all of this there is simply no recognition of the thought that what is good for me may conflict with what is good for my community, and I think it is fair to say that in general Aristotle simply does not *notice* that there is a potential problem here. It is perhaps something of an accident that he does notice it in IX.8, in the special context of friendship, but in any case the treatment that he there gives it is not one that we can endorse.

As for the claim of IX.9, that the good life must include friendships, I imagine that most people nowadays would be inclined to agree. Many of the more straight-forward points that Aristotle makes in that chapter (and elsewhere) do point to how, and why, we value friendship. (But the final and most abstract argument of the chapter, on which he presumably put most weight, is a complete failure; it seriously does rely, in a quite illegitimate way, on the metaphor that a friend is 'another self'.) What Aristotle is trying to prove, however, is the stronger claim that a life without friendships *cannot* be 'a good life', and when the arguments are considered in this light they do not seem to be strong enough. For example, suppose that we imagine ourselves back in the days when hermits were more common, and we consider one who says, 'it may be true that man is *born* a "political animal", but if so then that is a feature of man's nature—like several others—which he should learn to tran-scend'. Would you think that such a man ought to be convinced by what Aristotle says here? If so, you surely will *not* be convinced by what he himself will say later, in X.7, about the desirability of 'becoming immortal, so far as we can'.

---

[30] Pakaluk (1998: 229) gives a brief but useful table of the correspondences between *EN* and *EE* on friendship.

[31] This line of thought plays a central role in Irwin's discussion of the *Ethics* (and *Politics*) in chs. 16–21 of his 1988*a*. (For a succinct statement, see e.g. p. 439.)

Further reading

There is a full-length commentary on all of books VIII–IX in Pakaluk (1998), but most readers will probably want something shorter. So I mention here that Hardie (1968, ch. 15) gives a quick summary of almost all of these two books, though it is mainly without comment until he comes to IX.4, 8, 9. Concerning particular topics:

On the varieties of friendship it is useful to study the debate between Cooper (1980, pt. I) and Price (1989, ch. 5). (Cooper's views are not fully given here, and his 1977 might be consulted for greater detail; but the 1980 version is for most purposes more straightforward.)

Much has been written on whether Aristotle is committed to an unacceptable form of egoism, and discussion of IX.4 and IX.8 plays a large part in this. I suggest beginning with the debate between Annas (1988), and Kraut (1989, ch. 2). (Kraut offers a wide-ranging discussion, which uses parts of the *Politics* as well as the *Ethics*. But note that he sees no prima-facie commitment to egoism in the claim of I.7 that all men seek *eudaimonia*; on his account this means only that we all seek *someone's eudaimonia*, not necessarily our own, p. 145.) Annas offers a further response in her (1993, sect. 12.1), which modifies some previous claims, but it may be more helpful to consider a different view altogether, and I would recommend Madigan (1991). (The more adventurous may wish to try Kahn (1981), who attempts to find a genuine content in Aristotle's 'other self' metaphor by relating it to the notoriously obscure discussion of 'the active intellect' in *De Anima* III.5.) Pakaluk's commentary (1998), of course, pays attention to all the details of Aristotle's discussion. By way of a general orientation on what is to count as 'egoism', it may be helpful to consult Williams (1973), which is not addressed to Aristotle in particular.

The argument of IX.9 is discussed in an interesting way by Cooper (1980, pt. II). Some disagreements may be found in Price (1989, ch. 4) and in Kraut (1989, sects. 2.15–2.17).

Chapter IX

The good for man: second discussion (X.6–8)

1. Recapitulation

In his first discussion of the good life in book I, Aristotle says (in chapter 5) that by tradition there are three contenders for this title, namely the life of pleasure, the life of politics, and the life of contemplation (1095^b14–19). He rather quickly dismisses the life of pleasure as fit only for animals (1095^b19–22), but gives more of a discussion of the political life. He first suggests that its goal is honour, but then that it might be better to say that the goal is virtue, and it is clear that what he is thinking of here is the virtues which he calls virtues of character (1095^b22–31). We should evidently add to these the intellectual virtue of practical wisdom, which is closely connected with virtues of character, and which in VI.8 has been in a way identified with political knowledge (1141^b23–1142^a10). I shall call these collectively 'the practical virtues'.[1] They are what Aristotle thinks of as characteristic of 'the political life', both in book I and—as we shall see—in book X. It is true that in I.5 he concludes his discussion by saying that it cannot really be virtue that is the aim of the good life, and apparently dismissing the political life on this ground (1095^b31–1096^a2; cf. 1096^a7–10). But this is not a serious objection, for the response is that it is not virtue as such which is aimed for, but rather the activities in which it is expressed (1098^b30–1099^a7), and we find in book X that the political life is still viewed as a contender. (So too is the life of pleasure.) Finally there is the life of contemplation (*theōria*)[2] which is evidently associated with the one further virtue that Aristotle recognizes, namely the intellectual virtue of theoretical wisdom (*sophia*). In book I this is not discussed, but postponed for later treatment (1096^a4–5). To sum up: the tradition claims that there are three lives to consider, namely the life of pleasure (which involves no virtue), the life of politics (which involves all the many practical virtues), and the life of contemplation (which involves just one purely intellectual virtue). As is very clear in book X, Aristotle *accepts* this tradition. He does not seriously consider any other candidate for the title 'the good life' or 'the life of *eudaimonia*'.

As for the rest of book I, Aristotle proceeds in chapter 7 to give his own account of *eudaimonia*, but this is couched in general and abstract terms, and is intended (I believe) to be compatible both with the claims of the life of politics and with the claims of the life of contemplation. Seeking for a life that is specially distinctive of

---

[1] Aristotle himself uses this title once (1177^b6).

[2] Irwin translates 'study'. I discuss the translation below.

man, Aristotle says that it must be an activity of the rational part of the soul, but at the same time he indicates that this notion may be broadly construed, so as to include not only the part that has reason in itself, but also the part (containing desires and emotions) which can listen to reason (1098ᵃ3–5). So when he adds that the activity must be pursued 'with virtue' (i.e. 'with excellence'), we may understand this as including all the practical virtues, as well as the one purely intellectual virtue. Indeed, the main lines of Aristotle's argument at this point seem to require us to say that *all* the specifically human virtues should be included, and this *may* be what he means when he says 'and if there are many virtues, then in accordance with the best and most complete of them' (1098ᵃ17–18). For, in the context, it is not unreasonable to suppose that 'the most complete' virtue is to be understood as the virtue which includes all virtues (as 'justice', in the universal sense, includes all virtues of character). But evidently the phrase does not *have* to be understood in that way, for we may also understand it as meaning the virtue which is 'most final', or 'most end-like', and as intended to pick out just one virtue of all the many, in which case the virtue in question is evidently theoretical wisdom. As I suggested in Chapter I (p. 25), it may be that Aristotle means to play quite deliberately on this ambiguity, since at this stage he wishes to leave open the dispute between the life of politics and the life of contemplation.

There are other signs that this dispute is being left open in book I. In I.8 he says of 'the best activities' that

We say that *eudaimonia* is these activities, *or* the one of them that is the best. (1099ᵃ29–31)

There is a similar passage in each of the two discussions of pleasure. In VII.13 he says

If there are unimpeded activities of each disposition, then whether it is the activity of all³ of them *or* of some one of them that is *eudaimonia*, in either case . . . (1153ᵇ9–11)

And in X.5 he says

So, whether the activities of the completely blessed man are one *or* many, it is the pleasures that complete these that should properly be said to be human pleasures. (1176ᵃ26–8)

In each passage it is quite clearly being left as an open question whether the activity that is *eudaimonia* should be taken to be some one particular activity or a number of different ones. But it may be noted that, at least in the first two of these passages, the relevant opposition seems to be between *all* of the (best) activities or just one of them. We do not find here what at least *seems* to be the opposition in book X, between the many practical activities on the one hand (excluding contemplation) and the one purely theoretical activity (which is contemplation) on the other. But before I come to consider this, I insert here three quick comments on what we have had so far.

First, Aristotle gives us no justification besides 'tradition' for considering only the three lives that he does consider.⁴ These are thought of as the lives suitable for

---

³ 'All' should presumably be understood as meaning 'all that are virtuous', or something similar.
⁴ At 1096ᵇ5–7 he does briefly remark on 'the life of money-making' that money cannot *itself* be taken to be an ultimate goal.

an Athenian gentleman of leisure: he may cultivate things intellectual, or he can go into politics (for that is an honourable pursuit), or he can fritter away his time on what Aristotle clearly regards as the 'lower' pleasures. There is no suggestion that one might choose to be, for example, a doctor, or a lawyer, or an artist, or a businessman. There is certainly no suggestion that one might earn one's living as a cobbler, or a builder, or a scribe, or a flute-player. These lives are simply not regarded as candidates for being 'the good life', and one who is forced for financial reasons to engage in one of them is automatically debarred from *eudaimonia*. (Indeed, in the *Politics* Aristotle argues explicitly that the virtues require leisure, and that that is not available to those who have to work for their living. He appears to think that such people can only pursue pleasure.)[5] One can only note this point, and move on: Aristotle starts from a position that seems to us to be heavily blinkered.

Second, a word should be said about the meaning of *theōria*, which I (like Ross) translate as 'contemplation'. The root meaning of the word is to see, to watch, to spectate. But the word is commonly used of seeing 'with the eye of the mind' rather than that 'of the body', and that is clearly Aristotle's usage. When it is applied to 'intellectual seeing', the word can also cover intellectual enquiry, research, study, or indeed more or less any kind of thinking which falls under the general title 'the pursuit of knowledge'. So my vague phrase 'the intellectual life', or Irwin's phrase 'the life of study', are not obviously mistranslations. For the present, we may keep an open mind on what exactly Aristotle is thinking of when he speaks of *theōria*. But I think it will emerge quite clearly that in fact 'contemplation' is the best answer.

Third, Aristotle's phrase 'the life of such-and-such' is of course to be understood as referring to a life in which such-and-such is the *dominant* activity; it will not be the *only* activity. Thus even the contemplator must eat and drink, and no doubt he will enjoy doing these things (1154[a]17–18). He will also enjoy discussing his thoughts with his friends (1177[a]34); there is no reason to deny him the joys of family life; and so on. While contemplation may be, for him, the most valuable part of his life, it is surely absurd to think of it as the only thing he values. The same evidently applies to the life of political activity. But it then follows that Aristotle cannot consistently *identify eudaimonia* with the activity of contemplation, or with the activities that express the practical virtues, at least if he is to stick to his claims in I.7 that *eudaimonia* is an end which is 'most complete', or 'unconditionally complete', and 'self-sufficient'. For we saw (pp. 13–14, 21–5) that the most plausible interpretation of these claims requires us to understand *eudaimonia* as an end which includes within itself *all* the things that contribute to the goodness of the good life (though some, no doubt, are merely the means to others). Since Aristotle is about to argue that *eudaimonia* should be identified with contemplation, one supposes that he must now be abandoning the criteria for *eudaimonia* that he proposed in book I. But apparently he is not willing to admit this. At any rate, he repeats in book X what are verbally the same criteria as he

---

[5]   e.g. *Politics* 1278[a]20–1, 1319[a]26–8, 1328[b]39–1329[a]2, 1337[b]8–15. Cf. *EE* 1215[a]25–[b]1.

proposed in book I, though he is now interpreting them in a much less stringent fashion.[6]

With so much by way of preamble, let us now turn more directly to the arguments that he puts forward in book X.

2. Aristotle's arguments

Aristotle opens his discussion in chapter 6 by reminding us that *eudaimonia* is an activity, and one that is pursued for its own sake, not for some further end, since it is lacking in nothing and 'self-sufficient' ($1176^a30–^b7$). He at once adds that actions which express virtue would seem to be of this sort, for they are chosen for their own sakes ($^b7–9$). On the face of it, it appears that we should here stress the words 'would seem to be'. If the suggestion that he means to put forward is that the actions which express virtue satisfy *all* the points about *eudaimonia* that he has just recalled, then surely the words should be stressed. For it is not at all clear that such actions satisfy the criterion of self-sufficiency, in that they by themselves make one's life lacking in nothing, nor that they are not pursued for any further end. Indeed in I.7 he had said that they are pursued *both* for their own sakes *and* for the sake of a further end, namely *eudaimonia* ($1097^b2–5$). This apparently meant that they are pursued also as being *part* of a wider and more inclusive end that really is 'self-sufficient'. But perhaps all he meant to suggest at this point is that at least these actions are pursued for their own sake—for, after all, that is all he actually says—so they satisfy one of the criteria proposed for *eudaimonia*, though perhaps not the others. Even in this case, however, it appears that one should still stress the 'seems'. For in fact, as the discussion continues, Aristotle will argue that (at least in some cases) actions expressing the practical virtues are not really pursued for their own sakes, but rather for the sake of the state of affairs that will result from them ($1177^b6–18$). In either case, then, it appears that the 'seems' should be given due weight.

The word 'seems' should certainly be stressed with the next suggestion, namely that 'pleasant amusements' also seem to be pursued for their own sake ($1176^b9–10$), for Aristotle is going on to deny it. At any rate, he discusses these 'pleasant amusements' for the rest of the chapter, and certainly he claims that they are not what *eudaimonia* consists in. This discussion is best regarded, I think, as a further discussion of 'the life of pleasure', though in this case it would seem to be rather different pleasures that are envisaged. They evidently include social affairs where a ready wit is appreciated (1176^b14), and these could hardly be dismissed as 'fit only for animals'. Aristotle's arguments are, however, not particularly strong, and they surely would not convince a devotee of such amusements (for example, one who loves parties, or dancing, or watching football, or playing bridge). In broad outline he claims that though these activities may give pleasure to those who are keen on

---

[6] The extreme view (suggested by Nussbaum 1986: 377, and endorsed by Annas 1993: 216 n.) that X.6–8 is an independent essay which was not even intended to harmonize with book I seems to me not worth serious consideration. I.5 explicitly looks forward to a later discussion of the life of contemplation ($1096^a4–5$), and X.6 clearly opens with a reference back to book I ($1176^a32–^b6$).

them, still they are not the pleasures of the good man, and it is only these that are 'really' pleasures (1176^b12–27). And he adds that one *ought* to view such amusements simply as relaxation, which is not an end in itself but merely enables us to return refreshed to the more 'serious' activity which is the real goal (1176^b27–1177^a11). In any case, he ends the discussion by returning to the previous suggestion: *eudaimonia* is to be found not in such pursuits but in the activities which express virtue. This claim then functions as an unquestioned premiss in the discussion of the next two chapters.

Chapter 7 can only be understood, it seems to me, as offering arguments for the claim that *eudaimonia* is to be *identified* with the activity of the highest of the virtues, namely contemplation. A number of different arguments are offered. I briefly run through them.

(i) Pursuing the thought that *eudaimonia* is an activity expressing virtue, Aristotle claims it to be 'reasonable' to suppose that it must therefore be the activity of the highest virtue. This is at first vaguely described as 'thinking of some kind',[7] which seems by nature 'to rule and to lead, and to consider what is noble and divine', but is then said quite unambiguously to be contemplation[8] (1177^a12–21). The appropriate comment would seem to be that, even if we grant Aristotle his premisses, it is not at all clear why it is more 'reasonable' to pick on one virtue in particular, rather than including them all.

(ii) He claims that we can engage in contemplation 'more continuously' than in any action (1177^a21–2). Here one may wish to dispute the premiss, but I postpone discussion of this until we have seen more of what Aristotle actually means by 'contemplation'. In any case, one may also question its relevance. *Perhaps* the point is meant to be that *eudaimonia* is something that lasts for a whole life, so what we can do for longer, without a break, more approaches this. (But, if that is right, the conclusion to draw is that contemplation may perhaps *approach eudaimonia* in this way, but cannot *be eudaimonia*.)

(iii) He claims that we all think that *eudaimonia* involves pleasure, and that 'it is agreed' that the pleasure of contemplation is 'the greatest' of all pleasures of the activities expressing virtue (1177^a22–7). One can only say that there is no reason to suppose that someone who is committed to the political life would so agree.

(iv) He claims that self-sufficiency applies *more* to the activity of contemplation than to any other. It must be admitted that both the man of theoretical wisdom and the man of practical virtues need the necessities of life, but the latter also needs other people, to practise his virtue on, whereas the former does not strictly *need*

---

[7] More literally 'whether this is *nous* or something else' (1177^a13–14). Presumably *nous* here is to be understood in its general sense (cf. pp. 76–7), but in what follows it is clearly used for the part of the soul in which theoretical wisdom resides (1177^b19, $^b30$, 1178^a7, $^a22$, 1179^a23).

[8] Aristotle claims that it has already been said that the activity of the highest virtue is contemplative; I take this to be a general reference back to book VI, though the point is not stated there in any explicit way. He also claims that this agrees with what has been said earlier. I take this to be a reference to 1177^a3–6, which is then supported by 1177^a19–21.

them (though his life will, no doubt, be better if he has companions) (1177^a27-^b1). In much the same vein Aristotle adds later that to be generous, or to be able to repay debts, one must have money; to be brave one must have power;[9] to show oneself temperate one must have the resources to be intemperate; and so on. By contrast, none of these are needed for contemplation (1178^a23-^b5). We might be tempted to reply that the contemplator needs *other* resources (e.g. in our own day libraries), but in any case the main objection is as before: perhaps it is true that contemplation *more* approaches to the self-sufficiency that is a requirement on *eudaimonia* than other activities do, but it does not actually achieve it, as Aristotle's own remark about companions shows.

(v) He claims that only contemplation is pursued entirely for its own sake. At any rate, the most important of the practical virtues are shown in war and in political activity, but war is never undertaken for its own sake—rather, for the sake of the peace that will result—and the same applies to political activity too. For people engage in it hoping to secure positions of power, and honours, or at any rate hoping to bring *eudaimonia* to the citizens and to themselves, and this which they hope to bring about is not the same as the activity that they are engaged in. But none of this applies to contemplation (1177^b1-24).[10] This is a line of argument which one does not expect to find in Aristotle, for it certainly conflicts with what he usually says about virtuous actions. Let us briefly explore it.

His usual view is that the virtuous person undertakes the virtuous action 'for its own sake', which one may I think equate with 'just because it is the *right* thing to do (in the circumstances)'. As he also says, the action is done 'because it is noble (*kalon*)', which we need not take as conflicting with this, for his usual attitude seems to be that the noble action is *in itself* admirable, so long as it is undertaken 'because it is noble', or 'for its own sake'. It is true that the actions that he calls 'noble' are by and large those that we might regard as right because they benefit not the agent but others, and especially on important matters. (Thus facing death in battle is noble, and so also is a princely donation to charity; but refusing the tempting éclair is not noble, even though it is, in the circumstances, the right thing to do.) As I have noted, Aristotle seems at one point to offer this account of nobility himself—i.e. that what is noble is what conduces to the common good (IX.8, 1169^a6-13)—but there he thinks of the noble action as pursued for *two* reasons, i.e. both because it benefits others and because (being noble) it contributes to the agent's *eudaimonia* as well. (That is why to do such actions is to act from love of *oneself*.) But here he goes much further than one would expect, and claims that such actions are simply *not* chosen for their own sakes (1177^b18), but only for the

---

[9] It is not clear what kind of power (*dunamis*) is intended. The simplest suggestion seems to be that in order to exhibit bravery in battle one needs muscular strength.

[10] Aristotle interweaves this argument with a point about 'leisure', which I shall ignore, because it seems to have no independent force. (It is not clear why the activity of contemplating should be deemed 'more leisurely' than, say, the activity of devising, advocating, and putting into effect a piece of just legislation. And even if this is accepted, it is not clear why 'leisure' rather than 'seriousness' should be taken to be a mark of *eudaimonia*.)

good consequences that they bring.[11] Admittedly he offers only a couple of cases—courage in battle, and justice in politics—but it is obvious that he could have added many more. To take a much more trivial illustration, when you ask me the way to the station, I will give you a true answer. If I am an honest man, then I give you a true answer because I have been trained to tell the truth, regularly do so, and do so because I believe it is right. But *why* is it right? Well, the consequence is that you now know how to get to the station, rather than having no beliefs or false beliefs on the topic. And we might continue: why, in turn, is that state of affairs better than its alternatives? It will be tempting to answer in the same way, by spelling out its consequences, and so on. Pursuing this line of thought evidently leads to a 'consequentialist' system of ethics, of which the example best known to us is utilitarianism.

Now clearly Aristotle cannot be a utilitarian (if only because of his views on the value of pleasure, pp. 148–9). He of course accepts the initial move in consequentialist thought: some things are pursued *only* for the sake of others. But for the most part he is ready to offer a variety of things pursued for their own sakes, of which virtuous actions are standardly counted as one. (Others suggested are, for example, honour, pleasure, thought, 1097^b2; sight, memory, knowledge, 1174^a5–6.) But here he is saying that actions expressing the practical virtues do not actually qualify for this title, and—if we follow his argument through to its inevitable conclusion—there is only *one* thing that does, namely the activity of contemplation.[12] It follows that *everything* else that men (properly) pursue, they pursue for the sake of this. That certainly yields a consequentialist ethics, though it is one that astounds us. Of course one must at once add that men are often mistaken in what they pursue. Some do in fact assign an intrinsic value to pleasure, some to honour, and some to (the activities of) the practical virtues. But the truth, Aristotle claims, is that it is only contemplation that has intrinsic value. From this it really does follow, given Aristotle's other premises, that contemplation must *be eudaimonia*. For, as we have constantly been told, *eudaimonia* is an activity, and in particular an activity of the rational part of the soul, in accordance with virtue, and it is the only thing that is pursued only for its own sake and never for the sake of something else. If, then, it is also true that contemplation is the only such activity that is pursued only for its own sake, contemplation and *eudaimonia* have got to be the same thing.

To conclude, of all the five arguments that Aristotle first offers, at 1177^a12–$^b24$, it is

---

[11] Irwin clearly mistranslates two crucial sentences in this passage. At 1177^b1–2 the Greek clearly has 'It alone [i.e. study/contemplation] seems to be liked for its own sake.' But Irwin gives 'study seems to be liked because of itself alone'. At 1177^b17–18 the Greek clearly has, concerning actions which accord with the practical virtues, 'They lack leisure, and aim at a certain goal, and are *not* choiceworthy for their own sakes.' But Irwin gives, for the last clause, 'and are choiceworthy for something other than themselves'. Presumably Irwin cannot believe that Aristotle means what his text unambiguously says.

[12] More strictly, contemplation is the only *virtuous* activity that is valued for its own sake. Aristotle also holds that only activities are valued for their own sakes. But this allows us to add that there may be non-virtuous activities that are valued for their own sakes, the obvious candidate being perception. However, Aristotle has claimed in I.7 that perception is no part of the 'function' of man, so apparently it cannot contribute to his *eudaimonia*.

only this last that seems actually to yield the conclusion he is aiming for. It does so because of its very strong claim that contemplation is the *only* virtuous activity that is (properly) valued for its own sake alone. (No doubt some misguided people value it for other reasons, e.g. because of the rewards, honours, and reputations that a successful academic career can bring. But these people are, on Aristotle's account, misguided.) It cannot be said that Aristotle has offered a compelling argument for this claim. To be sure, he has offered a few indications—which could certainly be expanded—of how the activities which express other virtues could be valued for something else beyond themselves. But he certainly has not explained how, if we followed through the chain of consequences, they would all be found in the end to lead up to contemplation. (And one might reasonably ask: *whose* contemplation? It seems odd to suggest that the activities of the just legislator are good because they facilitate *his own* contemplation.) But when we follow up this line of thought, it is likely to lead to an objection to an earlier premiss: why should we suppose that it is only virtuous activities—or, indeed, only activities—that have intrinsic value? One has only to think for a moment about why it is that contemplation does *not* by itself satisfy the requirement of 'self-sufficiency' to see that there is a perfectly good question here. I do not mean to suggest, then, that Aristotle's fifth argument cannot be resisted; and surely we will wish to resist it, for the position that it aims to establish is one that strikes most of us as absurd. But resistance in this case must go deeper, whereas the first four arguments are easily turned, without any serious questioning of the premisses from which Aristotle is working.

But we have further arguments to come. Let us turn to these.

The next argument arises out of an objection, which is that the life of contemplation is too high to be a human life; one who lives thus is doing so not in so far as he is a man, but in so far as there is something divine in him, namely his intellect (*nous*). Its activity is superior to the activity of the rest of the virtues, so the life in accordance with it is a divine life in comparison with a human life ($1177^b26–31$). The objection is, then, that this life fails to secure the good *for man*, which is what we are seeking.

But Aristotle rejects this line of thought. We should not, he says, accept the proverbial instruction to think human thoughts because we are human, or to think mortal thoughts because we are mortal; rather, we should 'be immortal', so far as we can, and do everything possible to live in accordance with the best thing in us. Moreover, he claims that this harmonizes with what he has said before, which is presumably a reference back to the argument in I.7 about man's 'function', where it had been said that the good for man will be found in peculiarly human activity. It does so because each person *is* his intellect (or is so 'most'), since this is what is in control, and is what is best in him. In choosing to live the life of the intellect, then, he is choosing *his own* life, and not that of something else ($1177^b31–1178^a8$).[11]

---

[11] Everson (1998) appears to neglect this last point when he argues for the unusual view that, on Aristotle's account, theoretical activity *is* included in man's *eudaimonia* but *is not* included in man's 'function'.

Aristotle does not elaborate further on the claim that I *am* my intellect, and it is hard to make any suitable sense of it. I have allowed (pp. 000–0) that we can have at least some sympathy for the claim in book IX that a person may be identified with his *practical* reason, just because that is the part of him that is (or should be) 'in control'. But in this passage Aristotle must be changing his claim, for now it is *theoretical* reason that his argument concerns,[14] and I cannot see any ground for accepting that identification. Of course one may say that one who has already committed himself to the life of the intellect will in a way 'identify' himself with his intellectual activities, for they are the ones that he most values. But this is no use as an argument, for one can equally say that one who has committed himself to the political life will in the same way 'identify' himself with his political activities. We shall not reach any decision in this way on what 'part' of me is really *me*. In any case, the most obvious response is the tautology: all parts of me are parts of me. I simply am a composite being, in so far as I do have desires and emotions, as well as the capacity to reason, both practically and theoretically. We should conclude, I think, that the objection succeeds and Aristotle's reply fails: there are *many* activities that are peculiar to man, and the argument of I.7 directs us to include in the good life for man *all* of them that are pursued 'with excellence'.

Aristotle has two more arguments to offer in chapter 8. They are interwoven with his discussion of the other life, the life of practical activities, but it is convenient to consider them here and to turn later to what he has to say about the other life. Both these arguments take up and develop further the thought already introduced, that the intellect is something divine. The first of them is, I believe, an argument that weighed very heavily with Aristotle himself.

It begins from the premiss that we all suppose the gods to be supremely *eudaimōn*, and goes on to ask what their *eudaimonia* can consist in. It would be ridiculous, says Aristotle, to think that they engage in just actions, such as making contracts, returning what has been deposited with them, and so on. It would also be ridiculous to suppose that they display courage, by enduring what is frightening and dangerous, or again to suppose that they display generosity by giving gifts. As for temperance, it would be quite inappropriate to praise them for not having bodily desires that they should not have, for they do not have any bodily desires anyway. Quite generally, these human virtues and human actions are not fitted to the nature of the gods. But we suppose that the gods live, so they must be active in some way, and the only remaining possibility is that their activity is contemplation. For the gods, then, *eudaimonia* must be contemplation, and that is *eudaimonia* at its highest level. So for men too what is most akin to this is what holds most *eudaimonia* (1178^{b}8–23). After a brief remark about animals, Aristotle concludes: '*Eudaimonia* extends just as far as contemplation does, and those who have more of

---

[14] Broadie (1991) suggests that we can understand it to be *the same nous* that engages in both practical thinking (*qua* embodied) and theoretical thinking (independently of the body) (pp. 436–7 n. 56). I have indicated that this would in some ways be a welcome reform of Aristotle's doctrine (pp. 77–9), but it is quite clear that it is not what Aristotle's doctrine actually is.

contemplation thereby, and not coincidentally, have more of *eudaimonia*. For it is of value in itself. *Eudaimonia*, then, will be a form of contemplation' (1178[b]28–32).[15]

One thing that is quite clear about this argument is that Aristotle is relying on *his own* conception of what is fitting to the gods, which is not the usual conception of his own day, nor indeed of our day. The popular conception of his own day was, no doubt, not far removed from the conception we find in Homer. Homer's gods are, of course, immensely powerful by comparison with human beings—for example, Poseidon can always create a storm at sea if it will further his plans—but in other ways they are entirely anthropomorphic, and they do engage in all those activities which Aristotle says cannot be ascribed to them. The popular conception nowadays would be more sympathetic to *some* of Aristotle's points; for example, we would agree that it is not appropriate to credit God either with courage or with temperance, since nothing is a danger to him, and it is not in his nature to feel either fear or bodily desire. But many people would hesitate to say that God cannot be called just, and they would surely insist that he can be called loving. The basic difference between Aristotle's conception of God and ours is that in our view God cares about what happens in the world, and to the people in it, and he can interfere with what happens in the world in pursuit of this. But Aristotle's God[16] pays no attention to such things.

Aristotle's conception is elaborated elsewhere, most fully in *Metaphysics* Λ, 6–10, and this is not the place for a proper discussion of that often difficult text. But in broad outline Aristotle claims there that the only activity fitting to God is thought, and since this must be the best kind of thought it must be directed on the best objects of thought. At least this much is clear: the best objects of thought are those that cannot be otherwise, so (in *our* way of speaking) God's thought is of necessary truths, and not of contingent truths. He does not, therefore, think of human affairs, for they are all contingent. Aristotle *seems* to go on to say that since the *best* object of thought is God himself, God's only activity is thought *of himself*. But it is often suggested that he did not quite mean this, and that God may contemplate the necessary and universal truths of mathematics and of physics, as well as those of theology. I must here leave this question open, though certainly Aristotle's claim is that God's thinking falls under theoretical wisdom, as described in book VI, and not practical wisdom. (If I may add an anachronism: later theologians have sometimes claimed that God is 'outside time', from which it appears to follow that he is not aware of what happens in time. Though Aristotle himself does not say this, nevertheless it does suit his conception rather nicely.)

To return to the *Ethics*: one premiss to Aristotle's argument here is his own conception of the nature of the gods. This is clearly not the popular conception, but

---

[15] White (1988: 171–4) apparently ignores this passage when he argues that Aristotle is not recommending us to *maximize* contemplation.

[16] It is clear that Aristotle would wish to believe in just one god (*Metaphysics* Λ.10), and that he speaks here of gods in the plural as a concession to popular opinion, since it does not affect his present argument. (But one has to add that in *Metaphysics* Λ.8, which is presumably a later addition, he admits that the facts of astronomy seem to require a plurality of gods to explain them.)

a special and philosophical (or theological) conception of his own. But the other premiss is simply an appeal to popular opinion, namely that we all suppose the gods to possess *eudaimonia* in the highest degree. One may well doubt whether these two premisses can be held jointly. For if the popular conception of the gods is altogether mistaken, then the thought that the gods possess *eudaimonia* may surely be mistaken too, and this indeed seems a fair suggestion. For *eudaimonia* is the good life; it is living well and acting well; it connotes success and achievement, as well as something like happiness. Should we not say that if, as Aristotle claims, we should not think of the gods as performing any actions, then we equally should not attribute *eudaimonia* to them? They would seem to be 'above' such a description, just as they are also 'above' what count for men as the practical virtues. This, it seems to me, is quite a serious objection to the argument. There is also, of course, another and more obvious objection. Even if we grant that Aristotle is right about what *eudaimonia* is for the gods, nothing actually follows about what it is for man; men and gods are different 'species' (if we may talk in this way), just as men and asses are different, or men and fish. Now in fact Aristotle denies that *eudaimonia* applies to any animals other than man (1178^{b}24–5), though it is not obvious that he is right to do so. At any rate, he admits that men and asses each have their different pleasures (1176^{a}3–8), and that there is such a thing as what is good for fish, as well as what is good for man (1141^{a}22–3).[17] It is not clear why we should not apply to every species the notion of 'the good life' for it—according to the argument of I.7, this will be given by the 'function' of that species—but of course we shall get very different answers in different cases. Why should not the same be true for men and gods?

Aristotle's argument, then, is hardly compelling, even if we grant his premisses. Moreover, its conclusion is seen to be particularly unattractive, as we follow up the suggested analogy between men and gods, for this analogy shows that Aristotle really does mean that the *eudaimonia* of the contemplative life consists in contemplation *alone*, and not in the various other things that we associate with it. Those who devote themselves to academic pursuits, as Aristotle himself did, spend much of their time in what is nowadays called 'research' (and in writing up their results, and so on), and most of the rest of it in teaching. Clearly, the gods do not teach. But nor do they research. For research is a matter of *enquiry*, of trying to find the answers to problems which present themselves. But the gods, one must presume, already know all the answers, so that is not what their activity is. (Indeed, according to Aristotle's strict use of the notion of 'activity', research is not an *activity* at all, but a *process* that aims for a certain result, and is not complete until that result is reached.) So the activity of the gods is literally just contemplation, holding before the mind something that is already well known and entirely familiar. On Aristotle's account, they cannot do anything else, so that is what *eudaimonia* must be for them. According to his argument, then, it must be the same for us. But what is envisaged is an 'activity' which in fact virtually no one engages in for more than a few seconds at a time, and on rare occasions. A life devoted to it would appeal to no human being

---

[17] Both examples are taken from Heraclitus (fragments 9 and 61, Diels/Kranz).

at all. For what people who engage in the academic life do value about it is the opportunity to research, and (in many cases) to teach.

It may be replied that I am misconstruing Aristotle's position. For perhaps he did not mean to praise what I have called 'the academic life', which was *his own* life, but something quite different. This interpretation begins from the thought that God's activity is the contemplation of *himself*, and infers that *eudaimonia* for man is the contemplation *of God*. (It may be supported by noting that Aristotle's claim is that *eudaimonia* is a *kind* of contemplation (*theōria tis*, 1178ᵇ32), so perhaps he did have this special kind in mind.) Moreover, in the tantalizingly brief and obscure final paragraph of the *Eudemian Ethics*, where we might expect simply a reference to 'contemplation' in general, it is in fact only 'contemplation of God' that is mentioned (together with a mysterious 'service of God', which is nowhere explained) (1249 ᵇ16–21).[18] If this is right, then it can be added that there really are those who devote their lives to the contemplation of God, and we may think here of mystics, of hermits, or simply of nuns and monks.[19] I reply, first, that I doubt whether there were any such people known to Aristotle, who after all was writing well before the Christian era (and who would not have known of the Buddha). Second, that Aristotle's own god does not seem to be the kind of being that could evoke such continuous adoration. And third, that it really is improbable that his thinking on this issue simply *overlooks* the kind of life that he led himself—much of which was devoted to academic research—so that his discussion contrasts only the mystic and the politician. It is true that I am claiming that he is wrong about what it is that makes his own life worth pursuing, for what is specially enjoyable about it is research (including here the contemplation of a *new* result, not previously known), and perhaps teaching; while what makes it 'good' may perhaps be something else, for example the beneficial effects for society in general of an increase in knowledge, and of well-educated citizens. So even if it does share some features of the life of the gods, that is not what makes it a good life for man. But it seems to me overwhelmingly probable that Aristotle is trying to describe the academic life as he knew it, even if he is wrong about what is valuable in it.[20] As I have observed, the word *theōria can* cover all aspects of academic research, and this perhaps goes some way towards explaining why Aristotle failed to see that its use to mean 'contemplation' in particular gives altogether too narrow an account of what is actually attractive about that life.

The argument from the nature of the divine was, I think, influential in Aristotle's own way of thinking. It lies behind his misdescription here of the academic life, and

---

[18] With no manuscript authority, the present OCT text of the *EE* here reads 'of the divine' rather than 'of God'. For some discussion of this difficult text, see Kenny (1978: 173–8) and Woods (1982: 193–8).

[19] The thought is developed in a different and highly idiosyncratic way by Clark (1975, chs. V–VI; see esp. ch. V, sect. 3).

[20] I remark that Aristotle was familiar with the political life too, even though he was not a citizen of Athens (but a *metic*, or resident alien), and so could not take part himself in its political affairs. But obviously he would know those who did. Besides, he had himself been involved in political activities due to his involvement with first Philip and then Alexander of Macedon. (For a brief account of his life, see e.g. Ross 1923, ch. 1.)

it also underlies—I believe—his claim that it is only activities in the strict sense that are enjoyed, and that these are just perception and contemplation. But, as we have seen, the argument is not a good one. There is one further argument of this kind that Aristotle offers, at the end of chapter 8, but it is even worse. I comment on it only briefly.

If the gods do pay heed to human affairs, 'as it seems that they do', it will be 'reasonable' to suppose that they rejoice most in what is most akin to them, namely the intellect and its activity. Accordingly the one who most cultivates this, i.e. the man of theoretical wisdom, will be most loved by them. So it is likely that he will be most *eudaimōn* (1179^a22–32).

This argument again combines popular and Aristotelian premises, and this time clearly in an illegitimate way. It is a popular view that the gods pay heed to human affairs, love some men more than others, and reward them accordingly. But it is not Aristotle's own view, for his view is that their only activity is contemplation, and this evidently does not include loving and rewarding men. By contrast, it is an Aristotelian view that it is man's (theoretical) intellect that is most akin to the gods, but it is not the popular view. The premises, then, are not co-tenable (and, even if they were, the support that they would give to the conclusion would be tenuous at best). This final argument, then, can be ignored. It is a pity that Aristotle chose it to be the coping-stone of his discussion.

3. Aristotle's position

The most natural way of taking Aristotle's position is surely this. He means to compare and contrast two 'lives', one being the life of contemplation (or perhaps, more generally, of study), and the other the life of political activity. These are each two different ways of organizing and structuring one's whole life, and are distinguished by what one may fairly call the *dominant* activity in each.[21] Of course in any life there will be other activities too. Even the contemplator must also eat and drink, and do those other things that are necessary prerequisites to his main objective (which, according to 1176^b33–1177^a1, will include some relaxation). Perhaps we may also allow him some other enjoyments 'on the side', as it were. For example, he might be permitted, every now and again, to see a play when plays are being performed, or to hold a birthday party, or simply to play with his children, even if doing these things will not contribute to his contemplation.[22] But the idea is that contemplation is the major and overriding goal in his life, and he will do nothing that detracts from this in any serious way. The same will apply, *mutatis mutandis*, to the life of politics. No doubt these are ideals, and no one is likely to live up to them completely, but it is these ideals that Aristotle means to compare.

[21] White (1988) prefers to speak of the 'focus' of a life, but I think that what he has in mind is much the same as what I have in mind.

[22] But in *EE* VIII.3 Aristotle does seem very strict about this: he says that one should aim to acquire external goods—e.g. material possessions and money and friends—*only* to whatever extent will most produce contemplation (1249^b16–19).

He aims to compare them for *eudaimonia*, asking which life is the more *eudaimōn*, but his comparison focuses entirely on their respective dominant activities, and he proceeds by asking which of *them* 'is', or 'is most', *eudaimonia*. This is strictly an illegitimate way of phrasing the question. For if *eudaimonia* is supposed to be something 'complete' and 'self-sufficient' in the sense in which those criteria were used in I.7—i.e. if *eudaimonia* is something which 'by itself makes life lacking in nothing'—then it can only be identified with *all* that goes into making that life a good one, and not just its one dominant activity. Yet X.7 tries to argue that *eudaimonia* simply *is* contemplation. But I think we need not fuss too much about this discrepancy, for we may put the situation more neutrally in this way: Aristotle argues that contemplation has a higher value than political activity, and not unreasonably infers from this that a life which has contemplation as its dominant goal will have greater value than a life which has political activity as its dominant goal. This would seem to be enough to license his conclusion that the contemplative life is the one that is most *eudaimōn* (of those considered), whereas the political life is (at best)[23] only *eudaimōn* 'in a secondary way' (1178ª6–8).

How are we to understand what I have been calling 'the political life'? Aristotle himself uses this title when he introduces the idea in I.5 (1095ᵇ18, 23, 31), and repeats this title in X.7–8 (1177ᵇ12–18, 1178ª27).[24] But he also calls it the life 'in accordance with the rest of virtue', i.e. the practical virtues, excluding theoretical wisdom; and it is clear that he thinks of the particular practical virtues—e.g. justice, courage, temperance—as specially associated with this life (1177ª30–2, 1178ª9–14, ª28–34). These two descriptions do not seem to pick out the same idea, for while it is easy to conceive of a life that is organized round political activity as its dominant aim, it is altogether less easy to see the amalgamation of the practical virtues as providing a dominant aim for a life. This is a point that I shall return to, for it will be of some importance. But what I am calling the 'natural' way of construing Aristotle's position does call for two lives with different dominant goals, so I shall here take it that 'the political life' is the better title. Still, we need not construe the word 'political' in a narrow fashion. This word is of course derived from the Greek word *polis*, which is standardly translated as 'city' or 'state', but can perfectly well be regarded as the natural Greek word for 'the community' in a more general sense. (Thus Aristotle's well-known thesis that 'man is naturally a political animal' is best understood as claiming that 'man is by nature suited to living in communities'.) So we may not unfairly construe a 'political' life simply as a life that is, in one way or another, devoted to 'the service of the community'. No doubt the life of those whom we call 'politicians' may be so regarded, but the more general description would also include the life of the professional soldier, and even (say) the life of Mother Teresa, though no one would call *her* a 'politician'. But in any case, however the word

---

[23] I observe that book X does not provide arguments to show that political activity is in *any* way (a source of) *eudaimonia*, save for the claim that it involves (specifically human) virtues, and the general assumption of I.7 that every human virtue contributes to (human) *eudaimonia*.

[24] 1177ᵇ16–17 mentions not only politics but also war. This is in order to include courage, which Aristotle always thinks of as primarily displayed in battle.

'political' is understood—and I should myself be in favour of construing it broadly—my point is that we must see it as pointing to something that is the dominant aim of a certain kind of life, for it is only in this way that we can make sense of the 'two lives' that Aristotle means to compare.

We have, then, these two different lives, each with their own different goal. Apparently Aristotle does not consider any other lives to be worth discussion (apart from 'the life of pleasure', which he dismisses). Nor does he consider a 'mixed' life, which has both of these goals simultaneously,[25] and this is perhaps reasonable. For although it is in fact possible to try to combine an academic life with a career in politics, and some have done so, still it would be quite widely held that the combination is seldom successful. (One is likely to judge that one ingredient of this 'mixture' would have prospered more if it had been disencumbered of the other.) But in any case Aristotle simply does not consider such a 'mixed' life, just as he also does not consider any of the many other lives which might seem to us to be worthy of consideration. He considers only two, and undoubtedly he awards victory to the life of contemplation: it is the most *eudaimōn* whereas the political life is *eudaimōn* only in a secondary way. This is, as I say, the natural way of reading X.7–8, but it provokes many problems.

The first kind of problem is that, on this reading, X.7–8 seems to contradict much of what we are told elsewhere in the *Ethics*. Evidently, if we take I.7 to be proposing an 'inclusive' conception of *eudaimonia*, by which it includes *all* the (specifically human) virtues, and perhaps more besides—e.g. good birth, good looks, and good fortune—then we have a clear contradiction. But I have already discussed this question (in Chapter I), and will not reopen that debate now. Still, we may look at the issue more widely. Aristotle begins his discussion of the (human) virtues in I.13, and it continues until the end of book VI. Almost all of this discussion concerns the practical virtues (i.e. the virtues of character, together with practical wisdom). We are surely given to understand that they are worth discussing because they contribute to *eudaimonia*, for why else should they be discussed at all? Again, the various problems concerning *akrasia* which are discussed in book VII mostly concern its relation to the practical virtues, and these virtues clearly have a role to play also in the two discussions of pleasure. We may add that the account of friendship in books VIII–IX certainly does not suppose that friends are of value only because of the way in which they can contribute to contemplation. So when we are told in X.7–8 that true *eudaimonia* just is contemplation (or the life devoted to it), must we not say that it seems to have turned out that almost all of books II–IX is simply *irrelevant* to the question with which Aristotle began in book I, namely: what is the good life for man?[26] That is surely an unsettling result, and not much palliated by the concession—for which no good reason is given[27]—that the practical virtues do contribute to a 'secondary' kind of *eudaimonia*.

---

[25] Contrast *EE* 1214ᵃ30–ᵇ6.

[26] One may add that as soon as X.7–8 are finished, and Aristotle proceeds in X.9 to introduce the next phase of his discussion, i.e. the *Politics*, it is once again the *practical* virtues that he is thinking of.

[27] See n. 23 above.

There are also two particular points on which it is quite clear that X.7–8 contradict what is said elsewhere. The first is that everywhere else Aristotle says that the
activities which express the practical virtues are pursued for their own sake (or, for
the sake of 'the noble'), whereas in X.7 he says roundly that they are *not* pursued for
their own sake, but rather for the good consequences that they bring. There are no
doubt ways of attempting to water down or explain away this contradiction,[28] but
we cannot try to pretend that it simply is not there. Again, X.8 claims that a man is
(or 'is most') his theoretical intellect, a claim that is difficult to justify or even
understand, but anyway one that clearly conflicts with IX.8; for that says exactly the
same thing about a man and his practical intellect. No doubt one can try to fudge
the point,[29] but really it will not go away. However, one might not unreasonably
admit that in these two particular arguments Aristotle has overstated his case—in
Urmson's phrase (1988: 125) he 'has let his enthusiasm get the better of him'—and
not try seriously to defend them. But what has to be serious is that he clearly does
award victory to the life of contemplation, and yet it is hard not to see this as
conflicting with the general tenor of almost all of the rest of his *Ethics*.

The point has been put so far simply as one that concerns the internal consistency of Aristotle's *Ethics*, but we may also present it straightforwardly as a criticism
of the doctrine of X.7–8 as thus interpreted. For, on this account, Aristotle seems to
be saying that what we think of as *moral* behaviour just has nothing to do with what
is truly the good life. Apparently he *contrasts* the life of theoretical activity with that
which is 'in accordance with' the other virtues, as if neither had a part to play in the
other. So the devoted theoretician will secure the best kind of *eudaimonia*, even
though he pays no attention at all to the other virtues. His life may, then, be wholly
immoral, but it will still be the *best* life for man, so long as his theorizing is
successful. Can Aristotle really have meant this? To most interpreters it has seemed
to be a conclusion that must, somehow, be avoided.

4. Escape routes

One might suggest that we can accept the conclusion that the *most eudaimōn* life is
simply amoral, by stressing that it is the life of gods and not of men. So Aristotle
should be taken seriously when he says, 'But such a life would be above the human
level' (1177^b26–7), and when he emphasizes that it is his second life that is especially
human. For it is this second life that involves not only the rational part of the soul
but also the part that has desires and emotions. He mentions too, that these desires
and emotions would seem sometimes to arise from the body, and of course it is the
fact that human beings have bodies that most obviously distinguishes them from
gods (1178^a9–22). So the first life that he discusses is that of the gods, and not of
men. The second is *eudaimōn* only in a secondary way, but nevertheless it is the best
life *for man*. Man's *eudaimonia* is, after all, inferior to that of the gods.

---

[28] e.g. Keyt (1978: 151); Broadie (1991: 43–7).
[29] e.g. as Broadie does (n. 14 above), or as Reeve does (1992: 133–7) in a rather different way.

So far as I am aware, no commentator has seriously suggested this way out, for it is abundantly clear that the text will not permit it. Aristotle *rejects* the thought that the contemplative life is 'above the human level', and urges us to 'be immortal' so far as we can, since a human being *is* (or 'is most') his theoretical intellect (1177^{b}31–1178^{a}4). Nevertheless, a part of the view proposed is in a way endorsed by Cooper (1975, ch. III). For he accepts the comment that Aristotle's contemplative life is 'superhuman' (p. 179), and—more importantly—he thinks that Aristotle's second life genuinely does include *all* the human excellences, both practical *and* theoretical (pp. 165–8).[30] But there is simply no warrant for this last claim. The second life is described as one that is 'in accordance with the *rest* of virtue', and this description must exclude theoretical wisdom, which indeed is nowhere mentioned by Aristotle when he speaks of this life.

If we cannot get both kinds of human virtue into Aristotle's second life, can we perhaps find them both in his first? This is certainly a more popular suggestion, and it may be motivated in this way. What is called the life of contemplation must of course include whatever is needed for the efficient pursuit of contemplation, and then the suggestion is that these necessary conditions include the practical virtues (as they also include sufficient wealth to be free from concerns about one's daily bread, sufficiently comfortable surroundings, and so on). Aristotle himself seems to accept something along these lines when he says that the contemplator will need *some* external goods (though not many) and adds:

In so far as he is human, and lives together with many others, he will choose to do the actions that accord with [practical] virtue. So he will need such things [i.e. external goods] in order to be human. (1178^{b}5–7)

The thought seems to be that even the contemplator is bound to be living in a community, and his contemplation will not prosper unless he gets on well with other members of the community, and for this the practical virtues are required. They are necessary means to his primary objective.

The same moral may be obtained by reading rather between the lines of VI.12–13. There Aristotle is considering an objector who asks: what *use* are practical and theoretical wisdom? (1143^{b}18–33). His first reply is that both have value, just because both are virtues, even if neither produces anything further (1144^{a}1–3). His next reply is more informative:

In any case, they do produce something: theoretical wisdom produces *eudaimonia*, not in the way that medical skill produces health, but as health produces health. For it is a part of virtue as a whole, and, by possessing it and activating it, it makes [one who has it] *eudaimōn*. (1144^{a}3–6).[31]

This passage has sometimes been taken to say that theoretical wisdom produces *eudaimonia* by being a part of it (with the implication that there are other parts

---

[30] In his later (1987) Cooper revokes this view, and there adopts my next suggestion.

[31] I have given a literal but clumsy rendering, to make clear how I think that the received text can perfectly well be accepted, despite Bywater's obelisks in OCT. Both Ross and Irwin give smoother renderings, but in each case with the same sense.

too). But it does not say this. It does certainly say that theoretical wisdom is only a part *of virtue*, which indeed is obvious, for we know that there are other virtues too. But the way in which it 'produces' *eudaimonia* is 'as health produces health', which is most naturally seen as implying that it—or rather, the activity that expresses it— is the *whole* of *eudaimonia*. (The inference is not mandatory, but it is rather plausible.) Granted this interpretation, we naturally ask: then what is the use of the *other* virtues? I believe Aristotle means to answer this in the next line:

Further, the function [of man? of theoretical wisdom?] is fulfilled [only?] in accordance with practical wisdom and virtue of character. (1144^{a}6–7)

But his discussion then gets sidetracked into other questions, and this rather cryptic sentence is left without elucidation, until we come rather later to his response to his other problem (stated at 1143^{b}33–5). The elucidation when it comes is this:

Moreover, practical wisdom is not in control of theoretical wisdom, or of the better part [of the soul], as medical skill is not in control of health. For it does not make use of theoretical wisdom, but rather sees how it can come into being. So it issues commands not to it, but for its sake. (1145^{a}6–9)

Practical wisdom is not another part of *eudaimonia*, as medicine is not a part of health, but its use is that it gives directions on how theoretical wisdom can be achieved and exercised, and it is the exercise of theoretical wisdom that *is eudaimonia*. Moreover, practical wisdom is, here and throughout, strongly associated with the various virtues of character. So their role too is to ensure that theoretical wisdom (and its exercise, which is contemplation) can flourish. If indeed that is their only role, then perhaps Aristotle did really mean it when he said, in X.7, that the activities which express these practical virtues simply are *not* pursued for their own sakes, but because of their utility in promoting contemplation. It is true that Aristotle does not explicitly say, either in VI.12–13 or in X.7–8, that this is the *only* role of the practical virtues. But perhaps it is what he meant?

   This general line of interpretation is taken by many.[32] The main objections to it are (*a*) that it does contradict what Aristotle says elsewhere on what it is to have a practical virtue, and (*b*) that it is in itself a very implausible view. I take it that (*a*) is sufficiently obvious, and will here pursue only (*b*). It is notorious that those who are successful academically are not always virtuous, and there is no reason, even in theory, why they should be. Many kinds of immoral action do not affect one's research capacity (e.g. sexual indulgence), and some positively promote it (e.g. stealing someone else's ideas). More importantly, many good actions reduce one's research capacity, in particular those that involve paying heed to the needs and interests of other people, and so spending time and energy on them to the neglect of one's research. Perhaps our academic cannot ignore others' needs completely, for (in the *usual* case) he is a member of a community, and will not be able to

---

[32] I mention Ackrill (1981: 138–41), Cooper (both in 1975, ch. III, and in his reconsideration in 1987), Kraut (1989, *passim*, but esp. ch. III), Curzer (1991), Reeve (1992, ch. IV).

concentrate on his academic pursuits as he wishes unless he does remain on relatively easy terms with his colleagues. But, as Plato observed,[33] what this requires is merely the appearance of justice towards them, and not the real thing; or, in Aristotle's terms, such a man may act justly towards his colleagues, but he will not do so 'for its own sake' or 'because it is noble', so he cannot be accounted virtuous. In any case, this point affects only his relationship with his colleagues, and he has no such motive to act justly towards those who do not impinge on his own life. (So, if he is in need of money, and calculates that the simplest way to obtain some is by fraud, he will have no compunction in doing so, so long as he is confident that the fraud will not be detected.) The idea that practical virtue is a necessary condition of theoretical wisdom is very unconvincing, when one comes to think about it.[34] Kraut (1989: 178–82) does his best to defend the view that Aristotle may nevertheless have taken it for granted, but is forced to sum up in this way:

We can with good reason complain that X.7–8 does not spell out the way in which they [the practical virtues] make this contribution [to theoretical wisdom]. Aristotle should explain how temperance, courage, justice, and so on facilitate the intellectual life . . . The fact that he himself does not spell out these important details suggests that he expects no challenge from his audience: they do not need to be persuaded. (Kraut 1989: 195)

But the fact that he does not 'spell this out' may equally well suggest that he was *not* assuming it, and saw no need to. For at least it is clear that in X.7–8 he never even *asserts* the view in question; it is only those who wish to find some consistency in his overall position who feel the need to attribute it to him. But perhaps consistency can be restored in another way?

This brings me to my third 'escape route', which claims that everything that I have said so far has quite misconstrued the notion of the 'two lives'. I have been supposing that some people live the one life (of contemplation), other people live the other life (of politics, broadly construed), and that Aristotle does not consider someone who tries to live a mixed life, which somehow combines both. But the alternative suggestion is that these 'two lives' should rather be considered as but two aspects of a single life, so that what Aristotle is saying is that the *eudaimōn* life has *both* of these aspects (one of them providing a 'primary' kind of *eudaimonia*, and the other a 'secondary' kind, but both contributing to overall *eudaimonia*).[35] There is some dispute over whether Aristotle's word 'a life' (*bios*) can be taken to mean merely an aspect of one's whole life. (*We* certainly can use the word in this way. Thus we can talk about someone's 'sex life', as opposed (say) to their 'public life', or their 'life as a teacher (researcher, poet, etc.).' But the Greek word is not obviously so flexible.) I shall not enter this dispute, but simply concede for the sake of argument

---

[33] *Republic* II, 358b–367e.

[34] No doubt if the contemplator must choose between developing for its own sake either a practical virtue or one of its opposing vices, he would be well advised to do the former. But he is better advised to do neither (a possibility conspicuously omitted in the discussion by Curzer (1991: 60–5).

[35] I take Keyt (1978) as the representative of this view (but it can also be found elsewhere, e.g. in Gauthier and Jolif 1958–9: 860–6, 891–6; Engberg-Pedersen 1983, ch. 4).

that the word *could* be taken in this sense.[36] The question, then, is how it *should* be taken in the present context.

There can be little doubt that when Aristotle introduces the traditional 'three lives' in I.5 he construes them—as the tradition does—as three separate 'whole lives', and not three aspects of a single life. (For example, when he dismisses 'the life of pleasure' he is not saying that no respectable life has a 'pleasure-aspect'.) The main reason for saying that he continues to construe them in the same way in X.6–8 is just that that is how they were construed before, and he gives no sign that he is now changing the terms of that traditional debate. One may add that it is easy to see the contemplative life as one that has its own distinctive and dominant goal, and if this goal really does *dominate* then there is no room for a rival aspect of the same person's life in which something else dominates. But perhaps Aristotle does not mean to say that his second life has its own rival and dominating goal? I supplied such a goal by leaning on his description of this life as 'the political life', but if we pay no attention to this, and consider only that he describes it as 'the life in accordance with the rest of virtue', then (as I remarked) it is not easy to see it as a life with its own dominating goal. So we could perhaps see it as but one aspect of a total life which has theoretical activity as its major goal.

The best reason for supposing that this is how Aristotle himself must be thinking is his use of the locution *qua* or 'in so far as' (*hē(i)*). He tells us that a person pursues the theoretical life 'not in so far as he is a man, but in so far as there is something divine which belongs to him' (1177[b]27–8). On the other side, he never quite says that a person pursues his second life 'in so far as he is a man', but he does say things which strongly suggest this (e.g. that the practical virtues are specially 'human', 1178[a]9–14, [a]20–2; cf. 1178[b]5–7, quoted above). The thought obviously suggests itself that every person, all the time, both is a man and has something divine in him. So everyone should all the time pursue both 'lives' at once, one *qua* human (and composite) and the other *qua* having a divine part.

If we pursue this idea, then at once we are faced with the problem of how these two 'aspects' of a single life are to be combined with one another. Aristotle tells us that the one is 'most *eudaimōn*', and the other only '*eudaimōn* in a secondary way', but this does not obviously give guidance on how priorities are to be assigned between them. Keyt (1978: 142–3) usefully maps out some alternative views. One, that we have already discussed, is that the practical virtues have no independent value, and their purpose is simply to promote theoretical virtue. But if we assume that the practical virtues do have an independent value, then we might hold: (*a*) that theoretical virtue always overrides, so that practical activities are to be pursued (for their own sakes) only when purely intellectual activities make no demands; or (*b*) that the two are commensurable, and so may be 'traded off' against one another (for example, one might properly forgo some small amount of intellectual activity in order to perform a practical deed of real importance—say, saving one's

---

[36] Cooper (1975: 159–60) contended roundly that it could not. Keyt (1978: 145–6) replied with examples where this interpretation seems to be required. Cooper reconsiders the question in his (1987: 213–15), and is inclined to stick by his former contention.

neighbour from a burning house); or (*c*) that the practical virtues override, in the sense that theoretical pursuits should occupy one's attention only when all practical demands are satisfied. Keyt argues that all alternatives except the last would condone actions that we should regard as immoral (e.g. refusing to be interrupted by someone in need when one's theorizing is going well), and infers that Aristotle's view was, or should have been, the last.[37]

Perhaps he is right to say that this is what Aristotle's view should have been. (Though others have argued that if the practical virtues are taken seriously enough they leave *no* time for anything else.) However, there is absolutely no textual warrant for supposing that this is what his view actually was, for he simply has nothing to say on this question. Indeed he never, at any point, tells us how the different demands of the different virtues should be accommodated to one another.[38] Certainly he seems to suppose that practical wisdom will somehow harmonize the different practical virtues, but he says nothing useful on how it will do so. And when we come to the potential conflict—which seems even more important for him—between practical virtue on the one hand and theoretical virtue on the other, there is again simply no treatment of the problem. This may be because, though he has seen the problem, he does not think that anything of a general nature can usefully be said. But I think it more likely that he really has not seen how serious this problem is. He has to a large extent concealed it from himself by taking over the traditional conception of 'the three lives'—which he still thinks of as three separate lives, and not three aspects of a single life—and conducting his debate in terms of them. So he can assume that no one does try to live more than one of these lives at once, and the problem of apportioning a balance between them simply does not appear. But if he had been thinking as Keyt proposes, he surely could not have missed this question.

In all of X.7–8 we have only one relevant remark, and that is merely an aside. He has been claiming that a life of practical virtue will require many more external goods than does the life of contemplation (1178^a23-^b5). Then in one short sentence he concedes that since the contemplator is a man, living with other men, he will choose to act with [practical] virtue towards them, and this will require more by way of external goods than contemplation does by itself (1178^b5-7). He evidently does not think that this destroys his previous argument, presumably because he is taking it for granted that the contemplator will be *less* involved in the practical virtues than will one who lives the second (political?) life. But he gives it no further attention, and passes on at once to a different argument in favour of contemplation (1178^b7 ff.). We are left asking: what *reason* does the contemplator have for behaving virtuously towards his colleagues, and to what *extent* will he aim to do so? But Aristotle shows absolutely no interest in these questions. Indeed, he only mentioned the contemplator's virtuous action towards others because it has a bearing on what he had been saying about external goods.

---

[37] Similarly Engberg-Pedersen (1983: 108–21).

[38] Early on, in the discussion of mixed actions in III.1, he does seem to recognize that such decisions may be difficult (1110^a29-^b9).

I guess that Aristotle's thought is just this. The contemplator pursues theoretical excellence. But, at the same time, he is a man and not a god, so he will have to interact with other men, and there too he will pursue excellence. Why? Well, simply because good men just do pursue excellence in everything that they do. I do not imagine that he has taken the question any further than that. He has not pursued any of our 'escape routes', because he has not seen the problem that motivates them.

5. A comment

There is no such thing as *the* good life for man, and this can be argued even while accepting Aristotle's own premises.

Let us go back to his argument in I.7. The first claim is that what is special about man is that he can think. While *we* would no doubt wish to say that other animals can think too (though perhaps not quite so well), let us not stickle over this. And while *he* would wish to add that gods also think (indeed, they do nothing else), let us not stickle over this either. Let us accept, for the sake of argument, that thought is the distinctively human activity. Let us accept, too, Aristotle's next move, which claims that the good human being is therefore the one who thinks well, and infers from this that the life that is good for a man will be one that involves thinking well. (This is a fallacious inference, as has been noted, but nevertheless let us accept its conclusion.) Aristotle's thought then seems to be that the life that is best for a man will therefore be one that involves the best kind of thought, and indeed one that is devoted to this kind of thought as its dominant goal. Finally, since the best kind of thought is theoretical, rather than practical, it will be what he calls the theoretical life that is best for man. But we need not accept these last moves.

His arguments for saying that theoretical thinking is 'better' than practical thinking cut no ice with us. If we are asked what is valuable about theoretical thinking, we are hardly likely to reply that the main point is that it is useless for any purpose beyond itself (i.e. that it is pursued only for its own sake, and not for any further reason). On the contrary, a prevalent thought these days would be that it is theoretical thinking (in science) that has allowed man to advance far beyond other animals in the control of his environment. No doubt individual researchers are often motivated simply by the desire to know, but the result of their researches really does set man 'above' the other animals in all kinds of practical ways. This is a thought which would not have been so evident to Aristotle as it is to us, for it is only in the last few centuries that pure theorizing really has led to immense practical gains. Setting aside the obvious uses of pure mathematics (which even Aristotle could surely have appreciated), it must have seemed to him that man's superiority over other animals in practical matters is mostly due to the development of *technē* (craft, skill, expertise); and he very reasonably thinks of this as based upon experience and not theory, for so it was in his day. In his scheme of things, this makes it a kind of practical thinking, not connected with theorizing. So, if we try to put ourselves into his shoes, we should accept that pure theorizing is (mostly) useless for practical purposes. But why should that make it a 'better' kind of thinking? It seems to us an odd reason.

Perhaps the answer that best conforms to Aristotle's overall approach is to say that it is a kind of thinking that is furthest removed from any thought-processes that we can attribute to other animals; and perhaps we can take his own claim that it is most akin to the thought of the gods as a way of making this point. But we can still ask: why would this be a good reason for saying that it is 'the best' kind of human thinking? I do not see a feasible reply. No doubt, it *can* be pursued wholly for its own sake, but so too can many other kinds of thinking that human beings engage in (e.g. artistic creation, and the development of all kinds of skills—to take some trivial examples, the ability to solve crossword puzzles, or to discriminate fine wines, or to balance on a high wire). Moreover, these are just as much 'human' activities as is theorizing. We can run through Aristotle's other arguments in X.7 in a similar way. In each case there will be particular objections to be made, as we have noted, but there is also one that will be constant: why is this a reason for saying that theoretical thinking is 'the best' kind of human thinking? No argument seems satisfactory from this point of view. But if we drop this evaluative question, and ask instead whether there is reason to suppose that theoretical thinking is what specially distinguishes human beings, as a species, from all other animals, then the answer must surely be 'no'. As Aristotle would himself agree, theorizing is an occupation that is pursued only by very few human beings, whereas we *all* use practical thinking in working out what to do. It has a much better title, then, to be the distinctively human characteristic.

But it will not pick out some one life in particular as 'the good life for man'. Obviously, there are all kinds of occupations that people pursue, and Aristotle offers us no good reason for confining attention just to the academic life and the political life. One may set out to be an artist or a writer, a doctor or a lawyer, a designer of buildings or machinery or tableware, a singer or a footballer or an athlete, and so on and on without end. Or one's primary aim may simply be to care for one's household, one's spouse, and one's children. All of these are goals that may dominate a person's life, and all of them will call for practical thinking. (One may also, of course, try to combine several such goals simultaneously, e.g. to be both a doctor and a caring mother, and perhaps to do some painting on the side too.) I see no reason not to say that *any* such life may be 'a good life', and what makes it good (if it is) will be that it is pursued 'with excellence', in two different senses. First, to pursue the life of (say) a doctor 'with excellence' one must be a good doctor, successful at achieving the general aim of doctoring. This is a matter of technical excellence, but it surely counts as a specifically human excellence, even if it is not what Aristotle is thinking of as a 'human excellence' in I.7. The second condition is closer to what he is thinking of in I.7: a good life (in whatever career) is one that is pursued 'in accordance with the practical virtues', i.e. one that displays honesty, fairness, courage, temperance, generosity, and so on. These apply *whatever* one's career, and certainly are not confined (as Aristotle seems to suggest) to 'the political life'. (As we all know, politicians are as capable of immorality as all the rest of us, and this need not affect their (technical) excellence as politicians.) But the practical virtues are not themselves a career, an occupation, a goal that organizes and structures all the rest of one's life. One who sets out simply to cultivate the practical virtues has not yet chosen any dominant activity for his life, and has so far no definite direction in which to aim. That is not how the practical virtues function.

I should here add a qualification. *Some* occupations are closely connected with one practical virtue in particular. Aristotle would probably think of the soldier's life as one that calls specially for courage, though we might hesitate over this. (We do not associate courage with war so closely as he does, and we recognize other things too that are required of a good soldier, not to mention a good officer.) But we can surely agree that the occupation of a judge is specially connected with one kind of justice that he recognizes, namely justice in rectification. We might add that the virtue of munificence (i.e. generosity on a grand scale) is only to be expected of millionaires, and if we enlarged Aristotle's own list of virtues we could no doubt find further special connections of this sort. But still, most virtues are not tied to any one occupation or situation in particular. For example, *everyone* should be good-tempered, honest, trustworthy, and so on. These virtues are not specific to any particular occupations, and they do not themselves provide an occupation. One cannot, for example, embark upon a career of being honest, with everything else that one does chosen simply because it subserves that goal, for this by itself gives one no motive to *do* anything at all.

When Aristotle seeks for *the* good life (for man), he apparently thinks that the question is, or includes, the question of what occupation to follow. This is a mistake. The question seems feasible to him because right from the start he supposes that there are just two occupations to be seriously considered, namely the academic life and the life of politics, but in truth there is nothing specially 'good' or specially 'human' about these two occupations, and no good reason to ignore all others. He is mainly influenced, I imagine, by the tradition which he inherited, which contrasted just 'three lives' (the third being 'the life of pleasure'), but this tradition itself has no sound basis. I guess that it depends upon the thought that these three occupations are not *paid*, whereas all others can be thought of as pursued for the money that they bring. From our present perspective this is evidently unacceptable, for nowadays both academics and politicians are paid for what they do. But in any case a person may surely choose to pursue a particular activity, whether one of these two or many others (say, to be a sculptor, or a teacher, or an actress), even if they have no financial needs and are not interested in the money.

It is true that Aristotle does offer some reasons of his own, independent of the tradition, for preferring just his two lives. One is his suggestion that we may view all occupations as arranged in a hierarchy, with politics at the top, since the politician directs what other careers are to be pursued in his city, and to what extent (1094 ^{a}26–^{b}7).[39] But then, when we ask what the politician aims to promote, when issuing these directions, Aristotle's answer *seems* to be that he aims to maximize the pursuit

---

[39] It is never made clear quite how this supposed 'hierarchy' is actually organized, but I shall not fuss over that.

of theoretical wisdom (as is perhaps implied by 1145ᵃ6–9, cited earlier, p. 205). This is reinforced by his distinction in VI.4 between 'making' (*poiēsis*) and 'doing' (*praxis*), for apparently all other occupations can be seen as 'making' something, and it is only the theorizer who by his profession 'does' something (namely, he thinks) without 'making' anything. I have already said (Chapter IV, Section 3) that I do not think that Aristotle has this distinction quite right, but in any case it is not one that is likely to strike us as relevant here. For we could agree that a doctor 'makes' something (namely health), while an academic does not, without at all supposing that this puts the academic life above the doctor's.[40] As for the 'directing' role of politics, I have also offered some criticism of that idea (p. 10), but in any case it cannot be relevant here. For suppose that we accept Aristotle's highly idealized account of how the state 'directs' the various occupations; for example, it determines how many doctors there shall be. (This is surely more true of what happens in this country in this day than it ever was of what happened in any city in Aristotle's day.) Still, it cannot follow that this puts the politician's life 'above' the doctor's, for it is equally true that the state determines how many politicians there shall be (and, in our day, how many academics). This kind of 'direction', then, has no tendency to show that one kind of occupation yields a better life than another. As for the idea that the politician's aim is to maximize ('useless') theorizing, this should strike everyone as patently absurd, no less in Aristotle's day than in our own. I admit, then, that Aristotle does offer some arguments of his own, independent of the tradition, but they are quite without value.

We should conclude that the goodness of a life has very little to do with the particular activity (or activities) that are chosen as the dominant aim(s) of that life. It is true that we are apt to think of some occupations as 'more worthy' than others—for example, it is better to be a teacher than a barmaid—but such comparisons are difficult to justify without adopting some overall scheme of evaluation (such as utilitarianism) which has no basis in Aristotle's own thought. For all are equally 'human'. In any case, any reasonable scheme of evaluation will leave many different occupations on an equal footing. That is mainly why I say that there is no such thing as *the* good life for man, and that Aristotle should not have taken it for granted that there is such a thing, and have tried to arrange his arguments accordingly. On the matter of different occupations, the most that one can say is that some people are more suited to one, and some to another, but what makes a life a 'good life' is largely independent of this. My suggestion has been that it is one that is pursued 'with excellence' in both of the senses mentioned earlier: it is successful in pursuing its own aims, for it is informed by (technically) good practical thinking, and it takes due account of what Aristotle calls 'the practical virtues'. This suggestion seems to me to be broadly in line with what Aristotle himself says in I.7. But it will not yield any one life that is *the* good life.

---

[40]   This, in effect, simply returns to the point made earlier: we find it odd that what is 'useless' should be, just for that reason, ranked 'higher'.

Further reading

Aristotle's conception of God is outlined in his *Metaphysics*, book Λ, chapters 6–7 and 9–10. Since his conception is strange to us, this might perhaps be consulted, in order to fill in some of the background to his thought here in the *Ethics*. But for present purposes it is scarcely worth pursuing all the detailed problems that that text gives rise to. (A convenient translation and commentary is Judson, forthcoming.) Another issue that some commentators pay much attention to (e.g. Hardie 1968, ch. 16) is Aristotle's conception of the theorizing intellect, as that appears elsewhere in his work. (The most explicit discussion is *De Anima* III. 4–5, but that is desperately obscure.) My own feeling is that a student of the *Ethics* need not be too much concerned with that very awkward topic; what we are told in book VI of the *Ethics* itself (chs. 1, 3, 6, 7) is probably sufficient. (I add that Whiting 1986 has argued explicitly that the obscurities of the *De Anima* discussion are of no relevance to the *Ethics*.)

My selection of differing views on the doctrine of X.7–8 would be Cooper (1975, ch. 3), Keyt (1978), and Kraut (1989). Kraut's whole book is devoted to this one issue, but one might fairly concentrate just on his chapter 3, which gives the heart of his position. Alternatively one might see Curzer (1991), which presents a similar position clearly and more briefly. I should also note that Cooper has quite substantially revised his earlier views in his (1987). An account with which I have some sympathy is given by N. P. White (1988); a very different interpretation may be found in Everson (1998); a rather different kind of discussion, which pays less attention to problems of interpretation and is more concerned to extract an interesting moral, is Wilkes (1980).

Chapter X
Aristotle's methods in ethics

1. Dialectic

ONE of Aristotle's clearest and most explicit statements of method in the *Ethics* is at the beginning of book VII, introducing his discussion of *akrasia*. The passage runs:

> We must, as with other issues, first set out what appears to be the case and go through the puzzles that arise, and then demonstrate as far as is possible all the common opinions on what happens in this case, or if not then the majority of them and the most important. For if the difficulties are resolved, and the common opinions are left standing, it will have been demonstrated sufficiently. (1145^b2–7)[1]

There is a curiously modern ring to this proposal, reminding one of the approach of those philosophers earlier in this century who were called 'ordinary language philosophers' (e.g. Austin, Ryle, and in some ways Wittgenstein). The basic assumption is that what common people ordinarily say is likely to be broadly right, and yet it apparently gives rise to various puzzles. It may be that one thing that is commonly said seems to be in conflict with another, or that the two together seem to have an implication which is commonly rejected, and so on. The philosopher's task is then to resolve these puzzles by showing how they arise from misunderstanding. Typically it will turn out that each of the apparently conflicting views can be accepted, if it is taken in a suitably qualified way, and the puzzle arises only when the needed qualifications are overlooked. The philosopher's method, then, will be to distinguish various meanings of common terms, thereby revealing ambiguities in the common opinions, and showing how if these are resolved in the right way then the puzzles disappear.

There is, however, this difference between Aristotle's own approach and that of the 'ordinary language' philosophers just described. In the translation I have used, the passage just quoted speaks only of the *common* opinions, but in fact Aristotle casts his net more widely, and takes into account not only the opinions of 'the many' but also those of 'the wise', meaning by this the opinions of other philosophers. Of course the opinions of 'the wise' very often conflict with those of 'the many', but quite clearly Aristotle's general view is that we should consider both.[2]

---

[1] Note that the passage identifies what appears to be the case (the *phainomena*) with the commonly held opinions (the *endoxa*), and a little later Aristotle refers to these same things again as 'the things that are said' (the *legomena*, ^b20). (The translation 'the common opinions' is used by both Ross and Irwin. Barnes 1980 very fairly protests that the translation should be 'the reputable opinions'. The next paragraph explains why.)

[2] The *Topics* explains *endoxa* as 'what seems to be the case to all or to the majority or to the wise' (100^b21–2). For echoes of this in the *EN*, see e.g. 1095^a17–22, 1098^b27–9.

Our 'ordinary language' philosophers would here dissent, for in their view the philosophers' opinions are apt to arise from *mis*interpretation of the common opinions. Philosophers hanker for generality and system, so they fail to see that in ordinary language we constantly operate with tacitly understood qualifications and restrictions. Consequently they are apt to elevate into a general principle what should properly be understood only in a more limited way. Putting several such general principles together, and arguing rigorously from them, they are thus led into paradoxical positions which probably have no truth in them at all. So the modern 'ordinary language' philosopher will agree that the opinions of 'the many' should be taken very seriously indeed, but will view those of 'the wise' with great suspicion. Yet Aristotle *seems* to hold they each have an equal prima-facie claim to truth.

However, he is not in practice so even-handed. There certainly are occasions when he argues that this or that philosophical opinion has no truth in it at all. One example is his refutation of the Platonic view of 'the Good' in I.6; another is his rejection of the view that pleasure is a process in VII.12 and in X.3 (1152^b33–1153^a17 and 1173^a29–$^b20$). By contrast he is never, in the *Ethics*, willing to admit that a commonly held opinion has nothing whatever to be said in its favour, and where he does in effect reject such an opinion (as again with several common views on pleasure) he likes to explain how the view has arisen by hasty generalization from a perfectly good starting-point (VII.14, 1154^a22 ff.). We see how the wind sits from his much-quoted claim 'what seems to be the case *to everyone*, that we say is the case' (1172^b36–1173^a1). The method, of course, requires this principle, and on the surface it can apply only where 'the many' and 'the wise' are in agreement. But in fact the remark is used to reject the claim of Speusippus that pleasure is not a good at all, on the ground that everyone—intelligent or unintelligent—pursues pleasure (1172^b35–1173^a13; cf. 1153^b1–7). From this one might perhaps infer that it seems *to everyone except Speusippus* that there must be something good about pleasure, but Speusippus himself can hardly be included, since he explicitly denied it.[3] The moral is this: when speaking of what 'everyone' believes, Aristotle is prepared to ignore a philosophical opinion, but he evidently could not ignore a popular opinion in this way.

The method set out in VII.1 we may call the method of 'pure dialectic'.[4] When made fully explicit, it begins by stating what the relevant opinions are, and then stating what puzzles they give rise to. Subsequent discussion then aims to resolve these puzzles, and I shall understand the discussion as 'purely dialectical' if the only technique employed is that of drawing distinctions which allow the needed qualifications to be introduced, so that the appearance of contradiction is dispelled. It

[3] More strictly; in order to include Speusippus one would at least have to show that his actions belied his words, but Aristotle makes no attempt to show this.

[4] Dialectical reasoning is the subject of all of Aristotle's *Topics*. For a general description, see *Topics* I.1–2 and perhaps I.10–14. It is said to have three purposes, namely for training in argument, for success in conversations, and for philosophical knowledge (101^a26–8). The *Topics* itself is almost entirely devoted to the first of these, which envisages an argumentative debate between two people, one of whom is 'the questioner' and the other 'the answerer'. But Aristotle himself uses it for the third purpose in many of his philosophical writings, as is demonstrated by Owen (1961).

may fairly be said that the discussion of *akrasia* in VII.1–10 illustrates this method very nicely. As I have noted, the introduction does not quite say all that it should, for apparently it refers only to the opinions of the many, and it is only these that are then listed in the rest of chapter 1, at 1145^{b}8–20. But then when chapter 2 sets out the puzzles that arise, it in fact begins at 1145^{b}21–31 with a puzzle that arises from Socrates' opinion, and this is clearly not an opinion of 'the many', but an opinion of one of 'the wise'. Allowing for this minor correction, let us briefly review the structure of the whole discussion.

The Socratic opinion which creates a puzzle is his claim that knowledge cannot be dragged about like a slave, and to resolve this we must ask what kind of correct understanding a man has when he acts akratically. The puzzle is stated at 1145^{b}21–31, and some preliminary suggestions for disarming it are then briefly explored and rejected at 1145^{b}31–1146^{a}10. The resolution of the puzzle is given in the next chapter at 1146^{b}24–1147^{b}17, and it depends wholly on distinguishing different senses in which one may be said to 'know' that what one is doing is wrong. There are two main distinctions: one may know the universal but not the particular, and one may know in one sense (in that the knowledge is within one) but not in another (in that it is not before one's mind at the time of acting). Using both of these, Aristotle infers that Socrates was right at least to this extent: it is not universal knowledge that is 'dragged about', but only particular knowledge, and when this happens there is a sense in which it is not knowledge (i.e. it is not before the mind). But the common opinion is right too, in so far as there is also a sense in which it still is knowledge, even if it is not before the mind. This resolves the puzzle, showing that there is truth in both views, but no contradiction.[5]

In chapter 2 Aristotle has in fact enumerated six different puzzles, which may be summed up as posing these six questions:

1. In what sense does the akratic 'know' that he is doing wrong? (1145^{b}21–31)
2. What is the relation between *enkrateia* and temperance (*sōphrosunē*), and between *akrasia* and self-indulgence (*akolasia*)? (1146^{a}9–16)
3. What is the relation between *enkrateia* and obstinacy, or between *akrasia* and changing one's opinion? (1146^{a}16–21)
4. Can we call a man virtuous if he does the right thing because he holds wrong opinions but is also akratic, and so does not act in accordance with them? (1146^{a}21–31)
5. Should we say that *akrasia* is worse than self-indulgence, because *akrasia* cannot be cured while self-indulgence can? (1146^{a}31–^{b}2)
6. What is the sphere of *akrasia*? (1146^{b}2–5)

Chapter 3 opens by saying that we must discuss first (1) and then (6), and it adds a brief elaboration of (6) (1146^{b}9–24). Then, as we have seen, it proceeds to a resolution of problem (1), the Socratic problem. Problem (6) is then discussed at some length in chapters 4–7, problem (5) in chapter 8, and problems (4), (3), and (2) in chapter 9.

---

[5] Bolton (1991) gives a very detailed analysis of how the dialectical method is used in VII.3. He does not discuss its use elsewhere.

The discussion of the sphere of *akrasia* calls forth a wealth of distinctions. Chapter 4 claims that when we speak of *akrasia* without qualification we mean being overcome by the same bodily pleasures as were said in III.10 to be those relevant to temperance and self-indulgence, whereas when someone pursues other pleasures—e.g. victory, honour, wealth—more than he ought, and against his better judgement, he may by analogy be called akratic with respect to these things, but only if the relevant qualification is added. Chapter 5 distinguishes *akrasia* in its ordinary form from the extreme form, which is 'bestial', where the overmastering desire is one that is not natural to man at all, and adds some further distinctions of the same sort. Chapter 6 contrasts unqualified *akrasia* with *akrasia* with respect to anger (and argues that the latter is less disgraceful). Chapter 7 compares *akrasia* and *enkrateia* with softness and endurance, the latter being concerned with resisting (or failing to resist) pains, rather than yielding (or not yielding) to pleasures.[6] It also adds the useful distinction between impetuosity and weakness, as two contrasting forms of *akrasia*. Chapter 8 then turns to problem (5), and argues for a negative answer, on the ground that *akrasia* is not a permanent state, but something that occurs only every now and then, whereas self-indulgence is permanent. Moreover, the akratic does have the right principles, and knows that his action was wrong, and so can be exhorted or trained to change his ways. By contrast the self-indulgent man has the wrong principles, but no understanding that they are wrong, and so cannot be cured.[7] Finally, chapter 9 turns first to problems (3) and (4), which it not unreasonably runs together. The main point that Aristotle makes here is that *enkrateia* and *akrasia* are to be understood as sticking to, or failing to stick to, the *right* principles. This distinguishes *enkrateia* from mere obstinacy, which is a refusal to change one's mind on *any* principle, right or wrong, and at the same time it allows that one who changes his mind for good reason is not to be counted as akratic. For *akrasia* is a matter of being misled by one's desires, not of being persuaded otherwise by reasons. This resolves problem (3), and at the same time (though Aristotle fails to spell this out) problem (4). For, on this account, one who fails to live up to the wrong principles is not to be counted as akratic. Finally, problem (2) is explicitly addressed in the closing paragraph (1151^b32–1152^a6), though indeed the correct answer has been obvious all through the discussion: one who acts enkratically will act in the same way as the temperate man does, and one who acts akratically will act in the same way as the self-indulgent man does, but their motivations are quite different.[8]

During all of this long discussion there is virtually no appeal to what one thinks of as specifically Aristotelian principles of ethics or epistemology or metaphysics. In so far as there is any authority that is appealed to, to justify the various comparisons and distinctions that are drawn, it is simply common usage. (For example, what *we*

---

[6] That this is still part of the discussion of problem (6) is clear from 1146^b11–14.

[7] The Greek word for 'self-indulgent' (*akolastos*) means literally 'not punishable', i.e. 'not to be reformed by punishment' (cf. 1150^a21–2).

[8] Ch. 10 appears to be something of an afterthought, outside the original plan. (It perhaps aims to elaborate the earlier remark at 1146^a4–9.) But it introduces no change of method.

mean when we talk of *akrasia* without qualification is *akrasia* with respect to such bodily pleasures as food and drink and sex.) To be sure, one can find a few exceptions to this generalizing remark. For example, the presumption that knowledge of the universal is more properly counted as knowledge than knowledge of the particular (1147^{b}13–17) is a familiar Aristotelian (and Platonic) prejudice, but hardly a matter of common opinion. Or again, it is surely a specifically Aristotelian idea that the virtue of temperance lies between two opposing vices, one the familiar vice of self-indulgence, and the other nameless, and commonly not recognized, but a vice all the same (1150^{a}22–3). But such exceptions as these are few and far between, and play no important role in the main discussion. That is why I say that the whole discussion of *akrasia* in VII.1–10 is a very nice example of the 'purely dialectical' method which it describes at the outset: it draws numerous distinctions, providing a clear map of the 'logical geography' of the concepts concerned, all with a view simply to resolving the puzzles with which it began. And (apart from problem (1)) the resolution is entirely successful.

There is no other major discussion in the *Ethics* which so well exemplifies this purely dialectical technique,[9] though there are several which share some of its features. To illustrate, let us briefly consider the second part of book VII, which treats of pleasure. Here the discussion starts appropriately enough, by listing some opinions held about pleasure (1152^{b}8–12), though it soon becomes clear that they are mostly philosophers' opinions rather than common opinions. For the next thing that happens is that Aristotle gives arguments for these opinions, which are mostly philosophers' arguments (1152^{b}12–24). He does not pause to state explicitly the puzzles arising from the opinions, but instead launches into a critique of the arguments stated, which occupies the next two chapters (VII.12–13). One may agree that this procedure is very reasonable in the present case, for (*a*) each of the opinions listed is clearly a controversial opinion, and so itself provides a 'puzzle', and (*b*) when it is *philosophers'* opinions that are in question a critique of the arguments given for them is entirely appropriate. But the contrast with VII.1–2 should be noted. In the discussion of *akrasia* Aristotle never mentions or criticizes a philosopher's argument, not even where—as with Socrates' opinion—it would be very relevant to do so. Moreover in VII.1 the aim is clearly stated to be to *maximize* the truth in the opinions first listed, but there is no such aim in VII.11. On the contrary, these are philosophical opinions which Aristotle mostly disagrees with.

An equally important divergence in method is this. In the course of his critique of rival arguments (VII.12–13) Aristotle introduces a positive doctrine of his own, namely that pleasure is not a process but an activity, and this doctrine cannot be explained as motivated by purely dialectical considerations. The appropriate dialectical procedure would presumably be to observe that *some* pleasures (e.g. those of contemplation, 1153^{a}1) are activities and not processes, and to conclude that we shall maximize the truth in all views by allowing that pleasures may be either processes or activities. (The opponent's error can then be explained as due to a failure to distinguish processes from activities; cf. 1153^{a}15–17.) But plainly this is not

---

[9]   Some minor discussions may be claimed to be good examples, e.g. the discussion of equity in V.10.

the line that Aristotle in fact takes. While he does concede that one may be pleased while a process is occurring (obscurely in book VII, at 1152^b34–6 and 1154^b17–19; more plainly in book X, at 1173^b11–13), still he consistently denies that in such a case the process *is* the pleasure. For this denial he no doubt had what seemed to him to be good reasons, and I have tried to reconstruct them in my earlier discussion, but in any case it is quite clear that they are not 'purely dialectical'.

In fact I think that there is very little of the *Ethics* that can be explained as *simply* an application of the method set out in VII.1, though it is a reasonable generalization that most of what one may call its philosophy of mind and action does make *some* use of dialectical techniques. (For example, the discussions often begin by setting out various possibilities, and then giving arguments against each.) An exception is the discussion of what to count as voluntary or involuntary in III.1, but we may perhaps say that this discussion rides on the back of the earlier treatment in *EE* II.7–9, and that that earlier treatment was (in the way noted) dialectical.[10] Allowing for this, we can perhaps conclude that dialectical methods are given at least some role in all of III.1–5 and VII and X.1–5. But from now on I shall pay no further attention to this philosophy of mind and action, and shall concentrate on the specifically ethical part of the book, i.e. on what it has to say about what is and is not good. In this area we shall find that dialectic, as described in VII.1, has virtually no role to play.

2. Aristotle on first principles

An obscure passage in I.4 apparently tells us that there are 'first principles' of ethics, and that our task is to reach them. Unfortunately the Greek word (*archē*) which is often to be translated as 'first principle' has also the more general sense of a 'beginning' or 'starting-point' of any kind, and sometimes it is not clear just what kind of starting-point Aristotle has in mind when he uses this word. (A 'first principle' is a 'starting-point' for the orderly exposition of a completed science, for example the axioms of geometry, which deduction starts from.) I translate the passage using the ambiguous word 'starting-point' throughout.[11]

We must not fail to notice that there is a difference between arguments from the starting-points and arguments to them. Plato too would raise this question, quite rightly, and would ask whether our route was from or to the starting-points, as on a racecourse one may be going from the judges to the end or back again.[12] Now one must start from what is knowable,[13] but this can be taken in two ways, either as what is knowable to us or as what is knowable

---

[10] Meyer (1993, ch. 3), is a useful treatment of this point.

[11] Irwin prefers 'origin'; Ross uses sometimes 'first principle' and sometimes 'starting-point'.

[12] A Greek racecourse was shaped thus

Start

Finish

The judges would be at the end that was both starting-point and finishing-point.

[13] *Gnōrimon.* Irwin translates 'known' and Ross 'evident'. But it is odd to use either of these words of a starting-point (= first principle) that is not yet either known or evident.

without qualification. Presumably, then, *we* must start from what is knowable to us. That is why a person must have been brought up well in his habits if he is to be a satisfactory pupil on the topic of what things are noble or just, or more generally on political matters. For a starting-point is a 'that so-and-so', and if this is sufficiently clear there will be no need of an additional 'because'. Such a person will have or will easily get starting-points. But as for one to whom neither of these applies, let him hear what Hesiod says:

> The best of all is who knows all himself:
> Good too is who attends to sound advice.
> But who himself knows not, nor in his heart
> Can fix another's words, he's of no use.

$$(1095^a30-^b13)$$

Now it is fairly clear that the 'starting-points' mentioned in the first two sentences are first principles of a subject. This is confirmed by the reference to Plato, who in the *Republic* spoke of a 'route upwards' to a starting-point, and then a 'route downwards' from it, and clearly meant by this a first principle. (I shall say a little more of Plato's position shortly.) It is also confirmed by the ensuing distinction between what is knowable (i.e. known?) to us and what is knowable without qualification. For this is a distinction Aristotle also draws elsewhere,[14] and it is clear that he intends the latter to apply to the first principles of a subject. So *our* starting-point is one thing, and the starting-point which is a first principle is another. But presumably in either case a starting-point is a 'that so-and-so', for it is what is accepted to begin with, without further reason, whichever direction we are travelling in. It is natural to suppose that the pupil who has been well brought up either has or can easily get what are starting-points *for us*, i.e. starting-points on the upwards route to the first principles. It has to be admitted that the text could be taken differently, but it would be odd for Aristotle to suppose that genuine first principles lie readily to hand; he surely thinks—as Plato did—that some philosophical work is necessary to attain them. But what are these first principles of ethics, and by what method are we supposed to ascend to them?

The discussion of dialectical reasoning in the *Topics* very clearly claims that it is useful for the establishing of first principles. It says:

It is useful for philosophical knowledge, because if we are able to raise problems on both sides we shall more easily see the truth and the falsehood in each. Further, it is useful for the first things that each science is concerned with. For one cannot say anything about these that depends on the principles (*archai*) that are special to the science in question, since the principles are in every case what comes first. One must therefore treat of them through the reputable opinions (*endoxa*) held on each. This belongs peculiarly, or most specially, to dialectic. For, since it is investigative, it contains the path to the principles of all enquiries.
$$(101^a34-^b4)$$

This function of dialectic is not further explored in the *Topics*, since that book is almost entirely concerned with dialectic in a different role, and we are left to wonder how exactly it can achieve this task. Even if we grant the basic assumption

---

[14]   e.g. *Prior Analytics* 68^{b}35–7; *Posterior Analytics* 71^{b}33–72^{a}6; *Physics* 184^{a}16–23; *De Anima* 413^{a}11–16; *Metaphysics* 1029^{b}3–12.

that any reputable opinion is likely to have at least some truth in it, and that the procedure of resolving problems will clarify just how much truth there is in each, still there are at least three questions that arise. (i) It is not clear how such clarification can by itself tell us which truths are 'prior' to which, nor therefore which come 'first'. (ii) It is not even clear why we should expect that the genuinely 'first' principles must even be among the reputable opinions that we begin with, or among what results from their clarification. (iii) It is surely a possibility that *some* of the opinions that we begin from, even those very widely held, in fact contain no truth at all; if so, and if dialectic insists upon finding some truth in them, then it can only lead us astray. I shall have nothing more to say about the third question, but we can make some headway with the first two.[15]

Outside the *Topics* Aristotle does not usually refer to dialectic as providing the first principles of a subject, but speaks rather of *nous*, which works by induction (*epagōgē*). The *locus classicus* is the last chapter of the *Posterior Analytics*, but we have seen that he says the same in the *Ethics* too. (In VI.6 the task of discerning the first principles of the theoretical sciences is assigned to *nous*, though its *modus operandi* is not there described. But we can infer from VI.3 that it must operate by induction, for we are told at 1139^b24-31 that induction and deduction (i.e. syllogism) are the only two alternatives, and deduction is ruled out.) It may be suggested that this is not really a change of view from the *Topics*, since that work did count inductive reasoning as at least one variety of dialectical reasoning (105^a10-19). I think this would be over-generous, but the point is not worth debating.[16] Let us just say that Aristotle seems to recognize *two* methods of 'ascending' to first principles, one being the dialectical clarification of reputable opinions, and the other induction. The addition of induction goes some way towards answering our question (ii), for it would be perfectly reasonable to say that induction from the (clarified) opinions that reputable people have held may well lead us to more general principles, which in fact have never previously been recognized. But will that be enough?

A further resource which Aristotle surely should have recognized, though in fact he never does discuss it, is what Plato himself said about how to 'ascend' from what is known to us to what is genuinely a first principle. This is what Plato calls 'the method of hypotheses', which figures briefly in the *Meno* and is further discussed and elaborated in the *Phaedo* and the *Republic*.[17] If I may oversimplify very considerably, the essence of Plato's proposal is this. We start with some facts that are

---

[15] The third question is the mainspring of Irwin's major work (1988*a*). He hopes to distinguish between 'strong' and 'weak' dialectic, in such a way that the former can be seen both to be more reliable and to lie behind all of Aristotle's major theses. All that I shall say of this is that his attempt to apply this line of thought to the *Ethics* (chs. 16–21) seems to me quite unpersuasive. (Question (ii) is nicely elaborated by Barnes 1980.)

[16] When the *Topics* speaks of inductive arguments, it is thinking of how they may be used in a debate between 'questioner' and 'answerer'. There is no suggestion that they are especially useful in the discovery of 'first principles'.

[17] *Meno* 86d–87b; *Phaedo* 99e–101e; *Republic* 510b–511d. (Much of the *Parmenides* may also be construed as concerned with this method, but the interpretation of that dialogue is too controversial for me to discuss it here.)

known (or reasonably believed?), and we ask what hypothesis would best *explain* them. A proposed hypothesis is checked (by Socratic *elenchus*) for agreement with everything else already known, and if it passes this test, and really does explain what it was meant to explain, then it is tentatively accepted. We then seek for a yet higher hypothesis that will explain this in turn, and so on. To put the proposal in modern terms, it is that we employ the method known nowadays as 'inference to the best explanation', which is another way of reasoning (besides 'dialectic') that does not naturally fall under one or other of the two alternatives 'induction or deduction'. It is, moreover, much nearer to the mark as an account of how one does reach the first principles of the theoretical sciences (including mathematics) than is the rather naïve proposal 'by induction'. If we add this method of argument to those previously noted, then we have a response to our question (i) above, and a further response to question (ii). So far as (i) is concerned, it is precisely the question of what explains what that makes some facts 'prior' to others, and so gives us the hope of ascending to what is 'prior' to all else, i.e. 'first'. As for (ii), it is nowadays a commonplace that the search for explanations may lead us far beyond the 'data' that we initially began with.[18]

Reeve (1992, ch. 1) puts all these resources together in an attractive way. Dialectical resolution of problems will allow us to clarify our initial 'data' (i.e. the opinions of the many and the wise); induction will allow us to generalize from these; and by trying out tentative explanatory hypotheses we shall come to see what comes higher than what in the explanatory order.[19] Eventually, it will at last strike us that we must now have got it right, for everything fits. This is *nous*; it comes as the final crown of all our search. And that is how the ascent to first principles is accomplished.

I have put all this in completely general terms, as an account that Aristotle might accept of how this 'ascent' is achieved. But the evidence (such as it is) is almost entirely drawn from what he says of the first principles of the *theoretical* sciences. Would he say that the same applies too to the first principles of ethics? (Reeve thinks that the answer is 'yes', but nothing that I have cited so far would support this.) In any case, what does he take to *be* the first principles of ethics? The passage which I cited at the beginning of this section tells us that he thought that there are such principles, but it does not say either what they are or how we are to reach them.

The most relevant passage is from the end of I.7, which says,

Nor should one ask for an explanation in all cases alike; in some cases it is enough to have well demonstrated the 'that so-and-so', as indeed applies to the starting-points. The 'that so-and-so' is what comes first, and is a starting-point. Some starting-points are seen by induction, some by perception, some by a kind of habituation, and others in other ways. And one must try to pursue each in the way that accords with its own nature, and take pains to articulate it well, since it has a great influence on what follows. For the starting-point seems more than half of the whole, and through it much of what is sought will become clear. (1098^{a}33–^{b}8)

---

[18] It may also allow us to question the 'data', which gives some response to question (iii). (We may be able to explain why an opinion is held without assuming that it has *any* truth in it.)

[19] Reeve assumes that Aristotle too saw this point (1992: 63). I regard this as very charitable.

Again there is the awkward doubt over which kind of starting-point is under discussion, but there is some reason for saying that the whole passage should concern first principles. At any rate, it is certainly most natural to take the final sentence in this way, and in that case the previous two would seem to demand the same interpretation. Moreover, the context of the passage strongly suggests that it is all intended as a comment on the definition of *eudaimonia* just reached in 1097^{b}22–1098^{a}20, i.e. as an activity of the rational part of the soul, performed with excellence (virtue); and that would seem to be quite a good candidate for being a (or the?) first principle of ethics.[20] For as soon as this definition has been stated Aristotle proceeds to comment (*a*) that the definition is only an outline, which requires to be filled in further (1098^{a}20–6), and (*b*) that we should remember that ethics does not admit the same exactitude as mathematics (^{a}26–33);[21] (*c*) he then adds the passage just cited. Thus each of (*a*), (*b*), and (*c*) would seem to be intended as comments on this definition of *eudaimonia*. But if we accept this, then two very surprising consequences follow: (i) Aristotle is apparently saying that the definition of *eudaimonia* (or of 'the human good') that he has just given is not exactly stated here, and—it would seem—that no further elaboration could make it exact; and (ii) he certainly seems to be implying that *this* first principle, like other first principles, is reached either by induction or by perception or by habituation, yet none of these suggestions seems initially plausible.

On the contrary, one might very naturally suppose that starting-points reached by perception or habituation will be starting-points from which *we* start, rather than the first principles of a topic. In the case of perception this hardly needs further explanation; in the case of habituation further explanation is easily given. The kind of thing Aristotle will have in mind are such propositions as 'one ought to act virtuously', 'justice is a virtue', 'courage is a virtue', and so on. For these are the kind of starting-points that the well-brought-up young man accepts, without needing any 'because' in addition. Moreover, we may easily suggest why Aristotle might think of such propositions as 'inexact', namely because they do not give you clear guidance on how to act in all possible situations. On this account, when Aristotle talks of the 'lack of exactness' in ethics he will be thinking of the same thing as made him say, when defining virtue of character, that it is a middle 'relative to us' and 'not the same for all', namely that the situations in which one may find oneself are so many and so various that no simple rule can be laid down which will cover all cases.[22] Should we perhaps say, then, that while part (*a*) of 1098^{a}20–^{b}8 evidently is intended as a comment on the definition of *eudaimonia* just reached, nevertheless parts (*b*) and (*c*) have digressed into some other general points about ethics, and do not concern first principles at all?

We do not have to. First, we can accept quite straightforwardly that the definition

---

[20] That *eudaimonia* is an *archē* which we aim at in all that we do, is explicitly stated later in book I (1102^{a}2–4). It is probable, but not completely certain, that *archē* here means 'first principle', in the sense that I am now concerned with.

[21] The reference back is to 1094^{b}11–1095^{a}2. ('Exact' is Irwin's translation of *akribes*; Ross prefers 'precise'.)

[22] This interpretation fits the earlier passage at 1094^{b}11–1095^{a}2 well enough.

of *eudaimonia* given so far is merely an outline, for we know that Aristotle has more to tell us on this topic in X.6–8. Second, if Aristotle's main thought about the lack of exactness in ethics is as I have suggested, then we can see why he should say that this definition is 'inexact', and must always remain so, however much it is further filled out. For this does not mean that there is anything in the definition that is not entirely true, nor (despite 1094^{b}19–22) that it holds only 'for the most part'; that is something that applies to specific, but rough-and-ready, rules of conduct, such as 'keep your promises', but not to the definition of the key concept *eudaimonia*. What it does mean is just that, however full the definition, it will not by itself provide a guide to conduct that can be mechanically applied to all particular situations. For that task practical wisdom, based upon experience, will always be needed in addition. Finally, we need not suppose that genuine first principles have been lost sight of in the remark about how starting-points are to be attained. To be sure, it is difficult not to suppose that perception will only yield starting-points *for us*, but we should pause before saying the same of either habituation or induction.[23]

As I have said, induction is Aristotle's *usual* account of how first principles are reached, but there is also a reason for supposing that *in ethics* he means to assign this role to habituation instead. Indeed, in VII.8 he seems to say exactly this:

Virtue preserves the first principle (*archē*) and vice destroys it, and in actions it is the goal that is the first principle, as in mathematics it is the hypotheses [i.e. axioms]. In neither case is it reasoning (*logos*), but [in the case of actions] it is virtue, either natural or habituated, that teaches us right opinion about the first principle. (1151^{a}15–19)

But how could this apply to the present example of an ethical first principle, which is (by hypothesis) the definition of *eudaimonia* reached in I.7? The plain answer is that it could not.

One might say that this definition of *eudaimonia* consists of two parts: first Aristotle elaborates some interrelated formal or structural conditions on what the final good must be (i.e. it must be most complete (final), pursued only for its own sake, and 'self-sufficient'), and then he offers the argument based on man's 'function' to conclude that it is an activity of the rational soul pursued with excellence.[24] How does he think that the formal conditions are supported? Well, in fact he simply appeals to 'what seems to be so' (1097^{a}28, ^{b}8, ^{b}20), to 'what we say' (^{a}30, ^{b}8) and to 'what we think' (^{b}16). This is presumably an appeal to commonly held views, and as such it may be said to fall under dialectic, but is hardly to be credited to natural or habituated virtue. As for the argument about man's 'function', it is completely obvious that training and habituation in virtue makes no contribution to this. We must conclude that the starting-points that are attributed to virtue do not include the definition of *eudaimonia* that is reached in I.7. They are, if anything, the more

---

[23] We may surely leave out of account the throwaway remark 'and other starting-points in other ways'. If he is indeed meaning to comment on the definition of *eudaimonia* just reached, he cannot also be meaning this catch-all clause to cover it.

[24] As already observed (pp. 25–6), a major difficulty is to see how these two 'parts' are supposed to be related to one another.

concrete details which fill out this definition by specifying what does and does not count as a virtue.[25]

We are left with the view that in Aristotle's own opinion the definition of *eudaimonia* in I.7 is reached by induction, and this is perhaps comprehensible. Induction is basically a matter of generalizing from particular cases. Even today we are none too clear about how this works,[26] and Aristotle is certainly no clearer. But at least we can observe that the leading premiss to Aristotle's argument is that man has a function, and in support of this he does offer other examples of things that have functions (1097^{b}26–33); the second premiss is that where a thing has a function its good is to be found in its function, and here too he offers one example where this can be clearly seen (the good lyre-player, 1098^{a}9–12), and could easily have offered others.[27] So we may perhaps see these points as indicating two inductive arguments that lie at the basis of his discussion. (This strikes us, of course, as a very superficial account of how the argument is supposed to work, but it may, for all that, be Aristotle's own account.)

But if this account is right, then one can only conclude that Aristotle has not thought his position through. He believes that the first principles of any subject must be reached by induction, because he is apt to suppose that induction and deduction are the only two forms of reasoning, and that a *first* principle cannot be established by deduction. But in the present example he simply has not asked himself what the premisses to the induction are. For if he had so asked he must surely have seen that the premisses in question are *not* those rough-and-ready moral generalizations which are accepted without argument by those who have been well brought up. Anyone, however badly brought up, is in an equally good position to see that the flute-player has a function, and the eye has a function, and that the good eye or flute-player is the one that performs its function well. So the passage in I.4 with which this section began is wholly misleading, for it says that a good upbringing is required if the pupil is to grasp the starting-points for the route upwards to the first principles, yet on the reconstruction that I have offered this is clearly not the case. The role of good upbringing lies elsewhere.

I add one further observation before leaving Aristotle's own account of what his method is in reaching the definition of *eudaimonia* in I.7: he surely does not think that he is employing 'pure dialectic', as that was explained earlier. At one point it had looked as if he was going to. I.4 begins by noting that many different views on *eudaimonia* are held (1095^{a}14–28), and it goes on to say that, while it is not worth examining them all, it will be enough to examine 'those that are most widespread, or most seem to have something to be said for them' (^{a}28–30). If this promised

---

[25] It may be relevant to note here that in the *EN* it is only in books VI and VII that habituation in virtue is said to yield first principles. But these books are common books, and so may reflect the *EE* rather than the *EN*. *EE* II.1 does offer a general account of *eudaimonia* which is similar to that in *EN* I.7 (though it is reached by a very different route), but it never describes this as a 'first principle'.

[26] I do not believe that the so-called 'new riddle of induction', introduced by Goodman (1955), has yet been adequately resolved.

[27] In the somewhat similar argument in *EE* II.1 he does offer others, and he explicitly calls the inference an induction (1218^{b}37–1219^{a}6).

examination were aiming to be dialectical, it would seek a means of maximizing the truth in all of them. But in fact what happens is that a widespread view (concerning the 'three lives') is very briefly examined and dismissed in I.5, and then the Platonic view of 'the Good' is examined and dismissed in I.6. This ends the promised examination, with no truths established at all, and then I.7 puts forward Aristotle's own argument about the 'function' of man, which pays no attention whatever to the views of others. He was surely quite well aware that this argument does not use the dialectical procedure of VII.1. But he probably did think that I.8–12 does, for there he does try to show that the result that he has reached in I.7 is in harmony with (a selection of) views commonly held about *eudaimonia*. (With some charity, one might go a little further: perhaps his aim is to show how it explains those views.)

To sum up both this section and the last: Aristotle gives us two statements of the proper method to be followed, both in ethics and elsewhere, namely the purely dialectical method of VII.1 and the method obscurely indicated in I.4 of ascending to first principles 'by induction' from what is already known to us.[28] These methods do not distinguish ethics from other sciences. But he also in VII.8 mentions a method that is presumably peculiar to ethics, for here it is said to be training and habituation in virtue that provides us with first principles. (This is no doubt related to his point in I.4, that only those who have been well brought up are suitable students of ethics.) Yet none of these apply to the method that he actually uses in I.7 to reach his definition of *eudaimonia*. It certainly is not the dialectical method, nor something that relies on good upbringing, and though the function argument may use induction it obviously cannot be reduced just to this. (The premiss that man has a function is not adequately supported by the induction offered; the premiss that this function cannot include anything that man shares with other animals is not supported at all; nor is the premiss that only reason satisfies this condition; and so on. Aristotle no doubt has broader and more far-reaching views which make these assumptions plausible to him, but it would be over-simple to put these down just to 'induction'.) I conclude that what little Aristotle tells us about how to reach first principles in ethics does not actually fit what seems to be the most obvious candidate for being a first principle in his ethics, namely the account of what *eudaimonia* is.

It is not clear what other claims of his *Ethics* he would count as 'first principles', and I shall not speculate on this. Instead, I shall now ask what *we* might reasonably regard as its 'first principles', and go on to look at how they are in fact supported. This will at the same time serve as a summary of his argument, and as a reminder of some criticisms that I have already made.[29]

---

[28] On one interpretation (which I do not endorse) an obscure passage in VI.11 also says that *nous*, operating by induction from particulars, provides the first principles of ethics (1143ᵃ35–ᵇ5, discussed on pp. 91–3).

[29] As promised earlier, I shall not be paying any further attention to the associated philosophy of mind and action.

3. Aristotle's basic principles

A first attempt to list the basic principles of Aristotle's ethics, the main claims which form a framework for all the rest, might yield something like this:

1. The supreme good for man is *eudaimonia*.
2. *Eudaimonia* is an activity of the rational part of the soul, performed with virtue (excellence).
3. There are two relevant kinds of virtue (i.e. virtues of the rational part), namely virtues of character and virtues of intellect.
4. Virtue of character is a disposition to feel (and to act) that is 'in a middle relative to us', that involves choice, and is determined by practical reason.
5. There are two (relevant)[30] virtues of intellect, namely the virtues of practical reason and of theoretical reason.
6. The task of practical reason is to determine what actions will achieve the supreme good for man, *eudaimonia*.[31]
7. The task of theoretical reason is to discover the necessary and eternal truths.
8. The supreme good for man, *eudaimonia*, is the contemplation of the truths revealed by theoretical reason.

I shall, however, modify this list a little before discussing it.

Presumably a list of basic (or 'first') principles should not include those that are deduced from others. So presumably it should not include (8), since Aristotle certainly offers to deduce it. But when we consider the premises to his deduction, it is clear that they go beyond those already listed in (1)–(7). The more important of them are these:

(*a*) The highest virtue is that of theoretical reason.
(*b*) *Eudaimonia* is the only thing pursued for its own sake, and contemplation is the only thing pursued for its own sake.
(*c*) The good for man is special to what is really distinctive about man, and a man *is* his theoretical reason.
(*d*) The gods enjoy *eudaimonia*, and their only activity is contemplation.

Henceforth I shall simply replace (8) by (*a*) above. The first part of (*b*) is certainly part of Aristotle's argument for the thesis originally labelled (1), and I shall treat it as such. Similarly the first part of (*c*) is certainly part of his argument for the thesis originally labelled (2), and I shall treat it as such. The second parts of (*b*) and (*c*) I shall ignore in what follows, merely remarking here that we are given no good reason for them, and that they are not in themselves the slightest bit plausible, but seem to be adopted only to support a conclusion that Aristotle has already reached on other grounds. I shall similarly ignore (*d*), though in this case I suspect that it was in fact influential in Aristotle's own thinking. But I remark here that, in relation

[30] This sets aside as irrelevant the virtue of technical excellence (VI.4).
[31] Practical wisdom is defined as 'a true disposition, concerning action, involving reason, about what is good for man' (VI.5, 1140[b]20–1).

to Aristotle's *methods* in ethics, it displays an interesting incongruity that we shall also find elsewhere. The first claim in (*d*) is supported simply by an appeal to common opinion, but the second is certainly not a common opinion; it is rather a claim of Aristotle's own theology, which he has argued elsewhere (principally in *Metaphysics* Λ). The two make uneasy bedfellows. With so much by way of preamble, I now return to the initial list, as modified.

Let us begin with (1), which I now fill out thus:

> (1*a*) The good for man is pursued (by men) wholly for its own sake.[32]
> (1*b*) *Eudaimonia* is the only thing pursued (by men) wholly for its own sake.
> ∴ (1) The good for man is *eudaimonia*.

In our eyes (1) by itself is controversial, since it is understood as implying that each person ought to seek only *his own eudaimonia*. Later, in book IX, Aristotle will go some way towards recognizing this controversy, for (1) does appear to conflict with the commonly held view that I should do good to my friend for *his* sake, and not for mine. He does argue in reply that the conflict is only apparent and not real, but still the controversy is recognized. Nevertheless, in book I he simply treats (1) as a platitude, agreed both by 'the many' and by 'the wise', so his method here is simply an appeal to what he takes to be universal opinion. As for (1*a*) and (1*b*), he no doubt supposes that these are the obvious premises to supply in order to deduce (1), and hence that they should equally be accepted as platitudinous, though in so doing he overlooks a distinction which we are more aware of than he was. For one might surely accept that men *ought* always to pursue (their own) *eudaimonia* without supposing that they always *do*. One obvious counter-example to cite is the case of *akrasia*. Now again it is true that when Aristotle later comes to give his own account of *akrasia* he in effect disarms this counter-example, for on his account when someone acts akratically he does not actually have before his mind the knowledge that what he is doing is not what he ought to be doing. He claims, too, that even the wicked man fails to know that what he is doing is wrong (III.1, 1110^{b}29–30), apparently without seeing this as a controversial thesis. One may say, then, that Aristotle simply takes it to be obvious both that each man ought to pursue what is good (for him) and that each man does in fact pursue what (he thinks) is good (for him). The truth is that both claims are controversial (and, I would say, false), but since Aristotle does not recognize this he offers us no arguments in their favour.

Let us turn to thesis (2). This is now supplied with two separable premises, as follows

> (2*a*) The good for man is special to man, i.e. to what is really distinctive about man.
> (2*b*) The good for man is an activity, and one that is performed with virtue (excellence).
> ∴ (2) The good for man (i.e. *eudaimonia*) is an activity of the rational part of the soul, performed with virtue (excellence).

---

[32] I take (1*a*) to include the claims that the good for man is 'most complete (final)' and 'self-sufficient', but I shall not give any separate attention to these.

None of these claims could be regarded as platitudes, agreed by all, and Aristotle does not try to represent them as such. The truth is that (2a) relies largely on his own biological theories, and (2b) depends at least partly on his metaphysics, both developed elsewhere. It is his biology that supports the claim that there is a way of living that can be called the 'function' of man, a way of living special and peculiar to man, since it defines what a man is, so that the good (specimen of) man is one who displays this way of living to the full. Aristotle's real reasons for (2a) go way beyond the very superficial inductions indicated in the text. It is his metaphysics which supports the claim that what is good about this life must be its *activities*, and this claim is more or less controversial according as one gives a more or less restrictive meaning to the word 'activity' (*energeia*). Taking the word in the wide and general sense in which the English word is naturally taken, the claim may seem entirely reasonable, but when we recall that Aristotle distinguishes activities from processes, it becomes clear that the claim has much more force than at first appears. For an action which aims to produce a certain goal, and is not 'complete' until that goal is attained, is on his account a process and not an activity. So such actions are not, in the end, allowed to contribute to the value of the life they are part of, and nor are the goals thereby attained, unless such goals happen to be 'activities' in Aristotle's special sense. In this special sense, it appears that he will allow only perception and contemplation as genuine 'activities', and he has already dismissed perception as not specially distinctive of man. So there is actually very much more packed into this claim than is at all apparent when we first meet it in I.7. How much more Aristotle actually intends, when he first slips in this claim as if it were obvious, may be debated. But even if we take it in its weakest sense, as expressing merely a preference for activity (or actuality) over potentiality, still its justification lies elsewhere (most prominently in *Metaphysics* Θ).

In any case, as I have noted, the argument from (2a) and (2b) to (2) is a failure, since it rests on an equivocation between 'human goodness' (i.e. what is good about the good (specimen of) man) and 'the human good' (i.e. the good life for man, i.e. *eudaimonia*, i.e. what all men allegedly aim for). There is no reason to suppose that all men either do or should aim to be good specimens of their kind, and Aristotle's attempt to supply one does not work. We do best, then, to set this down as a real but unacknowledged 'first principle' of his ethics, not furnished with any justification. We may add that while it is no doubt quite fair to claim that the specifically human way of living will *include* the exercise of the rational part of the soul, Aristotle gives no good reason for *confining* it to this, and in this way too the deduction fails. (To put the point in traditional terms, man is both rational and an animal; it is his rationality that distinguishes him from other animals, but it is his animality that distinguishes him from the gods.) As for the addition 'with excellence (virtue)', we may no doubt agree that the good specimen of mankind will exercise his rationality *well*—and *we* may wish to add that he will also keep his animal body in good order—and if that is all that Aristotle means, we need not quibble. But one suspects that he is already drawing here on some preconceived list of 'the human virtues (excellences)', and if that is so then here is another step of the argument that is at this stage unwarranted. But this brings us to the next proposition.

Basic to the whole framework of the *Ethics* is this distinction introduced in I.13:

(3) The virtues of the rational part of the soul are divided into virtues of character and virtues of intellect.

The division is based upon a division of the rational part of the soul into two parts, one that has reason 'in itself' and one that contains desires and emotions, but can be said to partake in reason in a way, because it can listen to reason. The distinction between these two 'parts' of the soul Aristotle probably regards as at least a philosophical commonplace, if not exactly a popular one, but he does also offer to argue for it by citing the phenomena of *akrasia* and *enkrateia*. As ordinarily thought of,[33] these seem to involve an 'inner conflict' in which 'reason' pulls one way and 'desire' another, and this he not unreasonably takes to indicate that two distinct 'parts' are involved. This can certainly be regarded as an appeal to common sense, though obviously it falls far short of a compelling argument. (A more thorough investigation along these lines would evidently require one to ask whether it also happens that one 'reason' conflicts with another, or one 'desire' with another.) The thought that this distinction is commonly recognized by philosophers is surely more influential. That is why he refers us to the 'exoteric' discussions,[34] and does not attempt to spell out the distinction in any detail.[35] But I imagine that it is his own idea to classify the part that has desires and emotions as also rational in a way (since it can listen to reason), and by this means to include its virtues under the specifically human virtues. Though this may seem to be somewhat special pleading, we need not complain at the result.

But upon this division of the soul Aristotle bases his division of the virtues, and we may well ask what warrant he has for that. He appeals to common usage for the existence of a division between virtues of character (*ēthos*) and virtues of intellect (*dianoia*), but actually he offers no reason at all for his claim that this division corresponds to the division of the rational soul into its two parts. This makes for a neat scheme, of the kind that appeals to philosophers, but in fact it leads to several problems, as I have noted. In some cases, as with courage and temperance, it is relatively easy to accept that one virtue of character is specially associated with just one specific kind of desire or emotion, and this seems to have been Aristotle's original intention.[36] But in other cases, notably justice, no such specific association can be maintained. He in a way admits this himself when he characterizes a particu-

---

[33] Under Aristotle's own analysis, of course, the idea of a (conscious) inner conflict rather disappears.

[34] 'Exoteric' (1102ᵃ26) literally means 'outside', and it is not clear what Aristotle is referring to. Irwin takes it (as many scholars do) to be a reference to his own published works, which have not survived, and so translates 'our popular discussions'. Ross, with a different interpretation in mind, translates 'the discussions outside our school'.

[35] As a result, one can often be in doubt over whether a given 'appetition' (*orexis*) — say my wanting to win this game of chess — is to be classified as a 'wish' (*boulēsis*) belonging to the part that has reason in itself, or as a 'desire' (*epithumia*) belonging to the part that can listen to reason. (This point is developed by Anscombe, 1965: 144–8.)

[36] Note again 1117ᵇ23–4, which says that courage and temperance are *the* virtues of the unreasoning part of the soul. What, then, of other virtues of character?

lar virtue as associated with a particular range of actions, and not any particular desire or emotion (e.g. justice again, but also honesty, liberality, and several others). But a more pertinent criticism at the present stage is that almost all 'virtues of character' involve the intellect as well as the desires and emotions, and this is true both of those that do figure on his own list (for an extreme, but trivial, example, consider 'ready wit') and of those that do not (e.g. prudence). By what right, then, are they said to be *virtues of* the non-intellectual part? While one can perhaps supply some rationale for this (pp. 70–1), still I think it is clear that Aristotle's over-neat schematism is misleading, and the virtues do not clearly divide into two kinds as he suggests.

But all this presupposes that we do have some antecedent understanding of what a virtue of character is. Let us now turn to Aristotle's own elucidation.

(4) Virtue of character is a disposition to feel (and to act) that is 'in a middle relative to us', that involves choice, and is determined by practical reason.

I have sufficiently explored in my earlier discussion the difficulties in making sense of this definition, and will not recapitulate them now. The question I wish to raise now is: how does Aristotle reach this definition, and why should he expect us to accept it? In leading up to it, he gives very little indeed by way of supporting argument. There is a brief analogy with health and with strength (1104ª11–18, continued in 1106ᵇ36–ᵇ5), but this is clearly quite superficial. For example, there is no attempt to show that health and strength are good states of the body which are intermediate between two bad states, diverging from them in opposite directions. There is an equally brief analogy with the products of skill (1106ᵇ5–14), which again is equally superficial. And that is all. Certainly we are not offered a dialectical argument, which sets out the views of others and seeks to maximize the truth in them all. Nor, on the surface, are we offered an inductive argument either, for we are simply offered no (worthwhile) argument.

But one may not unreasonably suggest that it is actually an inductive argument that is at work. Aristotle has noticed that most of the states of character that are commonly recognized as virtues *do* lie between two opposing vices, and he infers by induction that all virtues of character will do so. On this account of the matter, his discussion in II.7, which aims to show how the general definition works out in particular cases, is actually his argument for the definition; the definition is shown to be correct because it does fit the particular cases. Accepting this for the sake of argument, three comments are in order. (i) Aristotle is expecting us to accept, without argument, that his own list of those dispositions that are virtues is correct, so it is here that he relies upon his audience having been well brought up, and thus sharing his own moral values. (ii) The induction is somewhat shaky, for, as Aristotle admits, he quite often has to invent vices and virtues that are not commonly recognized as such, in order to make his scheme fit. A clear example is the alleged vice of 'insensibility' (*anaisthēsia*, 1107ᵇ8), which characterizes one who takes too little enjoyment in bodily pleasures. While no doubt his well-brought-up audience might be expected to see that there is something amiss with one who is in this state, it is not at all obvious that they would be ready to call it a *vice* (i.e. a badness, *kakia*).

(iii) Worse still, the induction—if that is all it is—seems to be refuted by at least one clear counter-example, which is a virtue of some importance and admitted by Aristotle himself to be a counter-example, namely justice (1133^b32–3). One is inclined to feel that there should be something more behind Aristotle's definition than mere induction, if he can still maintain it in the face of this counter-example. Here there is a tempting speculation to offer. Aristotle begins his discussion, as we saw, by focusing too much on desires and emotions (and not actions). He perhaps holds that *no* desire or emotion which is natural to man can be wholly wrong.[37] So there must in all cases be a right way of feeling it (and acting accordingly), and from here it is easy to slip into the thought that wrong ways of feeling it will be characterizable as feeling it too much or too little. This is in many ways over-simple, as we have seen, but perhaps it is the (largely unstated) theory that explains Aristotle's confidence in his definition.

A further, and deeper, question should be raised before we leave this topic. What is the point of cultivating those firm and settled dispositions to feel and act in particular ways that Aristotle counts as 'the virtues'? After all, the overall aim, as given in thesis (2) is the good functioning of the rational part of the soul, including here both (practical) reason and the desires and emotions. Does the cultivation of settled dispositions in fact help or hinder this overall aim? In a pragmatic way, it may be said to help, because it reduces the amount of thinking required when considering how to act. But from a more theoretical viewpoint it may certainly be argued that it hinders, since it tends to close off options that are in fact worthy of consideration. So are 'the virtues' the best way of promoting 'excellence' in the rational part? Aristotle's own terminology tends to obscure the fact that this is a perfectly reasonable question to ask. His answer is, of course, affirmative, but this answer should be set down as a further unnoticed and unargued 'first principle' of his ethics.

The next three propositions we may reasonably take together:

(5) There are two (relevant) virtues of intellect, namely the virtues of practical reason and of theoretical reason.

(6) The task of practical reason is to determine what actions will achieve the supreme good for man, *eudaimonia*.

(7) The task of theoretical reason is to discover the necessary and eternal truths.

Aristotle's argument for making this distinction between the theoretical and the practical use of reason is that each deals with a different kind of proposition, the one with necessary truths and the other with contingent truths. This is sufficient ground, he says, for inferring that distinct 'parts' of the soul must be involved. This argument has a Platonic precedent. For Plato, though he did not distinguish practical and theoretical reasoning as Aristotle does, had argued in a similar way, but in the opposite direction, about knowledge and belief. In *Republic* V he begins from the thought that these must be two different 'faculties' (*dunameis*), since the one is liable to error and the other not, and infers from this that they must have different

---

[37] This is suggested by Urmson (1973).

objects. In his terminology, knowledge is of the Forms, and belief is of the perceptible particulars (476e–480a). But Aristotle would surely see an analogy—if not
an identity—between what Plato calls knowledge of Forms and what he calls knowledge of what could not be otherwise, and between what Plato calls belief concerning perceptible things and what he (usually[38]) calls knowledge of what could be
otherwise. Thus Plato had argued from difference of faculty to difference of object,
while Aristotle argues from difference of object to difference of faculty (or 'part of
the soul', as he prefers to say). But clearly each is making a similar connection. So if
one asks where Aristotle got his distinction from, and what led him to believe in its
importance, a probable conjecture is that he was quite strongly influenced by
Plato.[39]

But although Aristotle's distinction may be to some extent based on precedent,
still it is none the better for that. I have already observed that, as stated, it is neither
exclusive nor exhaustive. And if we widen its scope by saying that the main idea is to
distinguish reasoning with a view to action and reasoning with a view to acquiring
knowledge 'for its own sake', still we do not actually secure an exhaustive distinction, and it becomes even more obvious that exactly the same reasoning may be
employed for either purpose. But to understand Aristotle's own position we should
retain his own restricted account of 'theoretical' thinking, for it is when understood
in this narrow way that he sets a high value upon it, and thinks of it as, in a way,
divine. As for practical reasoning, I have described its task here in a way which is
surely uncontroversial, i.e. to determine what actions will achieve *eudaimonia*. But
this leaves room for much controversy over quite how much falls within its scope.
(Does it itself determine what *eudaimonia* consists in? Or is it something else,
namely *nous*, that spells out the content of *eudaimonia*? Or is it, as Aristotle seems
to say, neither *nous* nor practical wisdom but habituation in virtue that achieves
this task?) I shall here leave this question open, remarking only that it would seem
to be a criticism of the *Ethics* that it leaves us in so much doubt upon this apparently crucial point. Let us turn instead to X.6–8, where Aristotle gives us his own
views on what *eudaimonia* is.

For brevity I collapse Aristotle's claim simply to this:

(8) The highest virtue is that of theoretical reason.

Clearly Aristotle rejects the dictum that 'the proper study of mankind is man', and
bids us turn our attention to things that he regards as 'higher', i.e. to the study of
God and of nature (including mathematics). For man, he says in VI.7, is not the

---

[38] In VI.3 he defines knowledge as restricted to what could not be otherwise, but usually he does not
scruple to use the same word also of our grasp of the particular premisses employed in practical reason,
though recognizing that this is not equally 'knowledge-like' (*epistēmonikon*, 1147ᵇ13–17).

[39] There are other signs that Aristotle has Plato in mind when drawing this distinction. When he
introduces it he calls the two parts 'that concerned with knowledge' (*epistēmonikon*) and 'that concerned with calculation' (*logistikon*, 1139ᵃ12). Plato had used the latter title for the reasoning part in
general, but had specially associated it with the former function. Aristotle very probably intends this as
a criticism of Plato's failure to distinguish between practical and theoretical reasoning. Later, he slips
into calling the second part 'that concerned with belief' (*doxastikon*, 1140ᵇ26 and 1144ᵇ14), which
indicates in a different way the Platonic ancestry.

best thing in the universe, and these other objects of study have greater value (1141^a20-^b2). But for this interesting claim he offers us virtually nothing by way of argument, and one can only say that he is here giving us *his own* values, but no reason why we should share them. Certainly he could not here simply appeal to popular opinion, or even to the judgement of those who have been well brought up. Nor does he attempt to construct a dialectical argument, or an inductive argument, or any other kind of argument (e.g. inference to the best explanation) that one might suggest as an appropriate way of supporting a first principle. Admittedly in X.6–8 we do find a whole string of arguments in favour of the more specific proposition that true *eudaimonia* lies in the contemplation of the eternal and necessary truths that concern these allegedly more valuable objects. These arguments draw on a variety of premises, some with nothing to be said for them at all (e.g. that man *is* his theoretical soul), some that clearly conflict with what we have been told elsewhere (e.g. that actions in accordance with the practical virtues do not have their value in themselves), and some borrowed from Aristotle's own theories in other areas (e.g. on the nature of God). There is no one method that all these arguments follow, and anyway, as we have seen, they are all quite unconvincing. So here is another basic principle of Aristotle's ethics which one can hardly say is not argued at all, but for which the arguments offered are very clearly inadequate.

Moreover, in this case it is extremely difficult to believe that Aristotle really meant what he says. For the consequence appears to be that the virtues of character turn out to have *no* role to play in 'the good life for man', or at best an instrumental role, despite the fact that they have been the focus of interest for almost all the rest of the work. So here is a further way in which the *Ethics* turns out to be deeply unsatisfying. Though Aristotle does, in the end, give a relatively clear answer to his original question 'what is the good life for man?', one feels sure that he must have overstated his position, and there is no straightforward way of pulling all of his *Ethics* together into a coherent whole. The moral that I draw is that actually there is no such thing as *the* good life for man, so the question from which he began contains a mistaken presupposition.

Let us come back, briefly, to the question of *method*. This review of how Aristotle argues for the basic principles of his ethics shows well enough that there is no one method that he follows. In VII.1 he advocates the dialectical method, and proceeds to pursue it in VII.1–10. But despite the fact that he advocates it as the right method to pursue on all issues, he does not in fact rely upon it in any other major discussion, though one may reasonably say that dialectical techniques are quite prominent in his discussion of the philosophy of mind and action. However, in ethics proper the most that one can say is that he quite often shows some deference to common opinion, though at other times (e.g. in X.7) he ignores it altogether. In I.4 (taken together with the end of I.7) he says that first principles should be supported by induction from the views of those who are well brought up, and there are occasions when he can be seen as following this method. But he is never explicit about it, and in any case the occasions are few. The truth is that he simply employs whatever method of argument seems most likely to support the results he wishes to reach. Sometimes he invokes common opinion; sometimes he applies induction to it; sometimes he relies on what he takes to be established philosophical opinion; sometimes he imports results from his own theoretical enquiries elsewhere (i.e. from his biology, his metaphysics, and his theology); sometimes he simply insists upon the correctness of his own system of values. There is no one method that is systematically employed.

I do not regard this as a serious criticism of his ethics. Many philosophers have preached a particular method of philosophizing, but have not in practice followed it, and this is usually to their advantage. Aristotle is no exception. It is clear that if he really had relied on nothing but dialectic, or nothing but induction from what the well-educated young man already believes, his ethics would be much less interesting than it in fact is. There is no doubt room in philosophy for theories of method, but I am sceptical of their value. For example, philosophers have filled many pages on the question of what should properly be counted as 'scientific method', but scientists themselves have had the good sense to pay little attention to these pages, and have simply used any reasoning that comes to hand. I would say that the same applies to 'the method of ethics'. There is no good reason to suppose that philosophers can, in advance, set some boundary to the methods of reasoning that are appropriate in this subject.

Further reading

A good discussion of Aristotle's professed methods in philosophy, with special reference to the Ethics, is Reeve (1992, ch. 1). On the role of dialectic in particular, see Barnes (1980) and possibly Klein (1988). On 'first principles' in ethics there are useful points made in Irwin (1978). I am not aware of any helpful discussion which takes seriously the question of how Aristotle's actual methods diverge from those that he professes, save possibly Monan (1968, ch. 5). For a view which sees no serious divergence, see Irwin (1981), which puts in a brief space the arguments developed at length in his (1988a).

Concluding remarks

KENNY has said of the *Nicomachean Ethics* that 'no explanation succeeds in the three goals which most commentators have set themselves: (1) to give an interpretation of book I and book X which does justice to the texts severally; (2) to make the two books consistent with each other; (3) to make the resulting interpretation one which can be found morally acceptable by contemporary philosophers'.[1] But it is not just the apparent conflict between books I and X that creates this dilemma. There is also the question of how much Aristotle means to assign to 'reason', and to what kinds of 'reason', when he discusses practical wisdom in book VI. And there is too the question of what he took to be the methods of ethics, and in particular his own methods when studying the subject. All these are interrelated. I shall briefly set out two clearly contrasting approaches, which one might call a 'most intellectual' and a 'least intellectual' approach, and then ask whether there is any satisfying way of compromising between them.

As proponent of the most intellectual approach we may reasonably take Kraut (1989). On his reading, book I does not introduce an 'inclusive' conception of man's *eudaimonia*, but points forward to the claim of book X that *eudaimonia* simply *is* theoretical contemplation. Of course, this allows that a life devoted to contemplation, to the extent that that is possible for man, will also include the necessary (and facilitating) conditions for it, and that is where the practical virtues come in; they are means to the successful pursuit of contemplation. (It will also include suitable 'external goods', so far as they are needed both for contemplation and for the practical virtues which—on this conception—exist solely for its sake.) This approach evidently carries with it an answer to our question about practical wisdom: its task is to deliberate well on how to act so as best to achieve the desired end of contemplation. The practically wise man must of course know that the supreme end is contemplation, and this knowledge is achieved by the methods that Aristotle himself uses when arguing for this conclusion in book X. These methods are no doubt the same (in Aristotle's eyes) as are used when arguing for the first principles of any science, methods which he calls either dialectic or induction, and which he credits to *nous*. Since they are the same methods as are used in all other sciences, it is no doubt best not to credit them to practical wisdom, since that has a role only in ethics. But practical wisdom works from this first principle which *nous* supplies, when calculating how contemplation is best achieved. So, to go back to my earlier example, practical wisdom can in principle tell us whether, and to what extent, chastity is a virtue; this is simply a question of how, if at all, chastity

---

[1] Kenny (1992: p. 93). (Kenny's own 'solution' is to recommend us to turn to the *Eudemian Ethics* instead.)

conduces to the pursuit of contemplation. The same applies to all those other traits of character that are commonly called virtues, i.e. to courage, temperance, honesty, fairness, and so on.

There are clearly many aspects of the text which this approach must ignore or explain away. Here I mention just three: (i) the claim in VI.12 and VII.8 that the 'first principles' of ethics are given in a special way, peculiar to ethics, which is habituation in virtue; (ii) the stress in almost all of II–IX on the independent value of the practical virtues, and on the importance of doing what is 'noble' (*kalon*); (iii) connected to this, the fact that in X.8 the political life is admitted to be *eudaimōn*, if only in a secondary way. For why, on this account, should Aristotle suppose that it has any more value than the life of the doctor, or indeed the carpenter? (If the answer is that the politician works to promote a society in which theoretical contemplation can be maximized, then the same may be said also of the doctor and the carpenter.) But for many of us the chief objection will be the one that Kenny notes: the view that the practical virtues should be pursued *only* to the extent that they promote theoretical contemplation is simply immoral, and for that reason we should not accuse Aristotle of holding it.

Perhaps this last objection should be ignored. As I observed right at the beginning of my chapter I, when Aristotle speaks of 'the good life' he need not be taken as meaning what we might call 'the *morally* good life'. (There is no Greek word which has the same connotations as our word 'moral'.) Yet we have only to look at the bulk of his discussion of the virtues, in books II–IX, to see that he does share with us many of our views about what counts as 'good'. While there are no doubt several details which lead us to raise an eyebrow, still we do not come away feeling that his notion of 'goodness' has nothing in common with ours. Kraut has his own way out of this problem, which is I think special to him. While he accepts that *eudaimonia* (for Aristotle) just *is* theoretical contemplation, he does not suppose that in I.7 Aristotle is saying that we do (or should) pursue only *our own eudaimonia*. As he reads the passage (1097^a34–$^b6$), Aristotle claims that only *eudaimonia* is pursued for its own sake, but this allows us rightly to pursue anyone's *eudaimonia*, and not just our own. So, to take one of his examples, it may be right for me to curtail my own theoretical studies in order to look after an ageing father who is in need. Even if my father is quite incapable of theorizing, still Kraut argues that I am pursuing his *eudaimonia* when I so act, for I will be bringing him as near to theorizing as he can get (1989: 86–90). This is a sophistry which surely should not be taken seriously. If what we consider as a duty to help others is to be given any place in Aristotle's account, it can only be by taking seriously his claim that not only theoretical virtue but also the *other* human virtues are to be pursued 'for their own sake' or 'for the sake of what is noble'.

I turn therefore to my contrasting 'least intellectual' approach to Aristotle's ethics, for which one might fairly take Broadie (1991) as proponent.[2] This view takes I.7 to be proposing *eudaimonia* as an 'inclusive' end, and it pays little attention to the praise of theoretical contemplation in X.7–8. (In Broadie's own view, Aristotle

---

[2] I note that her earlier (1987) gives a very different view.

always means to treat a life of practical virtue as the *eudaimōn* life, and in X.7–8 he is led into somewhat extravagant praise of theoretical wisdom only because he fears that some might not see it as a virtue at all (pp. 388–92). The good life, then, is one which puts into practice all the human virtues (including theoretical wisdom), for all of them should be pursued for their own sakes.) When we turn to our questions about practical wisdom, asking how to determine what counts as a virtue, and how the virtues are supposed to fit together with one another, then—on this approach— all that can be said is that good training and long experience will provide the answers. The practically wise man has no 'Grand Conception' of the end, that could be articulated as one or more 'first principles' spelling out what is to be achieved, but only a kind of knack or instinct for getting things right. It is true that in VII.8 (and in VI.12) Aristotle speaks of habituation in virtue as providing 'first principles', but we should not suppose that he means principles that could be precisely formulated, for ethics is and must always be 'inexact'. The kind of thing that he has in mind are such general rules as 'it is usually right to keep one's promises', and the practically wise man is one who has developed an 'eye' for applying these to particular situations (1143^{b}13–14). That is why Aristotle so insists upon the point that practical wisdom is concerned with particulars, and shows almost no interest (in book VI) in the universal premises that are also involved. It is because the universals are commonplace, and well-known by all who have been well brought up, whereas what marks out practical wisdom is the ability to see morally relevant features of the particular situation, and thereby to say what reaction to it is the virtuous one.

On this account the arguments that Aristotle himself employs in the *Ethics*, when setting forth what I have called his 'basic principles' for the subject, are of absolutely no concern to the 'practically wise' man. He does not need to know that the good for man is an activity of the rational soul pursued with excellence, that the excellences divide into those of character and those of intellect, that the highest excellence is that of the theoretical intellect, and so on. These are no doubt 'first principles' in a way (for the first of them is explicitly said to be one), but they are principles of a different subject, which one might call 'meta-ethics'. For they cannot reasonably be represented as based upon training, habituation, and experience in virtue, which—on this account—is what the practically wise man relies upon when he engages in ethical reasoning. So we must posit a split, in Aristotle's own mind, between 'ethics' and 'meta-ethics', which is surely anachronistic, has no support in his text, and is anyway difficult to make sense of. The account is at odds with the text in another way, when it seeks to play down the force of what Aristotle says of theoretical contemplation in X.7–8. He does not say that theorizing is but *one* aspect of the good life; he says that it is the *only* thing about it that makes it good. ('So far, then, as contemplation extends, to that extent too does *eudaimonia*. And those who have more of contemplation thereby have more of *eudaimonia*, not coincidentally, but because it is contemplation', 1178^{b}28–31.)

We may conclude that Broadie's 'least intellectual' interpretation must conflict with some aspects of our text, as Kraut's 'most intellectual' interpretation does too. Broadie would claim that her interpretation gives Aristotle the more realistic, or more common-sensical, position, for the truth is that most of us do form our views on how to act simply on the basis of how we have been brought up. But we should hesitate before supposing that Aristotle thought the same of his ethical hero, the 'practically wise' man, who *knows* what to do. Aristotle's ethics is not an ethics for 'the common man'; he is unashamedly élitist, and he recognizes that *most* people have no chance to aim at the *eudaimonia* that he describes. (For example, all those who have to work for their living are automatically debarred.) I see no reason not to suppose that Aristotle thinks that his own discussion in the *Ethics is* of some relevance to practical wisdom, for he certainly does say that it is supposed to be of some practical use.

I cannot believe in either of these extreme positions. Against the first, I cannot believe that Aristotle really means that the practical virtues are to be pursued *only* so far as they contribute to theoretical contemplation. It is true that X.7, 1177^{b}1–26 seems to say this, but I set that down as an exaggeration. Despite 1177^{b}18, these virtues are to be pursued 'for their own sakes', or 'for the sake of what is noble', even if there are also other reasons for pursuing them too. Of course, this claim must at the same time point to a gap in Aristotle's discussion, for if there is more than one thing that should be pursued for its own sake then we need to know how priorities should be assigned: which is to take precedence over which, and why? This is not a question that Aristotle ever addresses. No doubt he supposes that 'practical wisdom' will somehow supply an answer, but he never tells us how it will do so. Indeed, he never tells us how 'practical wisdom' works at all, for—on the other side—I cannot believe that he really means to say that its role is limited to the calculation of means to ends that are given simply by training and habituation in virtue. No doubt it has such a role, but if that is *all* that it can do then it cannot criticize and appraise one type of moral education against another, and that leaves it powerless on substantive disputes. *Perhaps* that is Aristotle's intention, for *perhaps* he did not recognize that there are substantive moral disputes. But in that case we can only say that it is an odd and unexpected blindness on his part, for he certainly did recognize that there is dispute over what 'the good life' consists in. Indeed, he is committed to saying that one who has been brought up to believe that it lies in the pursuit of bodily pleasures, or that philosophical theorizing has nothing to do with it, has been *wrongly* brought up. Moreover, it is in a broad sense 'reason' that establishes, this, as the *Ethics* itself demonstrates. So it should apparently be similar 'reasoning' that shows which traits of character are virtuous, and if so then this too is a gap in Aristotle's overall discussion.

The gap is to some extent concealed by his emphasis upon another point, that ethics differs from other sciences in that its aim is not just truth *simpliciter* but 'practical truth', i.e. 'truth in agreement with right desire', and he does have things to say about right desire. He very fairly insists that desire is not automatically persuaded by argument, but needs training and habituation of a non-rational kind. Still, the object of this training is to get desire to 'listen to reason', so there must also be reasoning for it to listen to, and that is what we are missing. We may pinpoint the gap in this way: Aristotle thinks that virtues of character aim for what is 'noble' (*kalon*), but he gives us no analysis of this concept of nobility, and hence no account of how we know what it is and no general reason why it should be pursued. The argument about man's 'function' should presumably lead us to see the pursuit of nobility as a specifically human endeavour, but this gives precious little guidance on its content.[3] The same argument seems to promise us that one who 'reasons well' will be able to say what it is and why it should be pursued, just because it starts with the idea that what is really special to man is reason, but that is a step which one cannot see how to fill in on Aristotle's behalf. I think Aristotle is committed to the existence of such reasoning, but I also suspect that—if pressed—he would admit that he cannot himself supply it. Indeed, given only the resources that he permits himself, I do not see how he could supply it. (Other moral theories, e.g. Kantian or utilitarian, clearly call upon further resources.)

My own inclination, then, is to depart from both of the two 'extreme' interpretations of the *Ethics* sketched earlier, but I can do so only by supposing that it contains important gaps, which are in no way easy to fill. As for the method of ethics, if we may set aside the point about truth in agreement with desire, and concentrate just upon truth, I take it that Aristotle thinks of its method as being entirely similar to the method of other sciences. This is partly because he has a mistaken view about the *other* sciences. He thinks that in each case we begin from 'the appearances' (*phainomena*), and that these include not only what can be observed but also what people think. We would rather say that in other sciences what is crucial is what can be observed,[4] whereas in ethics observation has no such role, and if we have 'data' at all it can only be what people think. (That is because we have read Hume on 'is' and 'ought', and of course Aristotle had not.) But Aristotle draws no such distinction, and supposes that people's opinions are just as authoritative in either case, so naturally he does not see the great divide that we do (if we are well brought up).

I think we must conclude that in its overall structure Aristotle's *Ethics* is disappointing: either it is radically innovative but immoral, or it is a recipe for unthinking moral conservatism, or its argument contains gaps which it does not appear that he could fill. (I prefer the last alternative.) But the overall structure of the work is one thing, and the merits or demerits of its individual discussions are quite another. There are some of them that nobody will take very seriously today, for example his arguments in X.7–8 aiming to show that philosophical contemplation is the main ingredient of the good life for man. But others retain a perennial appeal. The thought that human goodness is specially connected with what it is to be human has refused to lie down and die; the idea that ethics should be based on virtues of character (rather than the rightness and wrongness of particular actions) is one

---

[3] Should we rule out, as not specifically human, the pursuit of athletic excellence (e.g. in running and jumping)? If so, should we also rule out skill and courage in battle? Or care for one's young? *Non liquet.*

[4] Whether mathematics is thus an 'empirical' science is, of course, a controversial question, which I cannot discuss here.

that is much in vogue today; one may well applaud his pioneering attempt to sort out our different conceptions of justice; and so on. If I may restore to the discussion his contributions to the philosophy of mind and action, then we may add: his account of voluntary action holds a special appeal to those who hope that this topic can be freed from that perennial chestnut 'the problem of free will'; there is surely something right about his claim that different pleasures are of different kinds from one another, and they cannot all be lumped together as just different degrees of the same 'feeling'; and at least he points us in the right direction when he speaks of action as a joint product of belief and desire, even if some details appear not to be quite right. There are more fundamental questions, too, which his *Ethics* certainly raises, even if we cannot accept his answer to them. Should we even try to seek for something that can be called *the* good life for man? And are we thinking of this in the right way when we characterize it as *eudaimonia*? Is the whole framework of his *Ethics* misconceived?

One would hardly expect to find general agreement on which features of the *Ethics* should be taken seriously today. Almost everyone will say that some bits can be ignored, but others cannot, and it must be unusual to find two people who have just the same lists as one another. Your task is to make your own list, and to be prepared to defend it and justify it. I hope that what I have said in this book will help you in this task.

References

Ackrill, J. L. (1965), 'Aristotle's Distinction between *Energeia* and *Kinesis*', in R. Bambrough (ed.), *New Essays on Plato and Aristotle* (London: Routledge & Kegan Paul), 121–41.

—— (1972), 'Aristotle on "Good" and the Categories', in S. M. Stern, A. Hourani, and V. Brown (eds.), *Islamic Philosophy and the Classical Tradition* (Oxford: Cassirer), 17–25; repr. in Barnes *et al.* (1977).

—— (1973), *Aristotle's Ethics* (London: Faber & Faber).

—— (1974), 'Aristotle on *Eudaimonia*', *Proceedings of the British Academy*, 60: 339–59; repr. in Rorty (1980).

—— (1978), 'An Aristotelian Argument about Virtue', *Paideia, Special Aristotle Issue*, 133–7.

—— (1980), 'Aristotle on Action', in Rorty (1980: 93–101).

—— (1981), *Aristotle the Philosopher* (Oxford: Oxford University Press).

—— *See also* Ross (1925).

Allan, D. J. (1953), 'Aristotle's Account of the Origin of Moral Principles', *Actes du XIe Congrès Internationale de Philosophie*, 12: 120–7; repr. in Barnes *et al.* (1977).

Annas, J. (1977), 'Plato and Aristotle on Friendship and Altruism', *Mind*, 86: 532–54.

—— (1980), 'Aristotle on Pleasure and Goodness', in Rorty (1980: 285–99).

—— (1988), 'Self-Love in Plato and Aristotle', *Southern Journal of Philosophy*, 27, suppl., 1–18.

—— (1993), *The Morality of Happiness* (Oxford: Oxford University Press).

Anscombe, G. E. M. (1957), *Intention* (Oxford: Blackwell).

—— (1965), 'Thought and Action in Aristotle', in R. Bambrough (ed.), *New Essays on Plato and Aristotle* (London: Routledge & Kegan Paul), 143–58; repr. in Walsh and Shapiro (1967) and in Barnes *et al.* (1977).

Austin, J. L. (1956–7), 'A Plea for Excuses', *Proceedings of the Aristotelian Society*, 57: 1–30.

Barnes, J. (1976), Introduction to *The Ethics of Aristotle*, trans. J. A. K. Thomson, rev. edn. (Harmondsworth: Penguin).

—— (1980), 'Aristotle and the Methods of Ethics', *Revue Internationale de Philosophie*, 34: 490–511.

—— (1984), (ed.) *The Complete Works of Aristotle: The Revised Oxford Translation*, 2 vols. (Princeton: Princeton University Press).

—— (1990), 'Aristotle and Political Liberty', in G. Patzig (ed.), *Aristoteles' Politik* (Göttingen: Vandenhoek & Ruprecht), 250–63.

—— Schofield, M., Sorabji, R. (eds.) (1977), *Articles on Aristotle*, ii: *Ethics and Politics* (London: Duckworth).

Bedford, E. (1956–7), 'Emotions', *Proceedings of the Aristotelian Society*, 57: 281–304.

Bekker, I. (1831), *Aristotelis Opera* (Berlin).

Bolton, R. (1991), 'The Objectivity of Ethics', in J. P. Anton and A. Preus (ed.), *Essays in Ancient Greek Philosophy*, iv: *Aristotle's Ethics* (Albany: State University of New York Press), 7–28.

Bostock, D. (1988), 'Pleasure and Activity in Aristotle's *Ethics*', *Phronesis*, 33: 251–72.

Bostock, D. (1991), 'Aristotle on Continuity in *Physics* VI', in L. Judson (ed.), *Aristotle's Physics: A Collection of Essays* (Oxford: Clarendon Press), 179–212.

—— (1994), *Aristotle, Metaphysics, Books Z and H* (Oxford: Clarendon Press).

Broadie, S. (1987), 'The Problem of Practical Intellect in Aristotle's *Ethics*', *Boston Area Colloquium in Ancient Philosophy*, 3: 229–52.

—— (1991), *Ethics with Aristotle* (Oxford: Oxford University Press).

Burnet, J. (1900), *The Ethics of Aristotle* (London: Methuen).

Burnyeat, M. F. (1980), 'Aristotle on Learning to be Good', in Rorty (1980: 69–92).

Bywater, I. (1894), *Aristotelis Ethica Nicomachea*, Oxford Classical Texts (Oxford: Clarendon Press).

Charles, D. (1984), *Aristotle's Philosophy of Action* (London: Duckworth).

Charlton, W. (1988), *Weakness of Will* (Oxford: Blackwell).

Clark, S. R. L. (1975), *Aristotle's Man* (Oxford: Clarendon Press).

Cooper, J. M. (1973), 'The *Magna Moralia* and Aristotle's Moral Philosophy', *American Journal of Philology*, 94: 327–49.

—— (1975), *Reason and Human Good in Aristotle* (Cambridge, Mass.: Harvard University Press).

—— (1977), 'Aristotle on the Forms of Friendship', *Review of Metaphysics*, 30: 619–48.

—— (1980), 'Aristotle on Friendship', in Rorty (1980: 301–40).

—— (1985), 'Aristotle on the Goods of Fortune', *Philosophical Review*, 94: 173–96.

—— (1987), 'Contemplation and Happiness: A Reconsideration', *Synthese*, 72: 187–216.

Crisp, R. (1992), 'White on Aristotelian Happiness', *Oxford Studies in Ancient Philosophy*, 10: 233–40.

—— (ed.) (1996), *How should One Live? Essays on Virtues* (Oxford: Clarendon Press).

—— (ed. and trans.) (2000), *Aristotle: Nicomachean Ethics* (Cambridge: Cambridge University Press).

Curren, R. (1989), 'The Contribution of *EN* III.5 to Aristotle's Theory of Responsibility', *History of Philosophy Quarterly*, 6: 261–77.

Curzer, H. J. (1991), 'The Supremely Happy Life in Aristotle's *Nicomachean Ethics*', *Apeiron*, 24: 47–69.

—— (1996), 'A Defense of Aristotle's Doctrine that Virtue is a Mean', *Ancient Philosophy*, 16: 129–38.

Dahl, N. O. (1984), *Practical Reason, Aristotle, and Weakness of Will* (Minneapolis: University of Minnesota Press).

Daube, D. (1969), *Roman Law: Linguistic, Social and Philosophical Aspects* (Edinburgh: Edinburgh University Press).

Davidson, D. (1963), 'Actions, Reasons, and Causes', *Journal of Philosophy*, 60: 685–700; repr. in his *Essays on Actions and Events* (Oxford: Clarendon Press, 1980).

—— (1970), 'How is Weakness of the Will Possible?', in J. Feinberg (ed.), *Moral Concepts* (Oxford: Oxford University Press); repr. in his *Essays on Actions and Events* (Oxford: Clarendon Press, 1980), 21–42.

Dawkins, R. (1976), *The Selfish Gene* (Oxford: Oxford University Press; 2nd rev. edn. 1989).

Devereux, D. (1981), 'Aristotle on the Essence of Happiness', in O'Meara (1981: 247–60).

Diels, H., rev. W. Kranz (1952), *Die Fragmente der Vorsokratiker* (Berlin).

Engberg-Pedersen, T. (1983), *Aristotle's Theory of Moral Insight* (Oxford: Clarendon Press).

Everson, S. (1998), 'Aristotle on Nature and Value', in S. Everson (ed.), *Companions to Ancient Thought*, iv: *Ethics* (Cambridge: Cambridge University Press), 77–106.

Fine, G. (1993), *On Ideas* (Oxford: Clarendon Press).

Finley, M. I. (1970), 'Aristotle and Economic Analysis', *Past and Present*, 47: 3–25; repr. in Barnes *et al.* (1977).

Foot, P. (1978), *Virtues and Vices* (Oxford: Blackwell).

Fortenbaugh, W. W. (1975), *Aristotle on Emotion* (London: Duckworth).

—— (1991), 'Aristotle's Distinction between Moral Virtue and Practical Wisdom', in J. P. Anton and A. Preus (eds.), *Essays in Ancient Greek Philosophy*, iv: *Aristotle's Ethics* (Albany: State University of New York Press), 97–106.

Furley, D. J. (1977), 'Aristotle on the Voluntary', in Barnes *et al.* (1977: 47–60); excerpted, with some revisions, from his *Two Studies in the Greek Atomists* (Princeton: Princeton University Press, 1967).

Gauthier, R. A. and Jolif, J. Y. (1958–9), *L'Éthique à Nicomaque* (Louvain: Publications Universitaires; 2nd edn. 1970).

Geach, P. T. (1956–7), 'Good and Evil', *Analysis*, 17: 30–42.

Glassen, P. (1957), 'A Fallacy in Aristotle's Argument about the Good', *Philosophical Quarterly*, 7: 319–22.

Glover, J. C. B. (1970), *Responsibility* (London: Routledge & Kegan Paul).

Gonzalez, F. J. (1991), 'Aristotle on Pleasure and Perfection', *Phronesis*, 36: 141–59.

Goodman, N. (1955), *Fact, Fiction and Forecast* (4th edn., Cambridge, Mass.: Harvard University Press, 1983).

Gosling, J. C. B. (1969), *Pleasure and Desire* (Oxford: Clarendon Press).

—— (1973–4), 'More Aristotelian Pleasures', *Proceedings of the Aristotelian Society*, 74: 15–34.

—— and Taylor, C. C. W. (1982), *The Greeks on Pleasure* (Oxford: Clarendon Press).

Gottlieb, P. (1993), 'Aristotle's Measure Doctrine and Pleasure', *Archiv für Geschichte der Philosophie*, 75: 31–46.

Grant, A. (1857), *The Ethics of Aristotle* (London: Longmans, Green; 4th edn. 1885).

Greenwood, L. H. G. (1909), *Aristotle: Nicomachean Ethics Book Six* (Cambridge: Cambridge University Press).

Haksar, V. (1964), 'Aristotle and the Punishment of Psychopaths', *Philosophy*, 39: 323–40; repr. in Walsh and Shapiro (1967).

Hardie, W. F. R. (1965), 'The Final Good in Aristotle's *Ethics*', *Philosophy*, 40: 277–95; repr. in Moravcsik (1967).

—— (1965–6), 'Aristotle's Doctrine that Virtue is a "Mean"', *Proceedings of the Aristotelian Society*, 66: 93–102; rev. in Barnes *et al.* (1977).

—— (1968), *Aristotle's Ethical Theory* (Oxford: Clarendon Press; 2nd edn. with new notes appended 1980).

Hare, R. M. (1952), *The Language of Morals* (Oxford: Clarendon Press).

—— (1956–7), 'Geach, Good and Evil', *Analysis*, 17: 101–11.

—— (1963), *Freedom and Reason* (Oxford: Clarendon Press).

—— (1971), *Practical Inferences* (London: Macmillan).

Heath, T. L. (trans. and comm.) (1908), *The Thirteen Books of Euclid's Elements* (Cambridge: Cambridge University Press; 2nd edn. 1925; repr. New York: Dover Publications, 1956).

Heinaman, R. (1988), '*Eudaimonia* and Self-Sufficiency in the *Nicomachean Ethics*', *Phronesis*, 33: 31–53.

—— (1993), 'Rationality, *Eudaimonia*, and *Kakodaimonia* in Aristotle', *Phronesis*, 38: 31–56.

—— (1995), 'Activity and Change in Aristotle', *Oxford Studies in Ancient Philosophy*, 13: 187–216.

Hume, D. (1978), *A Treatise of Human Nature*, ed. L. A. Selby-Bigge, rev. P. H. Nidditch (Oxford: Clarendon Press).

Hursthouse, R. (1980–1), 'A False Doctrine of the Mean', *Proceedings of the Aristotelian Society*, 81: 57–72.

—— (1984), 'Acting and Feeling in Character: *EN* III.1', *Phronesis*, 29: 252–66.

—— (1995), 'The Virtuous Agent's Reasons', in R. Heinaman (ed.), *Aristotle and Moral Realism* (London: UCL Press), 24–33.

Hutchinson, D. S. (1986), *The Virtues of Aristotle* (London: Routledge & Kegan Paul).

—— (1990), 'Aristotle on the Spheres of Motivation: *De Anima* III', *Dialogue*, 29: 7–20.

Irwin, T. H. (1975), 'Aristotle on Reason, Desire and Virtue', *Journal of Philosophy*, 72: 567–78.

—— (1978), 'First Principles in Aristotle's Ethics', *Midwest Studies in Philosophy*, 3: 252–72.

—— (1980), 'Reason and Responsibility in Aristotle', in Rorty (1980: 117–55).

—— (1981), 'Aristotle's Methods in Ethics', in O'Meara (1981: 193–224).

—— (trans. with notes) (1985a), *Aristotle: Nicomachean Ethics* (Indianapolis: Hackett).

—— (1985b), 'Aristotle's Conception of Morality', *Proceedings of the Boston Area Colloquium in Ancient Philosophy*, 1: 115–43.

—— (1988a), *Aristotle's First Principles* (Oxford: Clarendon Press).

—— (1988b), 'Disunity in the Aristotelian Virtues', *Oxford Studies in Ancient Philosophy*, suppl. vol., 61–78.

Jackson, H. (1879), *Peri Dikaiosunēs: Nicomachean Ethics Book Five* (Cambridge: Cambridge University Press).

Joachim, H. H. (1951), *Aristotle: The Nicomachean Ethics*, ed. D. A. Rees (Oxford: Clarendon Press; 2nd edn. 1955).

Jolif, J. Y. *See* Gauthier.

Judson, L. (1997), 'Aristotle on Fair Exchange', *Oxford Studies in Ancient Philosophy*, 15: 147–75.

—— (forthcoming), *Aristotle: Metaphysics Book Λ* (Oxford: Clarendon Press).

Kahn, C. H. (1981), 'Aristotle and Altruism', *Mind*, 90: 20–40.

Kant, I. (1948), *Groundwork of the Metaphysic of Morals*, trans. and ed. H. J. Paton (London: Hutchinson).

Kenny, A. (1963), *Action, Emotion and Will* (London: Routledge & Kegan Paul).

—— (1965–6a), 'Practical Inferences', *Analysis*, 26: 65–75.

—— (1965–6b), 'Aristotle on Happiness', *Proceedings of the Aristotelian Society*, 66: 93–102. rev. in Barnes *et al.* (1977).

—— (1966), 'The Practical Syllogism and Incontinence', *Phronesis*, 11: 163–84.

—— (1978), *The Aristotelian Ethics* (Oxford: Clarendon Press).

—— (1979), *Aristotle's Theory of the Will* (London: Duckworth).

—— (1992), *Aristotle on the Perfect Life* (Oxford: Clarendon Press).

Keyt, D. (1978), 'Intellectualism in Aristotle', *Paideia; Special Aristotle Issue*, 138–57.

—— (1991), 'Aristotle's Theory of Distributive Justice', in D. Keyt and F. D. Miller (eds.), *A Companion to Aristotle's Politics* (Oxford: Blackwell), 238–78.

Kirwan, C. A. (1990), 'Two Aristotelian Theses about *Eudaimonia*', in A. Alberti (ed.), *Studi sull' Etica di Aristotele* (Naples: Bibliopolis), 149–92.

Klein, S. (1988), 'An Analysis and Defense of Aristotle's Method in *EN* I and X', *Ancient Philosophy*, 8: 63–72.

Korsgaard, C. (1986), 'Aristotle and Kant on the Source of Value', *Ethics*, 96: 486–505.

Kosman, L. A. (1968), 'Predicating the Good', *Phronesis*, 13: 171–4.

—— (1980), 'Being Properly Affected: Virtues and Feelings in Aristotle's Ethics', in Rorty (1980: 103–16).

Kraut, R. (1988), 'Comments on Irwin, "Disunity in the Aristotelian Virtues"', *Oxford Studies in Ancient Philosophy*, suppl. vol., 79–86.

—— (1989), *Aristotle on the Human Good* (Princeton: Princeton University Press).

—— (1993), 'In Defense of the Grand End', *Ethics*, 103: 361–74.

Lear, J. (1988), *Aristotle: The Desire to Understand* (Cambridge: Cambridge University Press).

McDowell, J. (1979), 'Virtue and Reason', *Monist*, 62: 331–50.

—— (1980), 'The Rôle of *Eudaimonia* in Aristotle's Ethics', in Rorty (1980: 359–76).

—— (1998), 'Some Issues in Aristotle's Moral Psychology', in S. Everson (ed.), *Companions to Ancient Thought*, iv: *Ethics* (Cambridge: Cambridge University Press), 107–28.

MacIntyre, A. (1988), *Whose Justice? Which Rationality?* (London: Duckworth).

Mackie, J. L. (1977), *Ethics: Inventing Right and Wrong* (Harmondsworth: Penguin).

Madigan, A. (1991), '*EN* IX.8: Beyond Egoism and Altruism?', in J. P. Anton and A. Preus (eds.), *Essays in Ancient Greek Philosophy*, iv: *Aristotle's Ethics* (Albany: State University of New York Press), 73–94.

Mele, A. R. (1981), 'Aristotle on *Akrasia* and Knowledge', *Modern Schoolman*, 58: 137–59.

—— (1987), *Irrationality: An Essay on Akrasia, Self-Deception and Self-Control* (Oxford: Oxford University Press).

Meyer, S. S. (1993), *Aristotle on Moral Responsibility* (Oxford: Blackwell).

Mill, J. S. (1863), *Utilitarianism* (many editions since).

Miller, F. D. (1984), 'Aristotle on Rationality in Action', *Review of Metaphysics*, 37: 499–520.

Milo, R. D. (1966), *Aristotle on Practical Knowledge and Weakness of Will* (The Hague: Mouton).

Monan, J. D. (1968), *Moral Knowledge and its Methodology in Aristotle* (Oxford: Clarendon Press).

Moravcsik, J. M. E. (ed.) (1967), *Aristotle: A Collection of Critical Essays* (New York: Anchor Books).

Mothersill, M. (1962), 'Anscombe's Account of the Practical Syllogism', *Philosophical Review*, 70: 448–61.

Nozick, R. (1974), *Anarchy, State, and Utopia* (New York: Basic Books).

Nussbaum, M. C. (1985), *Aristotle's De Motu Animalium*, corr. edn. (first pub. 1978; Princeton: Princeton University Press).

—— (1986), *The Fragility of Goodness* (Cambridge: Cambridge University Press).

O'Connor, D. (1988), 'Aristotelian Justice as a Personal Virtue', *Midwest Studies in Philosophy*, 13: 417–27.

O'Meara, D. J. (ed.) (1981), *Studies in Aristotle* (Washington: Catholic University of America Press).

Ostwald, M. (trans. with notes) (1962), *Aristotle: Nicomachean Ethics* (Indianapolis: Bobbs-Merrill).

Owen, G. E. L. (1960), 'Logic and Metaphysics in Some Earlier Works of Aristotle', in I. Düring and G. E. L. Owen (eds.), *Aristotle and Plato in the Mid-Fourth Century* (Göteburg: Studia Graeca et Latina Gotheburgiensa), 163–90; repr. in Owen (1986).

—— (1961), '*Tithenai ta phainomena*', in *Aristote et les problèmes de la méthode* (Louvain, Éditions Nauwelaerts), 83–103; repr. in Moravcsik (1967), and in Owen (1986).

—— (1971–2), 'Aristotelian Pleasures', *Proceedings of the Aristotelian Society*, 72: 135–52; repr. in Barnes *et al.* (1977), and in Owen (1986).

—— (1986), *Logic, Science and Dialectic* (Ithaca: Cornell University Press).

Pakaluk, M. (1980), *Aristotle: Nicomachean Ethics, Books VIII and IX* (Oxford: Clarendon Press).

Pears, D. (1980), 'Courage as a Mean', in Rorty (1980: 171–88).

Price, A. W. (1980), 'Aristotle's Ethical Holism', *Mind*, 89: 338–52.

—— (1989), *Love and Friendship in Plato and Aristotle* (Oxford: Clarendon Press).

Rackham, H. (1926), *Aristotle: The Nicomachean Ethics* (Cambridge, Mass.: Loeb Classical Library; rev. edn. 1934).

Rawls, J. (1958), 'Justice as Fairness', *Philosophical Review*, 68: 164–94.

—— (1971), *A Theory of Justice* (Cambridge, Mass.: Harvard University Press).

Reeve, C. D. C. (1992), *Practices of Reason* (Oxford: Clarendon Press).

Roberts, J. (1989), 'Aristotle on Responsibility for Action and Character', *Ancient Philosophy*, 9: 23–36.

Robinson, R. (1969), 'Aristotle on *Akrasia*', in his *Essays in Greek Philosophy* (Oxford: Clarendon Press), 139–60; repr. in Barnes *et al.* (1977).

Rorty, A. O. (ed.) (1980), *Essays on Aristotle's Ethics* (Berkeley: University of California Press).

Ross, W. D. (1923), *Aristotle* (London: Methuen; 5th edn. 1949).

—— (1924), *Aristotle: Metaphysics* (Oxford: Clarendon Press).

—— (trans.) (1925), *Aristotle: Nicomachean Ethics*, vol. ix of W. D. Ross (ed.), *The Works of Aristotle Translated into English* (Oxford: Clarendon Press); rev. Ross (1954), Ackrill in Ackrill (1973), Urmson (1975 incorporated in Barnes 1984); further rev. Ackrill and Urmson for *Aristotle: The Nicomachean Ethics*, World's Classics (Oxford: Oxford University Press, 1980). I cite the last, except where indicated.

Rowe, C. J. (1975), 'A Reply to John Cooper on the *Magna Moralia*', *American Journal of Philology*, 96: 160–72.

Ryle, G. (1954), 'Pleasure', in his *Dilemmas* (Cambridge: Cambridge University Press).

Santas, G. (1969), 'Aristotle on Practical Inference, the Explanation of Action, and *Akrasia*', *Phronesis*, 14: 162–89.

Schofield, M. (1973), 'Aristotelian Mistakes', *Proceedings of the Cambridge Philological Society*, 19: 66–70.

—— *See* Barnes *et al.* (1977).

Shapiro, H. L. *See* Walsh and Shapiro (1967).

Siegler, F. A. (1967), 'Reason, Happiness and Goodness', in Walsh and Shapiro (1967: 30–46).

—— (1968), 'Voluntary and Involuntary', *Monist*, 52: 268–87.

Smith, A. D. (1996), 'Character and Intellect in Aristotle's Ethics', *Phronesis*, 41: 56–74.

Sorabji, R. (1973–4), 'Aristotle on the Rôle of Intellect in Virtue', *Proceedings of the Aristotelian Society*, 74: 107–29; repr. in Rorty (1980).

—— (1980), *Necessity, Cause and Blame: Perspectives on Aristotle's Theory* (London: Duckworth).

—— *See* Barnes *et al.* (1977).

Stewart, J. A. (1892), *Notes on the Nicomachean Ethics of Aristotle* (Oxford: Clarendon Press).

Taylor, A. E. (1928), *A Commentary on Plato's Timaeus* (Oxford: Oxford University Press).

Taylor, C. C. W. (1980), 'Plato, Hare, and Davidson on *Akrasia*', *Mind*, 89: 499–518.

—— *See* Gosling and Taylor (1982).

Telfer, E. (1970–1), 'Friendship', *Proceedings of the Aristotelian Society*, 71: 223–41.

—— (1989–90), 'The Unity of the Moral Virtues in Aristotle's *Nicomachean Ethics*', *Proceedings of the Aristotelian Society*, 90: 35–48.

Thomson, J. A. K. (trans.) (1955), *The Ethics of Aristotle* (Harmondsworth: Penguin; rev. H. Tredennick, with new introd. J. Barnes, 1976).

Urmson, J. O. (1967), 'Aristotle on Pleasure', in Moravcsik (1967: 323–33).

—— (1973), 'Aristotle's Doctrine of the Mean', *American Philosophical Quarterly*, 10: 223–30; repr. in Rorty (1980).

—— (1988), *Aristotle's Ethics* (Oxford: Blackwell).

—— *See also* Ross (1925).

von Wright, G. H. (1963), *The Varieties of Goodness* (London: Routledge & Kegan Paul).

Walsh, J. J. (1963), *Aristotle's Conception of Moral Weakness* (New York: Columbia University Press).

—— and Shapiro, H. L. (eds.) (1967), *Aristotle's Ethics: Issues and Interpretations* (Belmont, Calif.: Wadsworth).

Walzer, R. R. and Mingay, J. M. (1991), *Aristotelis Ethica Eudemia*, Oxford Classical Texts (Oxford: Clarendon Press).

Watson, G. (ed.) (1982), *Free Will* (Oxford: Oxford University Press).

Wedin, M. (1981), 'Aristotle on the Good for Man', *Mind*, 90: 243–62.

White, N. P. (1981), 'Goodness and Human Aims', in O'Meara (1981: 225–46).

—— (1988), 'Good as Goal', *Southern Journal of Philosophy*, 27, suppl., 169–93.

White, S. A. (1990), 'Is Aristotelian Happiness a Good Life or the Best Life?', *Oxford Studies in Ancient Philosophy*, 8: 103–43.

Whiting, J. (1986), 'Human Nature and Intellectualism in Aristotle', *Archiv für Geschichte der Philosophie*, 68: 70–95.

—— (1988), 'Aristotle's Function Argument: A Defense', *Ancient Philosophy*, 8: 33–48.

—— (1991), 'Impersonal Friends', *Monist*, 75: 3–29.

Wiggins, D. (1975–6), 'Deliberation and Practical Reason', *Proceedings of the Aristotelian Society*, 76: 29–51; rev. in Rorty (1980).

—— (1978–9), 'Weakness of Will, Commensurability, and the Objects of Deliberation and Desire', *Proceedings of the Aristotelian Society*, 79: 251–77; repr. in Rorty (1980).

—— (1995), 'Eudaimonism and Realism in Aristotle's Ethics', in R. Heinaman (ed.), *Aristotle and Moral Realism* (London: UCL Press), 219–31.

Wilkes, K. V. (1980), 'The Good Man and the Good for Man', *Mind*, 87 (1978), 553–71; repr. in Rorty (1980).

Williams, B. (1973), 'Egoism and Altruism', in his *Problems of the Self* (Cambridge: Cambridge University Press), 250–65.

—— (1980), 'Justice as a Virtue', in Rorty (1980: 189–99).

—— (1995), 'Acting as the Virtuous Person Acts', in R. Heinaman (ed.), *Aristotle and Moral Realism* (London: UCL Press), 13–23.

Woods, M. (1982), *Aristotle: Eudemian Ethics, Books I, II, and VIII* (Oxford: Clarendon Press).

—— (1986), 'Intuition and Perception in Aristotle's Ethics', *Oxford Studies in Ancient Philosophy*, 4: 145–66.

—— (1990), 'Aristotle on *Akrasia*', in A. Alberti (ed.), *Studi sull' Etica di Aristotele* (Naples: Bibliopolis), 227–61.

Yack, B. (1993), *The Problems of a Political Animal* (Berkeley: University of California Press).

Index of persons and places

Index of subjects

Index of passages

This index lists passages referred to, either generally by book and chapter or more specifically by page and line-number. Entries in boldface mark the main discussion of the passage.

Printed in the United States
By Bookmasters